Current Developments in Food and Nutrition Research

(Volume 1)

(Traditional Cheeses from Selected Regions in Asia, Europe, and South America)

Edited by

Celile Aylin Oluk

Eastern Mediterranean Agricultural Research Institute, Adana, Turkey

&

Oya Berkay Karaca

Karatas School of Tourism and Hotel Management,

Hotel Management Department,

Cukurova University, 01330 Adana,Turkey

Current Developments in Food and Nutrition Research

Volume # 1

Traditional Cheeses from Selected Regions in Asia, Europe, and South America

Editor: Celile Aylin Oluk & Oya Berkay Karaca

ISBN (Online): 978-981-14-3236-1

ISBN (Print): 978-981-14-3235-4

Bentham Science Publishers Pte. Ltd.
80 Robinson Road #02-00
Singapore 068898
Singapore
Email: subscriptions@benthamscience.net

BENTHAM SCIENCE

CONTENTS

PREFACE

'Farmhouse', 'traditional' and 'artisan' are more recognised as styles of cheese rather than as a method or scale of production. Cheese is manufactured in almost every country in the world and there exist more than 500 varieties of cheese. Cheese is generally classified into different types, such as ripened and unripened cheeses, low or high-fat content and soft or hard consistency. Recently, artisanal cheese consumption has changed in particular. Due to changes in consumer preferences, the concept of postmodern consumerism emerged. Artisanal cheese consumers focus on the unique characteristics of the products and their distinctive character in relation to mass-produced industrial cheeses. These cheeses are still being made in small family facilities. Artisanal cheeses differ depending on the type of production or milk used. Even if the milk source is the same, there may be different production methods in different parts of the world, but most of them are made only in the small processing units in a traditional simple way without good sanitary conditions as unstandardized products. Recently, there has been an increase in the production and consumption of traditional dairy products. With the increasing importance of geographic marking systems all over the world, production methods of traditional cheeses have started to be recorded in countries and transferred to the industry with more hygienic and standard production techniques. At the same time, the popularity of traditional cheeses as "niche products" in the gastronomy of countries is increasing day by day.

Cheese production is not only a craft that allows the milk to be stored and consumed longer than it is without spoiling. Thanks to this craft, tasty microbiological life forms can be produced in different regions of the world with different geographical and climatic features. These life forms also bear the cultural memory of producers who have developed countless traditional methods throughout the history of cheese production, which is thought to be 8000 years old. Today, rapidly developing dairy technologies are based on this cultural background. Defending, preserving and maintaining local-traditional cheeses means at the same time protecting the history and accumulation, cultural memory and the diversity of life.

The aim of this book is to describe the greatest traditional cheese types of country / region. This book also gives information to transfer regional cheese production methods from generation to generation to ensure the sustainability of cheese production and increase recognition of products. These cheeses were described by using the following context:

-Overview of cheese (History and origins, starter cultures, rennets, fundamentals of cheesemaking, ripening, packaging, nutrition, standards, economics and marketing, certification of type)
-The traditional technology of cheesemaking (types of milks, standardizing milk, heating or setting the milk, coagulation of milk, cutting the curd, syneresis of curd, cooking, draining the whey, milling, molding, salting, pressing, ripening, packaging, storage)
-Cheese characteristics (the chemical composition of cheese, texture, flavour, rheology, lipolysis, proteolysis, microbiological composition, physical, chemical and biological aspects of cheese ripening, geographical characteristics guaranteed by the "Appellation of origine").

The book has been written for a wide readership. The main target group of the book is primarily academicians (Department of Food, Dairy technology, Tourism, Gastronomy) and their students. Producers, cheese technologist and consumers can find some practical and interesting information. Since traditional foods are part of intangible cultural heritage, we think it will appeal to those who are interested in sustainability and geographic marking. It will also be a resource to protect traditional dairy products, to ensure their continuity and to increase their international recognition. For this reason, more valuable academics from Turkey and around the world met to explicate the local-traditional cheese. Our goal was to put together extensive detailed information on traditional cheeses from different countries. One can read the history, taste, components, geography and culture of local-traditional cheeses from researchers.

The appellation of origin for the sustainability of cheese production is important. The cheeses mentioned in the book are described by the typical technological and geographical characteristics guaranteed by the "Appellation of origine". In this book, the characteristics, technology of cheese making and artisan culture of the traditional cheese varieties in Republic of Macedonia (White brined cheese, Beaten and Kashkavalcheeses), Montenegrin (Njeguši, Pljevlja, Lisnati, Prljo and Sack cheeses), Brasil (Prato, Requeijão, Coalho, Manteiga, Minas, Serro, Canastra, Marajó, Paraíba, Reino, Serrano, Colonial and Caipira cheeses), Iran (Lighvan and Koozeh cheeses), Marmara Region of Turkey (Abaza, Beyaz, Çerkez fume, Ezine, Kaşar, Lor, Mihalic and Sepet cheeses), Blacksea region of Turkey (Civil, Imansiz, Kargi Tulum, Curry and Kolote cheeses), Eastern Anatolia Region of Turkey (Cacik, Civil, Gravyer, Herb, Kars Kashar, Tulum and Şavak cheeses), Southeastern Anatolia Region of Turkey (Adıyaman Şuji, Antep, Diyarbakır Örgü, Siirt Otlu and Urfa cheeses), Central Anatolia Region of Turkey (Konya Küflü, Divle Tulum, Gölbaşı Tulum, Çepni Tulum, Ereğli Cloth, Karaman Tulum, Yozgat Çanak (Pot), Çömlek, Kayseri Çömlek, Niğde Küp (Gödelek), Sivas Küp, Çankırı Küpecik and Ayaş Küpecik (Basma) cheeses), Aegean Region of Turkey (İzmir Tulum, Sepet, Kirlihanım, Kopanisti, Armola, Lor, Tire Çamur, Aydın Çörekotlu, Çoban, Köy and Peppered Lor cheeses), Mediterranean Region of Turkey (Surk, Tulum, Sünme, Carra, Künefe, Hatay Ezme, Hatay Dil, Isparta Küp, Isparta Dolaz, Burdur Akçakatık, Burdur Keş, Adana Lavaş, Karataş Basma, Dilim and Hellim cheeses) were discussed. This book also discusses the microbiological quality of traditional Turkish cheeses.

The editors wish to express their gratitude to the following scholars for their assistance in preparing this book: Erhan SULEJMANI, Tetova University; Seval Sevgi KIRDAR, Mehmet Akif Ersoy University; Slavko MIRECKI, Montenegro University; Jadranka FRECE, Zagreb University; Maria C. E. RIBEIRO, FATEC; Mostafa SOLTANI, Islamic Azad University; Nayil DINKCI and Cem KARAGOZLU, Ege University; Selin KALKAN, Giresun University; Aslı ÇELIKEL, Mardin Artuklu University; Mutlu Buket AKIN, Harran University; Ebru ŞENEL, Ankara University; Mehmet GUVEN and Zerrin ERGINKAYA, Cukurova University. We would like to thank our family members, Berke, Hakkı, Başar and Hüseyin for their great support.

Celile Aylin Oluk
Eastern Mediterranean Agricultural Research Institute,
Adana,
Turkey

&

Oya Berkay Karaca
Karatas School of Tourism and Hotel Management,
Hotel Management Department, Cukurova University,
01330 Adana,
Turkey

List of Contributors

Aslı Çelikel	Tourism Faculty, Gastronomy and Culinary Art, Mardin Artuklu University, Mardin, Turkey
Cem Karagözlü	Faculty of Agriculture, Department of Dairy Technology, Ege University, Bornova – İzmir, Turkey
Débora P. Baptista	Food Technology Department, Faculty of Food Engineering, University of Campinas, SP, Brazil
Deni Kostelac	Faculty of Food Technology and Biotechnology, University of Zagreb, Department of Biochemical Engineering, Pierottijeva 6, HR 10000 Zagreb, Croatia
Dushica Santa	Faculty of Agricultural Science and Food, Ss Cyril and Methodius University, 16-ta Makedonska brigada No 3, Skopje, Republic of Macedonia
Erhan Sulejmani	Faculty of Food Technology and Nutrition, University of Tetova, Tetovo, Republic of Macedonia
Ebru Şenel	Faculty of Agriculture, Department of Dairy Technology, Ankara University, Ankara, Turkey
Evrim Guneş Altuntaş	Biotechnology Institute 06135, Ankara University, Ankara, Turkey
Gözde Konuray	Faculty of Agriculture, Department of Food Engineering, Cukurova University, 01330 Adana, Turkey
Iva Čanak	Faculty of Food Technology and Biotechnology, Department of Biochemical Engineering, University of Zagreb, Pierottijeva 6, HR 10000 Zagreb, Croatia
İlyas Erdem Tonguç	Faculty of Agriculture, Department of Dairy Technology, Ege University, Bornova – İzmir, Turkey
Jadranka Frece	Faculty of Food Technology and Biotechnology, University of Zagreb, Department of Biochemical Engineering, Pierottijeva 6, HR 10000 Zagreb, Croatia
Kirdar Seval Sevgi	University Burdur Food Agriculture and Livestock Vocational Higher Education School, Burdur Mehmet Akif Ersoy, Burdur, Turkey
Ksenija Markov	Faculty of Food Technology and Biotechnology Department of Biochemical Engineering, University of Zagreb, Pierottijeva 6, HR 10000 Zagreb, Croatia
Karina S. Chaves	Federal University of Mato Grosso, UFMT, Campus Universitário do Araguaia, Barra doGarças, MT, Brazil
Marija Vrdoljak	University of Applied Sciences "Marko Marulić", Petra Krešimira IV 30, HR-22300 Knin, Croatia
Marko Jelić	University of Applied Sciences "Marko Marulić", Petra Krešimira IV 30, HR-22300 Knin, Croatia
Maria C.E. Ribeiro	Faculty of Technology of Capão Bonito, FATEC, Capão Bonito, SP, Brazil
Mayara S. Queirós	Food Technology Department, Faculty of Food Engineering, University of Campinas, SP, Brazil

Mostafa Soltani	Department of Food Sciences and Technology, Faculty of Pharmacy, Tehran Medical Sciences, Islamic Azad University, Tehran, Iran Nutrition and Food Sciences Research Center, Tehran Medical Sciences, Islamic Azad University, Tehran, Iran
Mehmet Güven	Faculty of Agriculture, Department of Food Engineering, Cukurova University, 01330 Adana, Turkey
Mihriban Korukluoğlu	Faculty of Agriculture, Department of Food Engineering, Uludag University, 16059 Bursa, Turkey
Musa Serdar Akin	Faculty of Engineering, Department of Food Engineering, Harran University, Sanliurfa, Turkey
Mutlu Buket Akin	Faculty of Engineering, Department of Food Engineering, Harran University, Sanliurfa, Turkey
Natasha Mateva	Institute of Animal Science, Ss Cyril and Methodius University, blv. Ilinden 92a, Skopje, Republic of Macedonia
Nayil Dinkci	Faculty of Agriculture, Department of Dairy Technology, Ege University, 35100 Bornova-Izmir, Turkey
Oya B. Karaca	Karatas School of Tourism and Hotel Management, Hotel Management Department, Cukurova University, 01330 Adana, Turkey
Sandra Mojsova	Faculty of Veterinary Medicine, Ss Cyril and Methodius University, Lazar Pop-Trajkov 5-7, Skopje, Republic of Macedonia
Selin KALKAN	Faculty of Engineering, Department of Food Engineering 28200, Giresun University, Giresun, Turkey
Semra Gürbüz	Tourism Faculty, Gastronomy and Culinary Art, Mardin Artuklu University, Mardin, Turkey
Slavko Mirecki	Biotechnical Faculty, University of Montenegro, Podgorica, Montenegro
Sonja Srbinovska	Faculty of Agricultural Science and Food, Ss Cyril and Methodius University, 16-ta Makedonska brigada No 3, Skopje, Republic of Macedonia
Tansu Taşpinar	Cukurova University, Agricultural Faculty, Department of Food Engineering, 01330 Adana, Turkey
Tuba Şanli	Faculty of Agriculture, Department of Dairy Technology, Ankara University, Ankara, Turkey
Vesna Levkov	Institute of Animal Science, Ss Cyril and Methodius University, blv. Ilinden 92a, Skopje, Republic of Macedonia
Zerrin Erginkaya	Faculty of Agriculture, Department of Food Engineering, Cukurova University, 01330 Adana, Turkey
Željko Jakopović	Faculty of Food Technology and Biotechnology, University of Zagreb, Department of Biochemical Engineering, Pierottijeva 6, HR 10000 Zagreb, Croatia

CHAPTER 1

Characteristics of Traditional Cheeses Produced in the Republic of North Macedonia

Natasha Mateva[1], **Vesna Levkov**[1], **Sonja Srbinovska**[2], **Dushica Santa**[2], **Sandra Mojsova**[3] and **Erhan Sulejmani**[4,*]

[1] *Institute of Animal Science, Ss Cyril and Methodius University, blv. Ilinden 92a, Skopje, Republic of North Macedonia*

[2] *Faculty of Agricultural Science and Food, Ss Cyril and Methodius University, 16-ta Makedonska brigada No 3, Skopje, Republic of North Macedonia*

[3] *Faculty of Veterinary Medicine, Ss Cyril and Methodius University, Lazar Pop-Trajkov 5-7, Skopje, Republic of North Macedonia*

[4] *Faculty of Food Technology and Nutrition, University of Tetova, Tetovo, Republic of North Macedonia*

Abstract: Traditional cheeses are specific products of the Republic of North Macedonia, mainly produced in small-scale farms or farm houses located in high mountains and rural areas. Some of the best-known cheese types are produced in almost all regions in North Macedonia: kashkaval, white brined cheese and beaten cheese (*"bieno sirenje"*). These types of cheese can also be found in other Balkan countries, only with different tastes and properties. Kashkaval is a type of hard yellow cheese with a natural rind and it belongs to the Pasta filata cheese group. Traditionally, it is made from sheep milk. White brined cheese is the most commonly consumed type of cheese that can be produced from sheep, cow and goat milk. Beaten cheese is a type of yellow hard cheese and the scalding procedure is a crucial step in its production. The great diversity in the manufacturing procedures results with variations in the physical, chemical and microbiological composition, as well as with variations in the proteolysis, the texture, and volatiles. In this chapter, the properties of cheese, the cheese-making technology and the artisan culture of the traditional cheese varieties in the Republic of North Macedonia are discussed.

Keywords: Beaten cheese (*"bieno sirenje"*), Kashkaval cheese, Proteolysis, Volatiles, White brined cheese.

INTRODUCTION

Beaten Cheese (*"bieno sirenje"*), Kashkaval from the Bistra mountain and White

---

* **Corresponding author Erhan Sulejmani**: Faculty of Food Technology and Nutrition, University of Tetova, Tetovo, Republic of North Macedonia; Tel/Fax: +389 76 466 139; E-mail: erhan.sulejmani@unite.edu.mk

Celile Aylin Oluk & Oya Berkay Karaca (Eds.)

Brined Cheese (“*belo salamureno sirenje*”) are our typical trademark products which vary in appearance, aroma, flavour, and texture. Artisanal cheeses are differentiated according to their strong bonds with the territory of their origin and so they represent a historical and cultural blueprint of the community they are produced by. The production of these cheeses takes place in limited geographical areas with the use of know-how techniques transferred from generation to generation and the use of milk that has undergone no treatment after milking. From milking to the end of the ripening process, these types of cheese pass through different surroundings where a variety of microorganisms have the opportunity to grow and develop. Raw milk is an example of an environment comprised of many complex communities and different taxonomical groups of microorganisms with approximately 300 species of bacteria, 70 species of yeast, and 40 species of moulds. The conducted research showed that traditionally made cheeses have unique properties in terms of palatable pleasure, richness, and diversity, as well as protection against pathogens. Undoubtedly, these properties are achieved due to the presence of unique indigenous microbiota, especially because of the use of raw milk, combined with the specific skills of the cheese-makers that result in the creation of their general properties and quality. Each traditional cheese originates from a complex system with its own biodiversity components such as environment, micro and macro climate, pasture, autochthonous breeds, use of untreated milk and its natural microbiota, natural coagulants, cheese-making procedures, ancient traditional tools and equipment, and natural ripening environment.

WHITE BRINED CHEESE

Macedonian white cheese is a brined variety of cheese with either soft or semi-hard texture. White colored, rindless, with close texture and salty acid to pickled flavour are the main sensory properties of this type of cheese. This cheese is most commonly made from cow milk, sheep milk and, less commonly, from goat milk, prepared in blocks, and ripened in brine for a period of 90 days (Fig. **9**). Traditionally, this type of cheese has been produced for decades by local farmers on a small-scale, using raw milk and traditional techniques handed down from generation to generation using only elementary equipment. Instead of using commercial starter cultures, artisan cheese-makers rely on the naturally present indigenous microorganisms in the raw non-pasteurized milk and the adventitious contaminants from the soil, the equipment, the surfaces, and the environment in general. According to Davis (1976) and Phelan (1993), the production of white brined cheese has mostly expanded in the regions of the Mediterranean and Balkan Peninsulas. The production range of white brined cheese is very wide, and it is therefore very difficult to determine the whereabouts of its first production.

Depending on the region provenance where it is produced, several modified methods, numerous varieties and sub-varieties of this type of cheese have been created. According to international standards, white cheeses in brine belong to the group of semi-solid and soft cheeses, with or without maturing phase. According to the Codex Standard for cheese A-6, 2000, this cheese has a distinctive white to yellowish colour, no crust, and it is ripened and stored in brine until repackaging or sale. According to the wide areas of origin of this type of cheeses in different countries on the Balkan Peninsula, they are included in the group of autochthonous dairy products, which, depending on the region, are characterized by a certain specificity in technology, quality, and name of the cheese. According to technology, this type of cheese is the so-called "Feta" in Greece (Anifamtakis, 1986), it is known as "White brined cheese" in Bulgaria (Dimov, 1963), and it is called "White soft cheese" in Macedonia (according to Srbinovska *et al.*, 1994). Dilanyan (1967) believes that the basic property and characteristic of the white brined cheese is the brine, which is the environment where the process of maturing and storage of the product takes place. Therefore, the duration of keeping the cheese in brine affects its taste and quality.

Production of Traditional White Brined Cheese

Sheep milk produced by autochthonous sheep breeds is exclusively used for the production of sheep white brined cheese (local sheep breeds, Pramenka, and crossbreeds). Accordingly, the milk used for the production of cheese should be of normal chemical composition in accordance with the Rulebook on the special requirements for safety and hygiene and the manner and procedure for performing official controls of milk and dairy products (Official Gazette of the Republic of Macedonia, No. 26, 21.2.2012), as well as the Rulebook on the requirements for the quality of raw milk, the quality standards for consumer milk, dairy products and the use of their names, the quality and activity of starter cultures, curdling and other specific substances and the manner of their use, the manner of additional labelling of the milk and dairy products, as well as the permitted deviation of weight in relation to the declared (Official Gazette of the Republic of Macedonia, No. 96/2011). The criteria for the chemical composition of raw sheep milk are the following parameters: minimum fat 6.09 (average 6.0-9.0%) and minimum proteins 5.2 (average 5.3-5.8%). The average of the chemical composition of raw sheep milk is given in Table **1** (Levkov, 2013; Talevski *et al.*, 2009)

Relatively, from the results, the research conducted by Levkov (2013) and Televski *et al.*, (2009), and the criteria provided in the Rulebook on raw sheep milk, it can be seen that the values for certain parameters are in accordance with the criteria of the Rulebook, but with certain deviations. That is, the values of all examined parameters of Talevski *et al.*, 2009, (Table **1**) are in accordance with the constitutional criteria of the Rulebook and in accordance with Simos *et al.* (1996), whereby normally sheep milk contains 6-9% milk fat, 4-7% protein, 4-6% lactose, and 17-21% dry matter. While according to Levkov (2013), there is a greater deviation of the increase in the value of milk fat (10.9%) and the value of milk proteins are below the prescribed (4.55%) (Table **1**). These obtained values of the chemical composition of raw sheep milk, according to Levkov (2013), are considered to be the result of the use of raw sheep milk at the end of lactation of the sheep, which is a practice among our producers, especially in the production of traditional authentic white cheese. The production of Macedonian White Brined Cheese (MWBC) depends on the hydrolysis of lactose by lactic acid bacteria to produce lactic acid (Sulejmani, 2010). The breakdown of the degradation of lactose in the curd has a major effect on the quality of the ripened cheese; for example, excessive lactic acid in cheese curd leads to a low pH, strong acidic harsh taste, and a brittle structure. Sulejmani (2010) studied the optimization of the production process of Macedonian white brined cow milk cheese, in which case the influence of the starter culture, the optimal temperature of milk curdling, and the optimum time of curd cutting on the quality and yield of the cheese were determined. The highest level of acidity (°SH), protein, and fat at day 1 of ripening (mean ± SE) were observed in the MWBC cheeses with a starter culture of 24.13%, 14.56% and 25.68%, respectively. The flavour and the total sensory scores were significantly influenced by the use of the starter culture, processing technology, and temperature of curdling. Cheeses made using the freeze-dried culture F–DVS YF–3331, curdling at 39°C MWCB2 had a more satisfactory appearance, texture, odour, and quality than other differently treated Macedonian white cheeses (Sulejmani *et al.*, 2011).

Table 1. Chemical composition of raw sheep milk.

Parameters	Levkov, 2013	Talevski *et al.*, 2009
Total dry matter, %	18.85	18.07
Proteins, %	4.55	5.89
Fat, %	10.09	7.07
Lactose, %	4.09	4.37
Dry matter without fat, %	10.09	11.0

The Cheese-making Process

Immediately after milking, which is done twice a day in the pen, the raw sheep milk is processed and transported to the cheese-makers, where it is filtered again and the production of cheese starts immediately. In the case when the milk is processed once a day, it should be cooled to a temperature below 6°C. For the production of traditional cheese, raw milk is used in the presence of natural microbiota. This means that pasteurization of the milk is not performed. For the production of white brined sheep cheese, the milk should have appropriate acidity and it is best to be from 9 to 10°SH, according to Srbinovska (2013).

Curdling

The drained sheep milk is placed in wooden barrels (Fig. **2**) with appropriate milking temperature and curdled, after cooling, *i.e.* T= 26-28°C. The duration of the curdling depends on the weather conditions, meaning that when the outdoor temperature is above 30°C the curdling lasts for 1 hour, while under 10°C the process is prolonged for 2 hours. The curdling is carried out at that temperature by adding a cheese enzyme (chymosin). Earlier, the rennet from the stomach of young lambs was used, while they were still fed with milk, that is while the enzyme chymosin was produced, which is the primary means for milk coagulation. The extracted rennet from the stomach of young lambs was dried in the sun, covered with the lamb's abomasum (the fourth compartment of the lamb's stomach where rennet is secreted) to prevent it from becoming contaminated by external factors and flies. Once dried, it was ground and then, in the form of powder, it was used as rennet. This way of rennet production was replaced by the production of various rennets of different origins, derived from certain microorganisms through synthetic ways. For the traditional way of production of these autochthonous white cheeses, the cheese enzyme used most commonly or recently is an organic production of 100% chymosin. Due to the high coagulation intensity of this type of rennets, they are added in amounts as small as 0.02%. It is recommended to use pure chymosin rennet, obtained synthetically, which is closest to the traditional in terms of intensity and activity of the enzyme, and because this enzyme has a precisely determined place of action and the production of creation of peptide chains of the same length, without the possibility of the appearance of bitter taste of cheese (Mateva, 2004).

In the production facility for renneting the milk, in addition to using wooden barrels (Fig. **2**), some cheese-makers use shallow wooden troughs (Fig. **1**) with dimensions 1m x 0.20-0.35m x 3-4m. Before putting the milk into these troughs, a cheesecloth or plastic foil is placed inside, in order to avoid major losses, especially for milk fats. When the curdling (coagulation) begins, the primary coagulation is measured, which usually begins in 10-15 minutes (once the cheese mass starts to detach from the vessel's wall), while the full coagulation is usually after 1 hour, depending on the temperature and weather conditions. The curds are considered curdled after obtaining a medium firm consistency, with porcelain cross-section and sweet taste, and the whey becomes yellowish-green (Kapac 1988).

Processing and Forming

The formed curd is bound on both sides with the cheesecloth (Fig. **3**), where the whey is separated. On the cheese mass above the cheesecloth, wooden weights are placed on both sides of the curd (Fig. **4**), so that it is nicely outspread and can be evenly divided. When the surface layer of the curd is cooled, it is turned over. After the curd rests for 3-4 minutes and is self-pouring from its own weight, the curd is cut into blocks with a large blade and is moved to a perforated frame, under which is a cheesecloth. Thus, through the cloth and the openings of the frame, the residual whey of the curd is separated, and the process of its separation or syneresis is finished. The separated whey is clear, and yellow and green in colour.

Fig. (1). Wooden troughs for renneting.

Fig. (2). Wooden barrels for renneting.

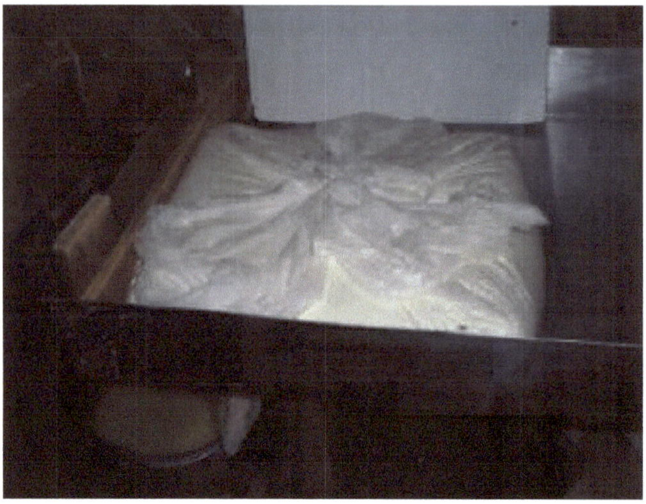

Fig. (3). The curd is bound with the cheesecloth.

Fig. (4). Added tools - wooden weights.

Pressing

After the self-pressing process has been completed from the very weight of the curd, some masters also practice the load-bearing phase in order to separate the remaining whey. In this stage of pressing under load, the temperature in the room is very important, and it should not be higher than 18°C, because higher temperature can strengthen the separation of the whey, which would cause strong decomposition of the milk sugar (lactose) and getting sour cheese with a larger number of cavities, considered as a cheese defect (Srbinovska, 2013). Pressing the load is carried out gradually so that it does not lead to forming a crust on the mass and its decomposition.

Cutting into Blocks

After finishing the pressing process and the separation of whey (syneresis), the curd, which should be 12 cm in size, is cut into blocks (Fig. **6**). The blocks are 11-12 x 11-12 cm in size because the diameter of the container is 22 cm. The salt, *i.e.* the brine, is prepared beforehand, and the concentration measured with a salinometer should be 18%. The brine is prepared in the production facility from sea salt and water used from natural resources. Troughs filled with pre-prepared souse are used for salting the brine and the cheese blocks are put inside (Fig. **5**). On the upper side of the surface layer, the cheese blocks above the brine are salted with coarse sea salt (which should meet the microbiological and hygienic requirements for its quality). Then, it remains in the brine for 12 hours, and during that time, the cheese blocks are overturned and salted on the surface layer to make the salt equal. By salting the cheese, the concentration and acidity of the brine changes. Each salting results in separation of the whey, which increases the acidity. On the other hand, with the reduction of salt in the cheese, the concentration of salt changes. Therefore, it is necessary to continuously monitor the concentration of salt and acidity of the brine (salt), which should be 8°SH (20^0T), and if it is higher, there will be a process of slow salting (Srbinovska, 2013). Therefore, the temperature of the brine should be 15 °C, while in warm weather, it should be up to 10°C. Ripening white brined cheese should contain 3-3.5% of salt. To achieve such salting, salt consumption in every 100 L of sheep milk should be about 3 kg.

Fig. (5). Salting in brine.

Fig. (6). Cutting the curd in blocks.

Packaging and Ripening

The packaging of white brined cheese is carried out in tin cans (Fig. **7**), and if there is a larger quantity, it is packed in wooden barrels and vats. The cheese blocks are lined up in barrels, and initially, on the bottom of the barrel, coarse sea salt is thrown, then cheese, then salt again, then cheese again, and so on until the barrel is filled (Fig. **8**). While arranging the cheese, regardless of the type of packaging used, it is important to keep an eye out so that there is no space left between the cheese blocks and they would not deform. Then, brine is poured, where the concentration of brine is 18%, but according to Kapac (1988), it should be 22-23%. The acidity of the brine where the cheese ripens should be 24-40°SH (60-100^{0}T). If the acidity is lower, conditions for developing rotting bacteria are created, while higher acidity prevents the ripening and the cheese gets a crushing consistency. The quantity of brine in the barrels should be 10-15% of the total amount of cheese and it should be controlled, it should be clear with a pleasant smell and no mild liquid. Packed cheese in barrels is carried to a special room for ripening, with controlled temperature and humidity conditions for that purpose. Most often the ripening in the pens is done in a special room, an excavated underground basement with good insulation. The walls are made of stone and they are tinned and thatched for better insulation. The ripening is carried out at a temperature of 15-17°C, with humidity of 75-80%, and in the duration of 2 to 3 months. According to Kapac (1988), the ripening of white brined cheese is carried out at a lower temperature of 13-15°C for 30 to 45 days.

Fig. (7). Ripening in cans.

Fig. (8). Stacking in cans.

A major highlight of the white brined sheep cheese ripening process is the formation of milk acid, which begins with the pressing stage of the cheese so that after 15-20 days of ripening, the amount of milk sugar (lactose) is reduced to a minimum or is not present at all. Therefore, the quality of this cheese depends on the dynamics of ripening, which, if faster or slower, inhibits the development of the desired lactic acid bacteria and the creation of their enzymes (ferments). According to Penev (1972), the basic indicator is the pH value which, at the end of the ripening period, should be between 5.8 and 6.0 in order to achieve this acidity of cheese and proper dynamics during the ripening. The same authors consider that for the first ten days, the ripening should be carried out at a temperature of 15-18°C, and afterwards, the cheese should be carried to basements with a lower temperature regime of 10-15°C, until the final stage of ripening.

Fig. (9). White brined cheese.

Ripening and Cheese Yield

Until a certain fermentation stage, the cheese ripens in a warehouse at a temperature below 10°C, where the process continues, which consists of supplementing the brine (as needed). The yield of sheep's white cheese is different, depending on the sheep's lactation period. For the production of 1 kg white brined sheep cheese, an average of 3.3 L sheep milk is needed.

Chemical Composition of White Brined Cheese

The average chemical composition of the traditional white brined cheese is presented in Table **2**. The research by Mojsova (2013), regarding the use of the same raw material (sheep milk), is comparable to Kapac's (1988). While regarding the final product – white brined cheese, the results of Kostova-Mojsova (2013) and Kapac (1988), for the production of traditionally white brined cheese, are comparatively in line with small deviations in the industrial production of white brined cow cheese, according to Mateva (2012). According to Mojsova (2013), in traditional white sheep cheese, the lowest pH value was 4.04, and the highest value was 5.05. The obtained results were consistent with the results of

Turantas *et al.*, (1989) which, for the Turkish white cheese, received pH values of 4.11-5.65.

Table 2. Chemical composition of white brined cheese.

Parameters, %	Kostova-Mojsova, 2013 Traditional Sheep Cheese			Kapac, 1988 Traditional Sheep Cheese	Mateva, 2012 Industrial Cow Cheese		
	average	*min*	*max.*	*average*	*average*	*min*	*max.*
Moisture	49.05	46.97	51.58	52.95	50.43	48.91	52.54
Dry matter	49.98	48.85	53.07	47.05	49.57	47.46	51.09
Fat	28.50	28.00	30.00	26.42	24.02	22.14	25.12
Fat in dry matter	57.62	52.40	60.00	56.21	48.47	45.31	51.87
Nonfat dry matter (NFDM)	21.48	18.85	25.07	20.63	25.55	25.32	25.97
Moisture in NFDM	70.25	64.73	73.07	71.96	66.37	65.31	69.70
Proteins	19.65	18.21	21.37	23.79	17.26	14.42	19.24
Soluble Nitrogen	0.8458	0.6700	0.9213	/	0.2978	0.2677	0.3211
Ripening coefficient	27.19	19.64	31.65	/	11.28	10.67	11.85
Salt	4.90	4.38	5.43	3.26	3.75	3.05	4.01
PH	4.46	4.04	5.05	4.65	4.51	4.26	4.61
Titratable Acidity °T	216.25	200.00	240.00	220-260	222.21	216.93	225.15

Titratable acidity in fresh cheese ranged from 100-120°T, while in mature cheese, the lowest measured titration acidity was 200°T, and the highest value obtained for titratable acidity was 240°T. The derived results completely coincide with the results obtained for Sjenicko cheese, obtained from sheep milk, where titratable acidity in mature cheese ranged from 143.85-244.32°T (Ruzic-Muslic *et al.*, 2011). According to the moisture content in nonfat dry matter (NFDM) and rheological properties, this type of cheese (Codex standard for cheese, CODEX STAN 283-197) is classified as soft cheese ripened in brine, since this parameter is above 67%, *i.e.* it ranges from 64.73% to 73.07% and is in accordance with the trials of Kapac (1988), with representation of 71.96%, (Srbinovska 2013). According to Mateva (2012), the results for NFDM in cow cheese obtained by industrial production are slightly lower than 67%, *i.e.* by an average of 66.37%, which is the result of using cow milk and other production methods. The content of dry matter in cheese is the most important parameter that affects the ripening process. As a rule, cheeses with a higher moisture content go through a faster ripening process. Therefore, processing the crude as a technological operation aims to regulate the moisture content of the cheese. According to the results from the analyses carried out by the producers themselves and from literature data

(Mojsova, 2013; Mateva, 2012; Kapac, 1988) it can be concluded that the dry matter ranges from 47.05% to 53.07%. The values for fat in dry matter in traditional cheeses examined by Anifantakis (1998) ranged from 49.38-58.05%, which are somewhat lower compared to our results. On the other hand, Ruzic-Muslic *et al.* (2011) published higher values than our lowest value, more precisely with Sjenicko cheese - the values for MCM ranged from 53.32 to 62.50%. Milk fat ranged from 27.51% to 30%, compared to Kapac, where there was low content of fat 26.42%. Based on the content of milk fat, this type of cheese meets the criteria of CODEX STAN 283-197 and belongs to the group of full-cream cheeses. The obtained values coincide with the values for Sjenicko cheese where fats ranged from 24.50-29.50% (Ruzic-Muslic *et al.*, 2011). The results, according to Mateva (2012), for the value of milk fat are justifiably lower, because it is industrially produced white brined cow cheese. According to Mojsova (2013), the lowest values for total protein content are from 18.21% to 23.79%, which is the average according to Kapac (1988). The values for total proteins, according to Mateva (2012), are in accordance with the values of the previous authors. As a basic indicator to determine the coefficient and dynamics of ripening, larger variations in this parameter can be seen between traditional and industrial white brined cheeses. From the results of the value of the ripening coefficient, we can see a large difference of 11.28% (Mateva 2013) to 27.19 (Mojsova 2013) in the final ripening stage of white brined cheese. The lower value of the ripening coefficient in industrial cow cheese is due to the lower intensity of proteolytic decomposition, and at the same time, due to the lower value for soluble nitrogen in this type of white brined cheese. The data for the value of 11.28% (Mateva 2012) for the ripening coefficient in the final stage of ripening are somewhat smaller and in relation to the studies of Maćej (1989), who found a ripening coefficient of 12.63% in Sjenicko cheese. According to authors Maćej (1989) and Jovanović (2001), who compared certain parameters in white brined cheeses produced traditionally and industrially, in relation to cheeses produced on the basis of co-aggregates, they found insignificant differences in the content of dry matter and percentage of milk fat, while in relation to the total nitrogen content, they found an increase of 50% for traditional cheeses compared to industrially produced cheeses and on the basis of co-aggregates. From here arises and is standardized the research in which, according to Stević (1962), biochemical transformations in the ripening process can be divided into primary (basic) ripening – in all types of cheese with special emphasis in solid and semi-hard cheeses with formed crust, and secondary (additional) ripening which is characteristic for cheeses with soft consistency and without crust. Typical for softer cheeses are the changes that occur on their surface, forming compounds that give it a spicy taste and smell. These compounds have the ability to diffuse inside the cheese mass during ripening and form the specific aroma of cheese.

Microbiological Properties of White Brined Cheese

Bacterial biodiversity coming from untreated milk and the surrounding contamination constitute the primary source of microorganisms essential for the peculiar taste, unique flavour and typical consistency and appearance of this traditional cheese. Indigenous microbiota from the raw milk and microorganisms originating from environmental exposure during the cheese-making process are also involved in the fermentation process, the biochemical properties and the final development of the cheese product (Kostova-Mojsova, 2013). Microorganisms, including bacteria, yeasts, and moulds are present in the whole process of ripening and make their contribution to it through their metabolic and enzymatic activity. The quality of the cheese largely depends on the composition of microbiota. When cheese is traditionally made, environmental microbiota plays a fundamental role in the fermentation process and is one of the most essential parameters affecting the quality of the cheese (Demaringny *et al.*, 1997). Microbial populations are numerous and diverse. Microorganisms vary in terms of their abundance and diversity during the cheese ripening process. The major microbial groups, isolated from white brined cheese in the first days of the ripening process were *Lactococcus, Lactobacillus, Enterobacteriaceae, Leuconostoc, Enterococcus,* and yeasts. As the process evolved, these populations changed and at the end of the ripening process, the most prevalent ones were *Lactobacillus* and *Lactococcus* and yeasts groups (Mojsova *et al.*, 2013). Lactic acid bacteria (LAB) constituted the predominant bacterial group during the ripening process with a population higher than 7 log cfu/g. Among LAB, lactococci were found in higher numbers at day 10, ranging from 5.44 to 7.94 log cfu/g when compared with leuconostoc (4.11 – 6.58 log cfu/g) and lactobacilli from 5.17 in sample A to 6.9 log cfu/g in sample B (Fig. **10**). These results indicate that lactococci constituted the predominant bacterial group in the beginning of the ripening period. At the end of the ripening, lactobacilli were the most prevalent group and their number was within the range of 5.35 – 7.43 log cfu/g. The decline in lactococci was probably due to the inhibitory action of the low pH and high salt in moisture values in the maturing cheese (Vafopoulou-Mastrogianaki *et al.*, 1990). The presumptive *Leuconostoc* at day 10 were found within the range from 4.11 – 6.50 log cfu/g. The obtained results are slightly higher than the ones reported by Manolopoulou *et al.* (2003). At the end of the ripening process, at day 90, there was only one cheese sample where *Leuconostoc* was detected. These groups of microorganisms may influence the ripening process through the production of lactic acid, the decrease in oxidation/reduction potential and their proteolytic and lipolytic activities (Steele, 1995). Enterococci were also found in high values at the beginning of the ripening process (from log 2.6 – 6.03 cfu/g) (Mojsova *et al.*, 2013). These high numbers were in accordance with counts of enterococci found in Turkish white cheese (Oner *et al.*, 2006) where the mean log was 5.34 cfu/g. At day 90, they rapidly decreased and enterococci were not detected in some cheese samples (Fig. **10**). The variations in their number in different types of cheeses depend on the type of cheese, the initial contamination of the milk, their survival and growth under certain circumstances in the production and ripening processes (Macedo *et al.*, 1995; Litopoulou-Tzantaki 1990). Enterococci may influence the ripening process due to their proteolytic activity, citrate activity, and lipolytic activity, as well as the production of diacetyl and other aromatic compounds (Kostova-Mojsova 2013; Folquie *et al.*, 2006). The most prevalent species of *Lactobacillus* at the initial point of the ripening process were: *Lb. plantarum, Lb acidophilus, Lb. curvatus, Lb. lidneri, Lb. helveticus, Lb. salivarius. Lb. delbrueckii* subsp. *delbrueckii*. The most prevalent species of Lactobacillus at the final point of the ripening were: *Lb. paracasei* subsp. *paracasei*, *Lb plantarum, Lb. brevis, Lb. fermentum, (*Kostova-Mojsova 2013*)*. Among the *Leuconostoc* genus, the most frequently detected species were *Leuconostoc mesenteroides* subsp. *mesenteroides dextranicum, Leuconostoc lactis,* and *Leuconostoc mesenteroides* subsp. *cremoris*.

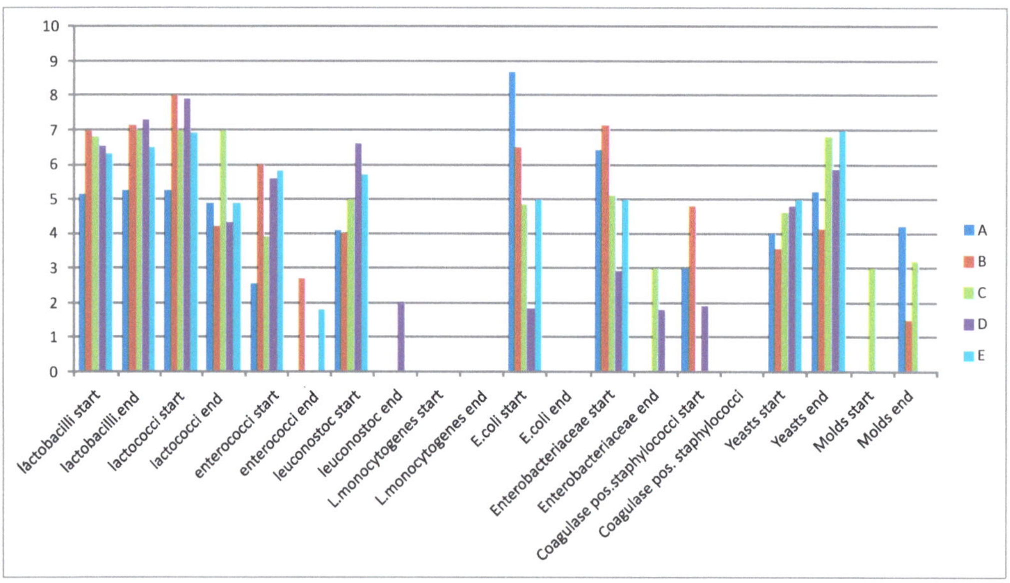

Fig. (10). Log count of microbial groups during ripening of the white brined cheese (beginning and ending point).

Enterococci are part of the lactic acid bacteria (LAB) and present a complex, diverse and important group of bacteria in terms of their interaction with food and humans. Probably because of their usual presence as contaminants of raw milk, thermotolerance, low toxicity and the ability for acidification, enterococci became an essential part of fermented food (Franz *et al.*, 1999). Enterococci are omnipresent bacteria that harbour the digestive tract of humans and animals, and are also part of the environment including the soil, surface waters, plants, and vegetables. Furthermore, they can be found in food, and specifically in cheese (Giraffa, 2002). They are often present in artisanal cheeses made in the southern regions of Europe, mostly from raw sheep or goat milk, where they occur as contaminants from animal faeces, water or milking equipment, and storage tanks. (Folque-Moreno *et al.*, 2006). As an important component of artisanal cultures, they play a key role in the ripening process, (Manolopoulou *et al.*, 2003) giving the authentic taste and aroma (Kostova-Mojsova 2013). According to Mojsova (2016), the most dominant species of enterococci in traditionally made cheeses are *E. faecium* and *E. Faecalis,* and less dominant are *E. durans, E. hirae* and *E. gallinarum.* These results are completely in accordance with the results found in other Mediterranean cheeses (Gelsomino *et al.*, 2001; Aarestrup *et al.*, 2002; Manolopoulou *et al.*, 2003; Foulquie *et al.*, 2006). Enterobacteriaceae are part of the normal microbiota of many dairy products. In fact, contamination of raw milk inhabited with enterobacteriaceae is naturally unavoidable, and even additionally during inappropriate cooling after milking and improper transport of the milk in terms of failing to meet the requirements for cold chain transport, as well as during handling when processing. The growth of enterobacteriaceae can be influenced by the pH value during the process of fermentation, stress factors along the production stages, salt concentration, and the presence of bacteriocins. As a result of all of the abovementioned factors, these groups of microbiota are not commonly present in the final product. At the beginning of the ripening process, a high level of contamination with *Enterobacteriaceae* from 7.24 log cfu/g and *E. coli* 8.69 log cfu/g occurred, which was expected due to the use of raw milk and artisanal rennet, as well as the high moisture and low salt content. All of the factors mentioned above could promote the growth of these bacteria. Furthermore, during the maturation process, the number of *Enterobacteriaceae* was decreased to 3.07 and 1.77 log cfu/g, and in some cheese samples, they were not present at all. *E. coli* was not detected in the final stage of ripening (Mojsova *et al.*, 2013). Similar results were obtained in Feta cheese, where at day 4, *E. coli* reached 5.3 log cfu/g, after which they are significantly dropped and at the end of the ripening they are not detected at all (Manoloupoulou *et al.*, 2003). According to Kongo (2008), the number of *Enterobacteriaceae* is very important for evaluating the quality of raw milk and hygiene during the process of making, handling, and storage. On the contrary, their presence is very important in creating the flavour, aroma, and texture in some of the traditional cheeses (Dahl *et al.*, 2002). In addition, the presence of *Enterobacteriaceae* in all stages of the cheese production could be related to their role of ripening and creating the typical aroma. (Mojsova *et al.*, 2013). The presence of coagulase positive staphylococci at the beginning of the ripening process was within the range from 3.11-4.77 log cfu/g, and at the end of the production, they were not detected in any of the samples (Mojsova *et al.*, 2013). Even though staphylococci tolerate salt, their growth is not favoured by the cheese environment, because of the combined inhibitory effect of the pH and NaCl concentration (Litopolou-Tzanetaki 1997). In addition, some bacteriocin-producing enterococci isolated from the cheese samples showed inhibitory activity against them (Mojsova *et al.*, 2015).

Although there is evidence of survival of the *L. monocytogenes* in brine, Papageorgiou & Marth (1989) and other authors, after their conducted researches concluded that Listeria can grow and survive in mature Feta cheese, while *L. monocytogenes* was not detected in the Macedonian white brined cheese samples (Mojsova *et al.*, 2013). This data could be explained by the presence of bacteriocin producing enterococci in the cheese that showed the activity predominantly against *L. monocytogenes*. (Mojsova *et al.*, 2015).

Yeasts are inevitable microorganisms in traditional cheeses and they were present through the whole ripening process within the interval from 2.74- 5.30 cfu/g (Mojsova *et al.*, 2013). High yeast numbers can be attributed to their tolerance to low pH values, low water activity, ability to grow at low temperatures, ability to use the lactose and organic acids and tolerance to high concentrations of salt and cleaning agents and sanitation. (Bouton *et al.*, 1998). Although the presence of yeasts may have adverse effects in terms of spoilage of cheese (Pereira-Dias *et al.*, 2000, Minervinu *et al.*, 2001), according to some authors they play an important role in creating the aroma and the nutritional quality as a result of the producing essential amino acids, vitamins, aromatic compounds such as diacetyl, acetaldehyde, methyl ketone, and ethyl alcohol. The increased number of yeasts found in the study of Mojsova *et al.*, (2013), may contribute to the typical organoleptic characteristics of our traditional cheese, since recent examinations have shown that some lipolytic and proteolytic enzymes produced by these microorganisms contribute to the development of the aroma and flavour compounds (Marino *et al.*, 2003). Moulds were found at day 90 in some of the samples and could probably be due to contamination occurred during the ripening process (Mojsova *et al.*, 2013).

TRADITIONAL BEATEN CHEESE

Traditional cheeses are characteristic products of Balkan and Mediterranean countries, one of which is the Republic of Macedonia. Their production is typical of the rural areas, but also of the mountain regions where the livestock is kept and pastured in summer. The consumers' increased interest in traditional cheeses due to their unique organoleptic properties (Montel *et al.*, 2014; Franciosi *et al.*, 2009) is positive from many aspects, such as: preservation of the microbial biodiversity, support to the rural regions' economic development, creation of new jobs, etc. (Zeppa *et al.*, 2004; Talevski 2012). It is considered that the production of traditional Beaten cheese dates back as early as the age of the Ottoman Empire (Sulejmani *et al.*, 2014b), and more precisely – that Beaten cheese was invented by the Vlach sheep breeders under the name Vlashko Cheese (Talevski 2012). According to Micev (1969), Beaten cheese was produced in rural households in the region of Mariovo from raw sheep milk. Over the course of time, its production spread almost on the whole territory of today's Republic of North Macedonia and, besides sheep milk, it has also been produced from cow and goat milk or their mixture. According to Dubrova-Mateva *et al.* (2016), very often for the production of Beaten cheese, milk from the end phase of the lactation is used, since this milk is richer in milk fat. The small changes in the production technology, in combination with climatic and geographic properties, the milk quality and the animals' nutrition (Micari *et al.*, 2007) influence the quality and the properties of the cheese. A common feature of all the variants is the specific production process, namely the stage of churning (beating the cheese, wherefrom comes its name) when the curd is crushed into tiny particles like larger grains of sand in size. Beaten cheese is distinguished by its hard consistency, has a yellowish colour, and when cut, a lot of cavities of different sizes and unequal disposition can be seen. A very important feature of this type of cheese is the high concentration of salt. According to Kapac-Parkačeva (1974), by its physical and chemical properties and the technology of production, the Beaten cheese belongs in the group of hard cheeses, while by the way of ripening and keeping (in brine), it belongs in the group of sour-salty cheeses.

The Cheese-making Process

The milk for making Beaten cheese is drained through a cheesecloth (not obligatory) and poured into a curdling vessel. The curdling is most often done by using an enzymatic rennet with the strength of 1:5000, or the rennet chymosin CHY-MAX (2080 mica/g) at the temperature of the milk of 25–35°C. In the past, for curdling, a home-made rennet was used, obtained from the lamb's stomach (Talevski 2012). The curdling process lasts 30–50 minutes. After that, the curd is submitted to processing (churning or beating) by using a wooden tool (the so-called "kyurkalo") (Fig. **11 a**). The process of churning (beating) is done in three series of 50 strokes (150 strokes in total), and after each series, the curd is left to "rest" for 5–10 minutes. In this process, it may come to the separation of a part of the milk fat, in which case the fat is skimmed and removed from the vessel. When the beating process ends, the curdled mass is warmed up by adding warm water at a temperature of 53-90°C, depending on the particular manner of production. Then the curdled mass is left to stay with warm water for some 5–20 minutes, whereupon the curdled mass slowly settles down at the bottom of the vessel. After that, the water and the whey are removed from the vessel and the curdled mass is shaped into a ball by hand (Fig. **11 b**). The curd is put into a cheesecloth, hung up and left for some 16–20 hours so that the mass can drain and self-press (Fig. **11 c**). After that, the stage of ripening follows, which differs depending on the traditions in the particular region where the cheese is produced. The cheese is laid on a wooden table covered with cloth, to ripen at room temperature for 2–7 days (Fig. **11 d**), or it can be exposed to sun ("sunning") for 2-3 days, wherein the cheese gets waxy yellow colour caused by the milk fat emerged to the surface. When the stage of ripening is finished, the cheese is cut into pieces 5–6 cm thick and it is salted with coarse salt (Fig. **11 e**). The amount of salt depends on the cheesemaker's estimate. After salting, the pieces are put into tin or plastic cans and left to stay for 1-2 days (Fig. **11 f**). Then the cans are filled with brine (20-22%) till the pieces are fully covered, and the cheese is left to ripen for 30-45 days.

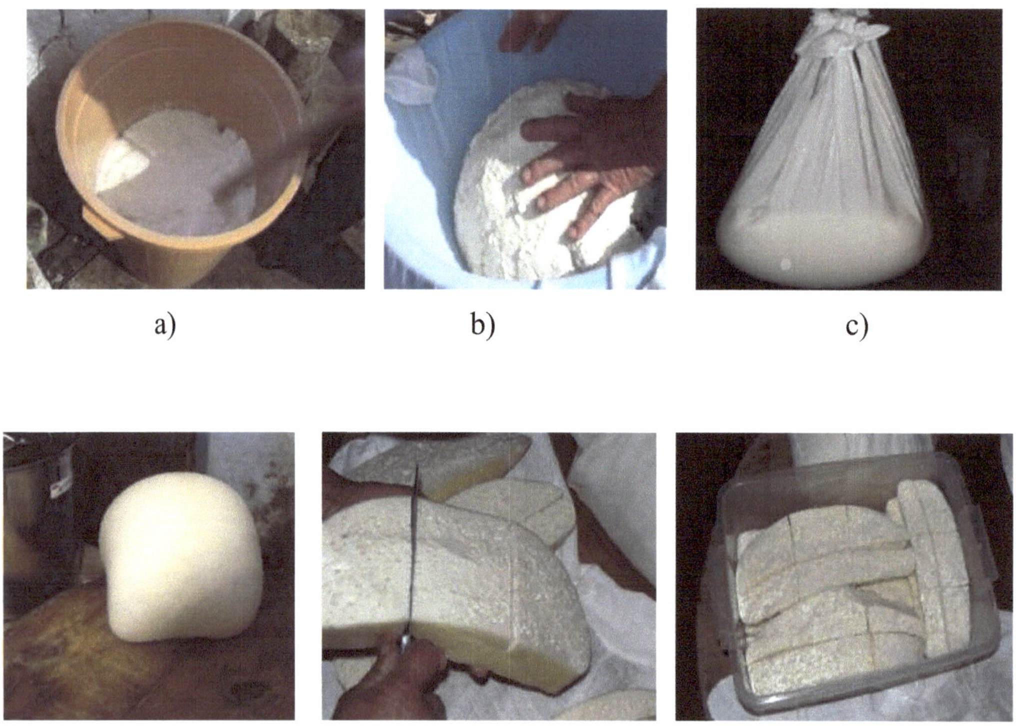

a) b) c) d) e) f)

Fig. (11). Technology of beaten cheese (Bieno sirenje) production; a) cheese processing (churning of curd); b) shaping of curdled mass; c) curd draining and self-pressing; d) dry ripening; e) preparing for salting with coarse salt; f) cheese stacking in can.

Physical and Chemical Properties of Traditional Beaten Cheese Made of Raw Sheep Milk

As can be seen from Table **3**, the pH value changes during the process of cheese ripening. In the beginning, there is a steady decrease, whereas during the period of ripening in brine (from day 10 to day 45) there is a slight increase. The differences in the pH values between the cheeses made in different households are not statistically significant ($p<0.05$). The differences in the production technology, the milk quality and the season of the year when the cheese is made, have an influence on the active acidity of the cheese which resulted in lower pH values compared to the pH values obtained by Sulejmani *et al.* (2014a, 2014b) and Talevski (2012). The amount of moisture in the cheese decreases during the period of ripening when a statistically significant difference ($p<0.05$) was found. This is due to the unequal amounts of salt added and the speed of its diffusion in the cheese as well as the expulsion of the water from the curdled mass. The results obtained are in line with those obtained by Kapac-Parkačeva *et al.* (1974) and Dimitrovska *et al.* (2017). In the process of ripening, the amount of total proteins and fats gradually increases (Table **3**). The comparative research shows that statistically significant differences ($p<0.05$) in the presence of total proteins and fats have been found not only in the fully ripened cheese but also after the first and the tenth day of ripening in brine. The results obtained on the amounts of proteins and fats in the Beaten cheese are in line with those obtained by Sulejmani *et al.* (2014a, 2014b), Dimitrovska *et al.* (2017) and Talevski (2012). The dynamic of the amounts of fat in dry matter in both variants of Beaten cheese is similar, although certain statistically significant differences ($p<0.05$) were found after 10 days of ripening in brine. Some similarities have been found with the results obtained by Mijačević and Bulajić (2008), Mijačević *et al.* (2005) and Caridi *et al.* (2003). The main feature of the Beaten cheese is the high concentration of salt added in the stage of salting and in the stage of preparing the brine, put by the cheesemakers at a rough estimate. Despite that, the differences in the percentages of salt in the examined samples of beaten cheese are statistically significant ($p<0.05$) only in the stage of ripening, after 30 days (Table **3**). These results are in line with those obtained by Sulejmani *et al.* (2014a, 2014b), Dimitrovska *et al.* (2017), Talevski (2012) and Kapac-Parkačeva *et al.* (1974). The non-standardized way of salting also influences the amount of salt in moisture (S/M) which shows similarities with the results obtained by Sulejmani *et al.* (2014a), but it is higher than the results obtained by Arenas *et al.* (2004). The percentage of ash in the Beaten cheese increases in the process of ripening as a consequence of the increase in the amount of dry matter (Abdel-Razig & Al Gamry, 2009). Statistically significant differences ($p<0.05$) in the amounts of ash between both variants of Beaten cheese were noticed in the stages of ripening in brine, namely, after 10 and 45 days of ripening. The higher amount of ash is also a result of the high concentration of salt in the cheese, whereas the variations in the percentage of ash are most probably due to the higher diffusion of inorganic substances in the brine caused by the higher percentage of moisture in the cheese (Abdalla & Ahmed, 2010). The results obtained regarding the concentrations of salt in this research are similar to those obtained by Kapac-Parkačeva *et al.*, (1974), Sulejmani *et al.* (2014a, 2014b) and Dimitrovska *et al.*, (2017).

Table 3. Physical and chemical properties of Traditional Beaten Cheese ("bieno sirenje").

Ripening Days	House Hold	pH	Titratable Acidity	Moisture %	Proteins %	Fat %	MFDM %	Salt %	S/M %	Ash %
dry ripening	A	5.09-5.13 Av 5.11±0.02	60.00-64.00 Av 62.13±2.01^{a}	43.37-47.98 Av 45.54±2.31	26.33-26.51 Av 26.41±0.09	21.82-23.66 Av 22,78±0.92	40.34-43.23 Av 41.84±1.44	nd	nd	2.39-2.47 Av 2,43±0.04
	B	5.13-5.21 Av 5.17±0.04	69.60-75.20 Av 75.20±4.87^{b}	45.38-48.12 Av 46.43±1.47	24.29-25.63 Av 24.82±0.71	23.66-28.57 Av 25,62±2.60	45.43-52.30 Av 47.78±3.91	nd	nd	2.21-2.28 Av 2,25±0.03
ripening with salt	A	5.06-5.11 Av 5.09±0.03	62.40-69.60 Av 65.07±3.95^{a}	36.48-38.65 Av 37.86±1.19	26.67-29.17 Av 27.69±1.31	28.66-30.86 Av 29,73±1.10	46.56-48.58 Av 47.84±1.11	2.43-4.49 Av 3,75±1.14	6.66-11.61 Av 9.85±2.77	5.13-6.51 Av 5,94±0.72
	B	5.02-5.12 Av 5.08±0.05	77.60-80.00 Av 79.20±1.39^{b}	29.69-35.22 Av 33.36±3.18	27.99-29.91 Av 28.93±0.96	28.49-33.58 Av 30,31±2.84	43.95-47.75 Av 45.42±2.04	3.71-3.79 Av 3,76±0.04	10.53-12.76 Av 11.34±1.23	6.39-7.09 Av 6,85±0.39
1 day brining	A	5.09-5.10 Av 5.10±0.01	72.80-76.00 Av 74.77±1.72	37.26-40.31 Av 38.80±1.52	25-96-26.98 Av 26.63±0.58	26.22-28.29 Av 27,09±1.07^{a}	43.74-45.09 Av 44.25±0.72	2.87-5.24 Av 3,92±1.20	7.39-14.06 Av 10.17±3.46	6.98-7.87 Av 7,39±0.44
	B	5.05-5.14 Av 5.10±0.05	72.80-87.20 Av 78.13±7.90	36.48-38.42 Av 37.70±1.06	26.37-26.93 Av 26.59±0.29	28.64-30.38 Av 29,35±0.91^{b}	45.08-49.16 Av 47.13±2.03	4.36-5.05 Av 4,71±0.34	11.95-13.21 Av 12.48±0.65	6.72-7.20 Av 6,92±0.24
10 days brining	A	5.09-5.17 Av 5.15±0.07	62.70-68.80 Av 66.23±3.16^{a}	38.26-40.16 Av 39.08±0.97	24.22-25.90 Av 24.82±0.93	26.31-27.71 Av 26,96±0.70^{a}	43.89-44.88 Av 44.25±0.54^{a}	5.41-6.68 Av 6,01±0.64	13.90-16.63 Av 15.35±1.35	7.39-7.86 Av 7,56±0.25^{a}
	B	5.10-5.16 Av 5.13±0.03	80.00-84.80 Av 81.87±2.57^{b}	36.93-38.82 Av 37.91±0.94	24.54-28.09 Av 26.81±1.96	28.38-30.04 Av 29,31±0.85^{b}	45.76-48.21 Av 47.20±1.27^{b}	5.04-6.27 Av 5,59±0.62	13.26-16.97 Av 14.76±1.95	7.95-8.29 Av 8,13±0.17^{b}
30 days brining	A	5.14-5.20 Av 5.17±0.03	67.20-69.60 Av 68.27±1.22^{a}	37.39-38.79 Av 37.93±0.74	25.60-28.00 Av 26.81±1.20	27.32-30.07 Av 28,72±1.37	46.14-53.34 Av 49.17±3.73	530-5.86 Av 5,60±0.28^{a}	13.66-15.67 Av 14.77±1.02^{a}	7.93-8.44 Av 8,25±0.28
	B	4.96-5.18 Av 5.10±0.12	81.60-88.00 Av 85.33±3.33^{b}	37.53-39.48 Av 38.38±0.99	24.59-27.96 Av 26.80±1.91	26.61-28.26 Av 27,61±0.88	43.00-46.69 Av 44.82±1.84	6.90-7.34 Av 7,14±0.22^{b}	18.16-19.24 Av 18.59±0.57^{b}	8.62-8.82 Av 8,69±0.11
45 days brining	A	5.18-5.29 Av 5.22±0.06	74.40-78.40 Av 76.27±2.01	37.08-40.66 Av 38.86±1.79	25.70-26.24 Av 25.90±0.29^{a}	26.72-30.34 Av 28,17±1.91	43.61-49.59 Av 46.08±3.12	5.20-6.51 Av 6,07±0.75	14.02-16.76 Av 15.59±1.41	6.99-7.47 Av 7,19±0.25^{a}
	B	5.03-5.18 Av 5.13±0.06	73.60-83.20 Av 78.93±4.84	37.00-39.07 Av 38.36±1.18	27.31-28.26 Av 27.78±0.47^{b}	26.73-30.83 Av 28,11±2.35	43.82-48.94 Av 45.57±2.91	6.52-6.72 Av 6,62±0.10	16.99-17.62 Av 17.27±0.31	8.92-9.94 Av 9,28±0.57^{b}

Average values±SD within the column with no common superscript letters differ ($p<0.05$)
nd – not determined

Microbiological Properties of Beaten Cheese

The ripening of Beaten cheese, like in all other traditional types of cheese, is a result of the interaction between the rennet enzyme, the natural milk enzymes, and the naturally present lactic-acid bacteria. The dynamics of the amounts of the examined groups of micro-organisms in the sheep Beaten cheese produced in two households in the region of Mariovo is shown in Fig. (**12**).

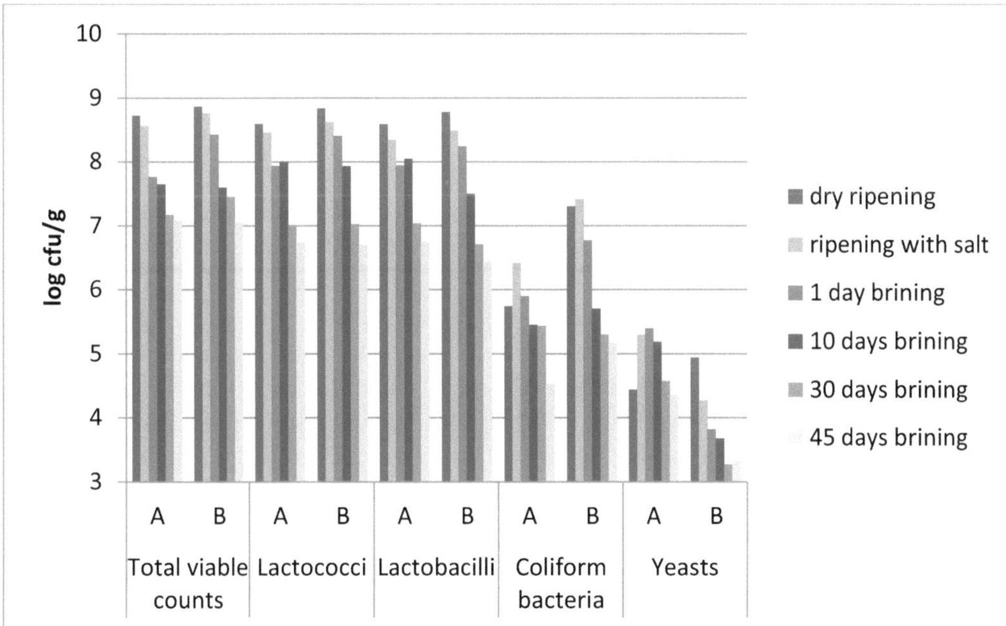

Fig. (12). Dynamic of examined microorganisms in sheep Beaten cheese.

From Fig. (**12**), it can be seen that there is a higher presence of total viable bacteria, especially in the stage of dry ripening. The traditional sheep Beaten cheese is most often made of combined milk (obtained from morning and evening milking) and ripens at room temperature (20-24°C), so the higher presence of total viable counts (TVC) is in some way within the expected. According to Levkov & Kakurinov (2012), the sheep milk used for making Beaten cheese contains a higher number of aerobic mesophilic bacteria (6.72-7.10 log cfu/g), and this, in combination with the climatic factors (the cheese is produced in summer) results in the high number of TVC in the cheese (Bonetta *et al.* 2008). A significant decrease in the total viable counts in the examined samples was noticed after the 30th and the 45th day of ripening in brine when their number was in the range of 7.05-7.45 log cfu/g. The examinations conducted by Levkov & Kakurinov (2012) show that the TVC is most probably influenced by the high concentrations of salt and the percentage of salt in moisture which after day 30 and day 45 of ripening ranged in the scope of 5.60-7.14% and 14.77-18.58%, respectively. According to Levkov *et al.* (2014), the negative coefficients of correlation ($r=-0.84$ and $r=-0.80$, $p<0.01$) for NaCl and TVC or S/M and TVC, respectively, indicates the inhibitory effect of these factors. The lower value of the coefficient of correlation between the pH values and TVC ($r=0.61$, $p<0.01$) is explained by the same authors as a consequence of the microorganisms' ability for adaptation to the lower pH values. The higher presence of TVC in the traditional cheeses was also found by Enriquez *et al.* (2016), Gaglio *et al.* (2014) and Şengül (2006). The lactic acid bacteria (LAB), as the most important for the process of ripening (Parente & Kogan 2004; Özkalp *et al.*, 2007) are present in the largest number during the stage of dry ripening of the Beaten cheese. The number of presumptive lactococci in this stage ranges from 8.59 to 8.83 log cfu/g, and of lactobacilli from 8.58 to 8.77 log cfu/g. After the stage of dry salting, a slight decrease of their number has been noticed (8.45–8.62 log cfu/g for lactococci and 8.34–8.48 log cfu/g for lactobacilli), while after 45 days of ripening in brine, the number of lactococci and lactobacilli was 6.70–6.74 log cfu/g and 6.45–6.74 log cfu/g, respectively (Fig. **12**). The obtained results are similar to those found in the examinations done by Andalla & Omer (2017), Picon *et al.* (2016), Mrkonjić-Fuka *et al.* (2013), Levkov *et al.* (2013) and Kirdar & Korsun (2011). From Fig. (**12**), it can be seen that the dynamics of the number of lactobacilli is very similar to that of lactococci, though it is well-known that the lactobacilli are more resistant to the unfavourable conditions that occur during the cheese's ripening, such as low pH, reduced amount of loose water, reduced concentration of oxygen, and high concentration of salt (Fitzsimons *et al.*, 2001; Marino *et al.*, 2003). Nevertheless, their number starts decreasing after day 30 and day 45 of the stage of brining. The study done by Levkov *et al.* (2014) shows that the number of LAB is influenced in the largest amount by the high concentration of salt, one of the main characteristics of Beaten cheese, as well as the high percentage of S/M. The authors support these conclusions with the statistical analyses regarding the coefficient of correlation between lactococci or lactobacilli and the percentage of salt or S/M. According to the analyses, the coefficient of correlation between the salt or S/M and the lactococci is $r=-0.83$ or $r=-0.81$ ($p<0.01$), respectively, while regarding lactobacilli, this coefficient is $r=-0.87$ or $r=-0.85$ ($p<0.01$), respectively. The negative coefficient of correlation between lactococci and pH in the cheese ($r=-0.46$, $p<0.05$) indicates that the lower pH value may have an influence on the decrease of the number of lactococci. Anyway, the percentages of salt and salt in moisture have a key role in the process of inhibition of these bacteria's growth. According to the examinations done by Levkov *et al.* (2014, 2017), mainly mesophilic lactic acid bacteria were isolated from the traditional Beaten cheese. Also, variations in their composition were noticed depending on the particular household where the cheese had been made.

The determination of LAB was made on the basis of their morphological, physiological and biochemical characteristics, using API 50 CHL system (BioMérieux) patterns. The largest percentage of the isolated cocci-shaped LAB belongs to the species *Lactococcus lactis* ssp. *lactis* (35%). Of the isolates, 16.7% belongs to the genus *Pediococcus*, and only 1% to *Leuconostoc*. According to Levkov *et al.* (2017), *L. lactis* ssp. *lactis* were isolated from the cheese only in the initial stages of its production (after draining and dry ripening, and after dry salting), while during the stage of ripening in brine, this species was not found in the samples taken. Members of the genus *Pediococcus* sp. were isolated from the cheese in all the stages of its ripening, whereas members of *Leuconostoc* sp. were isolated only from the curd prepared in the household A. Of the isolated lactobacilli, the species *Lactobacillus paracasei* ssp. *paracasei* (22.5%), then *L. planatarum* (15.8%) were present in the largest percentage, while the species *L. brevis* (9.2%) was isolated only from the cheese from the household A. *Lactobacillus paracasei* ssp. *paracasei* and *L. planatarum* were isolated from the curd as well as from the cheese in the stages of ripening, dry salting and ripening in brine, while *L. brevis* was isolated from the cheese in the stages of dry salting and ripening in brine. The lactic-acid bacteria's composition by species during the production of the Beaten cheese, and all the other cheeses produced in a traditional way, is subject to change. The changes in the composition by species occur because of the microorganisms themselves, namely – because of their potential for certain metabolic activities. On the other hand, the expression of this potential depends on the conditions in which the microorganisms live. These conditions are, in the first place, the properties of the milk, then the biochemical changes that occur with the formation of the curd, as well as the technology implemented in the production and the conditions that occur during the process of ripening (Montel *et al.*, 2014). According to the authors, an additional source of microorganisms in the milk and the cheese is the equipment for making the cheese, the environment where the animals are kept (*e.g.* the sheep pens, the rooms for milking), as well as the food, the silage and the hay used for feeding the animals. The lactic-acid bacteria's composition by species in the Beaten cheese corresponds with the data in the literature about traditional cheeses (Terzić-Vidojević *et al.*, 2015; Domingos-Lopes *et al.*, 2017; Li *et al.*, 2017; Enriquez *et al.*, 2016; Hayaloglu *et al.*, 2002). The coliform bacteria in both variants of sheep Beaten cheese were present in a lesser number compared to the lactic-acid bacteria (Fig. **12**). The lower pH value of the cheese during ripening, as well as the increased number of lactic acid bacteria and their higher competitiveness, affect the number of coliform bacteria in relation to its decrease (Manolopoulou *etal.*, 2003; Caridi *et al.*, 2003). It is also evident from Fig. (**12**) that there is a higher presence of this group of bacteria in the Beaten cheese variant from household B as compared with the variant from household A in the stage of ripening in brine. The number of coliform bacteria in the cheese depends in a large amount on their presence in the milk, the hygiene of the vessels and the personnel engaged in making the cheese, the temperature of ripening and keeping, the season when the production takes place, and the possibility of additional contamination especially when the cheese is produced in poor hygienic conditions (Manolopoulou *et al.*, 2003; Psoni *et al.*, 2003; Bonetta *et al.*, 2008). Although, in the process of ripening in brine, a reduction in the number of coliform bacteria was noticed, nevertheless, after 45 days of ripening their number was higher compared with the data from the examinations done by Turkoglu *et al.* (2003), Kamber & Çelik (2007), Kirdar & Korsun (2011), and Mrkonjić-Fuka *et al.* (2013). The high presence of coliform bacteria was found by Pešić-Mikulec & Jovanović (2017) in Serbian traditional cheeses, as well as by Abdalla & Omer (2107) in the cheese Gybna Bayda. The higher presence of coliform bacteria indicates that the combination of stress factors will not lead to the disappearance of these microorganisms from the Beaten cheese. Anyway, the present coliform bacteria should not be considered pathogenic microorganisms, but their presence can be considered an indicator of contamination with more detrimental bacteria (Arenas *et al.*, 2004). The number of coliform bacteria is, in the largest extent, influenced by the high concentration of salt in the cheese and salt in moisture (Levkov *et al.*, 2014). In support of this conclusion are the statistical analyses of the coefficients of correlation of the salt and of S/M, which are $r=-0.82$ and $r=-0.76$ ($p<0.01$), respectively. The presence of coliform bacteria in high amounts may result in the occurrence of defects in the cheese or it can be detrimental to the consumers' health. In order to eliminate or at least reduce this risk, it is recommendable the ripening of traditional cheeses to last for a longer period of time (60 days). The examinations for the presence of *Salmonella* sp. in sheep Beaten cheese showed results of its absence in all the production stages. Similar are the data reported by Dimitrovska *et al.* (2017) and Talevski (2012), who found no members of *Salmonella* sp. and *Listeria monocytogenes* in traditionally produced ripened cow Beaten cheese. According to Dimitrovska *et al.* (2017), in the samples of cheese produced in five different households, the presence of *Enterobacteriaceae* was found in the scope of 20–2,000 cfu/g, of *E. coli* in the scope of 10–1,200 cfu/g, and of coagulase-positive *Staphylococcus* 36–180 cfu/g. The examinations of cow Beaten cheese done by Talevski (2012) showed very similar results. Namely, in 10 examined traditionally produced cheeses, the presence of *Enterobacteriaceae* was in the range of 260-3,000 cfu/g, of *E. coli* in the range of 0–1,300 cfu/g, and of coagulase-positive *Staphylococcus* 5–110 cfu/g. Yeasts are present in the sheep Beaten cheese during the entire production process. Their presence was also found in the milk used for producing the cheese, in the amount of 4.09-4.38 log cfu/g (Levkov & Kakurinov, 2012). Their presence was higher in the variant produced in household A than in the one produced in household B (Fig. **12**), a phenomenon which, according to Manolopoulou *et al.* (2003), is nothing unusual. According to the authors, the number of yeasts can differ between the cheeses coming from two dairies, but also between the batches coming from the same dairy. This is a result of the variations in the efficiency of pasteurization, the efficiency of the cleaning agents, the additional contamination with yeasts and the hygiene during the production process itself. The analysis of the samples taken in the stage of ripening in brine shows the presence of yeasts in the cheese, which is owed to their ability to grow in extreme conditions such as lower pH, high concentration of salt, and low content of moisture (Alonso-Calleja *et a.*l, 2002; Tarakci *et al.*, 2004). Similar dynamics of the yeasts was found in the research done by Özdemir *et al.* (2010), Pešić-Mikulec & Jovanović (2005), Abdalla & Omer (2017) and Golić *et al.* (2013). The yeasts isolated from the sheep Beaten cheese in this study have been determined on the basis of their morphological, physiological and biochemical properties. The isolates belong to the genus *Trichosporon* sp., *Torulopsis* sp., *Debariomyces* sp., *Rhodotorula* sp., *Pichia* sp., *Kluyveromyces* sp. and *Candida* sp. The isolated yeasts belong to genera most often present in traditional cheeses (Montel *et al.*, 2014). Their composition by species is in line with the results obtained by Bintzis and Papademas (2002), Gardini *et al.* (2006) and Golić *et al.* (2013). In the beaten cheese, moulds have occasionally been found in various phases of the ripening stage, which is a consequence of subsequent pollution caused either from the premises where the cheese is made or the equipment used (Levkov & Kakurinov, 2013). From both variants of cheese, 11 isolates of moulds have been isolated, belonging to the genus: *Aspergillus* sp., *Fusarium* sp., *Alternaria* sp., *Curvularia* sp. and *Penicillium* sp.

TRADITIONAL KASHKAVAL CHEESE

In recent years, we have been witnessing increased consumption of traditional sheep dairy products, especially cheese produced with traditional technology (Santa & Srbinovska, 2017). Cheese made using sheep milk is very popular among consumers, primarily due to its specific sensory characteristics (Kalantzopoulos, 1993). Kashkaval is one of the most popular hard cheeses in many Mediterranean countries and its production dates back to the eleventh and twelfth century (Kindstedt *et al.*, 2004). It belongs in the group of Pasta filata cheeses, which are characterized by a unique texture which is malleable, smooth, fibrous, and sliceable. These qualities arise mainly from the cooking/stretching step which is common to all these varieties, whether they are soft, semi-hard, or hard (Fox, 2017). Pasta filata cheeses have a long tradition in most East-Mediterranean and neighbouring countries as well. In many of them, these cheeses are highly consumed and in some are the second most popular cheese group after white brined cheeses. Kashkaval-type cheeses are the best known

Pasta filata cheeses, made for centuries in this region (Alichanidis & Polychroniadou, 2008). According to Bertozzi & Panari (1993), geographical origin indication and the traditional character of the cheeses are the two most important factors which influence the selection of cheese by the consumers.

One of the traditional cheeses with a long history is the kashkaval produced on the Bistra mountain in Macedonia. The flora in this country is among the richest, not only in the Balkans but also on the entire European continent (Niketić, 2014). The Bistra mountain belongs to the group of mountains which constitute the most interesting mountain-pasture area of the Balkan Peninsula, with the most beautiful hilly pastures (Srbinovska, 2013). The dominating plants in this region are from the family of Poaceae grasses with 48.84%, then leguminous plants with 11.63%, aromatic plants with 18.60% and other species with 20.93% (Micevski, 1994). The specific climatic conditions, the hilly mountain, as well as the sheep breeding tradition are altogether excellent preconditions for the production of cheese in this region. For the production of cheese on the Bistra mountain, the milk originates from local breeds, which are adapted to the specific geographical and climatic conditions (Santa & Srbinovska, 2017). Although traditional cheeses are recognized as important for cheese producers, they are losing the battle against the industry and they might become extinct (Mijacevic *et al.*, 2005). In Macedonia, the number of traditional cheese producers is continuously decreasing (Santa & Srbinovska, 2014). Furthermore, there is not much information in the literature about the Macedonian kashkaval. Traditional kashkaval is produced in only two regions in Macedonia - Galichnik and Lazaropole, both located on the Bistra mountain. In the region where it is produced, local sheep breeds (Pramenka and cross breed) are dominant and characterized by low milk productivity which is 60-80 litres during lactation. In the summer period, sheep graze on pastures at 1500 m above sea level. The kashkaval from this region has a long history (Fig. **13**). The produced cheese was sold in all of the markets in the region and abroad. In 1927, this cheese was even exported to the US market, and this was documented in the Yugoslavian newspapers as a significant event.

The Cheese-making Process

The production process of the traditional kashkaval for both regions where it is produced is shown in the flowchart (Fig. **14**), (Santa & Srbinovska, 2014; Santa, 2015). Some of the steps are different between the two types of cheese, depending on the region.

Fig. (13). Old facility for production of kashkaval on the Bistra mountain.

Raw sheep milk

↓

Clarification - filtration

↓

Renneting (at 33-37 °C, coagulation complete in 45 min)

↓

Cutting and stirring the coagulum

↓

Whey draining

↓

Wrapping the curd with cheese cloth

↓

Pressing (3-6 hours)

↓

Cutting the curd in blocks

↓

Curd ripening (cheddaring) Galichnik- 18h 12-15C, Lazaropole 3h-30°C

↓

Cutting the curd ("baskija") in pieces (62-65°SH and pH 5,2-5,4)
size of the pieces (20x5-10x0,5-1 cm)

↓

Stretching in hot brine, plasticization (5%NaCl, 72-75°C, 5 min)

↓

Hand shaping, dry salting and moulding

↓

Removing the moulds (after 24 hours)

↓

Ripening (3months, RH 75-80%)
Galichnik – 12-15 °C, Lazaropole 15-18°C

↓

Dry salting, cleaning with cloth, wheels rotation

↓

Expose the cheese to the sun (10-15 min for each side)

↓

Packaging

↓

Storage (5 °C)

Fig. (14). Flow chart for traditional kashkaval production.

Milk Reception and Renneting

After milking on the farm, raw milk is transported to the production facility, where, again, it is filtered. In cases when the milk is processed once a day, it is cooled to a temperature lower than 6°C. The renneting is carried out by adding powdered chymosin in an amount which enables coagulation of the milk for 45-60 minutes at 33-38°C and the initial coagulation should occur 10 minutes after adding the enzyme. In cases when the milk is cooled, a second heating treatment is performed. This is usually done when a certain amount of milk is heated at a temperature of 65 to 75°C, and then the heated milk is well mixed with the remaining amount of milk. The cheesemaker assesses the curd if it has enough firmness and elasticity and then starts the production. The curd is cut and stirred for around 30 min with a special wooden tool called "krstach" and then rested for about 10 minutes (Fig. **15**). Once the required size of the curd is uniform and given some time to rest, the curd falls to the bottom and the whey is separated on the surface and drained from the vat (Fig. **16**). Then the curd is placed on the cheesecloth on the table, and the whey is drained from the table.

Fig. (15). Cutting and stirring the coagulum with "krstach".

Fig. (16). Draining of the whey from the vat.

Then the fresh curd is chopped by hand and this is very quick (5 min) so that the curd does not cool. The aim of the chopping is to remove the remaining whey from the curd. The curd is wrapped in a cheesecloth and pressed using 12-15 kg per 1 kg of curd. The duration of the pressing process is 3-6 hours, depending on the room temperature. The pressed curd is called "baskija" or "teleme" which is usually in blocks of 25 cm width (Fig. **17**). For the cheese from Lazaropole, the cheesemaker places hot water over the pressed "baskija" so it does not cool.

Fig. (17). Baskija.

Curd Ripening (Cheddaring)

During the pressing, the whey continues to be drained both from the surface and from the inside of the "baskija". This whey contains a higher percentage of milk fat. After the process of pressing, the curd ("baskija") is cut into smaller pieces weighing 5-8 kg, with width and height of 10-15 cm and length of 25-30 cm. The curd pieces are covered with a cheesecloth and a polyethylene cloth and the process of curd ripening called "cheddaring" starts. The process is usually extended at lower temperatures (not lower than 10°C) for around 20 hours (Galichnik), or at warm temperatures of 30-35°C for 3-4 hours.

Stretching the Curd in Hot Brine (Plasticization)

Cheddared curd contains small holes, an elastic consistency, a light yellow colour and a pleasant smell with optimum pH value between 5.2-5.4 or 62-66 °SH. According to the cheese maker's decision, the production continues when a small piece of cheese is soaked in hot brine and stretched well. Then the curd ("baskija") is sliced into pieces (20 × 5-10 × 0.5-1 cm) (Fig. **18**) and placed in a wooden basket by immersing the basket in the hot brine (5% NaCl, 73-75 °C) for about 5 min (Fig. **19**). The cheesemaker manually conducts the mixing and stretching of the curd, and according to Kosikowski (1982), this is the hardest and most intense work required in the cheese-making process. The most characteristic result of mixing and moulding is the removal of some water from the fresh curd and the transformation of the rough, short-textured curd into a smooth, malleable, long-grained, satiny plastic mass.

Fig. (18). Baskija sliced into pieces.

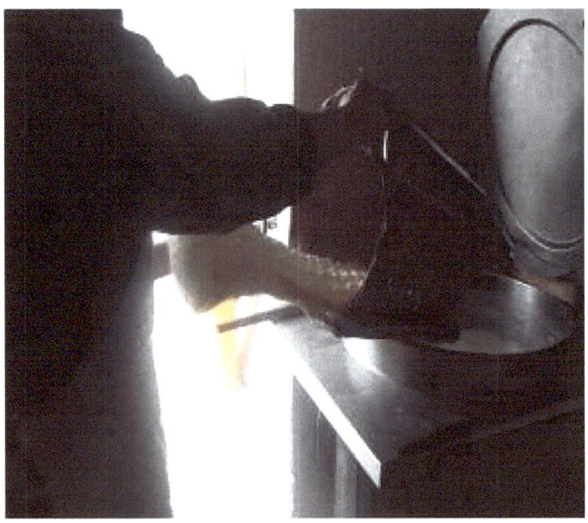

Fig. (19). Stretching the baskija.

Moulding

The cheese dough is placed on the table and kneaded carefully for about 10-15 minutes until a compact elastic mass is obtained and air bubbles are eliminated.

According to Kindstedt *et al.* (2004), texturing has an additional advantage, *i.e.* a pasteurizing effect which suppresses undesirable microbial growth and encourages desirable fermentation and ripening, resulting in high-quality cheese. In the process of kneading the curd (around 5-10 minutes), the cheesemaker applies dry salt. During moulding, the cheese is carefully closed like a mushroom and a small piece of cheese called "navel" ("papok") is pulled out in order for the dough to completely fill the mould (Fig. **20** and Fig. **21**).

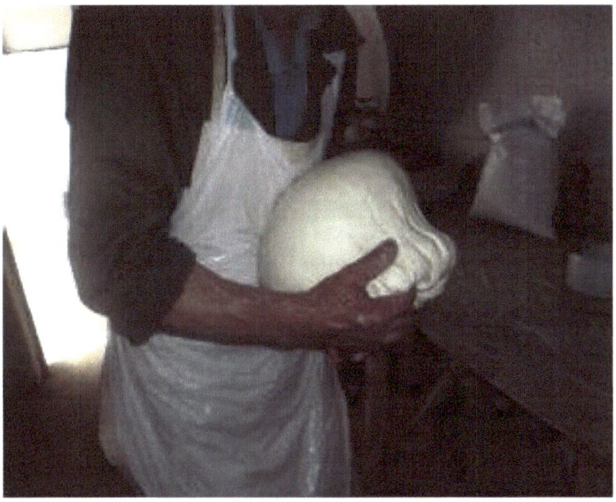

Fig. (20). Molding the cheese.

Fig. (21). Cheese in mold.

Ripening

In order to remove the air from the cheese, the cheesemaker carefully pinches the cheese with metal needles of 2 to 3 mm diameter. Within 24 hours, the mould with cheese has been turned over several times and then it is removed from the cheese. The cheese is dry salted several times (5-7 times) during the period of 20-30 days. The dynamic of the salting depends on the cheesemaker.

Ten days after the final salting, each wheel is brushed in warm water (35-40 ° C) (Fig. **22**) and then dried for seven to eight days (Simonovska, 2018). The drying temperature should not exceed 22 ° C, and the optimum temperature is 15-18 °C. One of the specific characteristics of the Bistra mountain kashkaval is that the cheese is exposed to the sun. The process of "sunning" takes around 10-15 minutes on each side where the surface softens and fat globules appear on the cheese (Figs. **21** & **23**) (Simonovska, 2018). The cheese is placed on the shelves for ripening one over another to a maximum of four wheels and they are each rotated periodically (Fig. **24**). The ripening period is for the duration of three months at a temperature of 12-15 °C (Galichnik) and 15-18°C (Lazaropole). The ripened kashkaval acquires the shape of a flattened cylinder with a weight of 3-3.5kg, 5-6kg or 8-9 kg. It is usually packed in vacuum polymer foil.

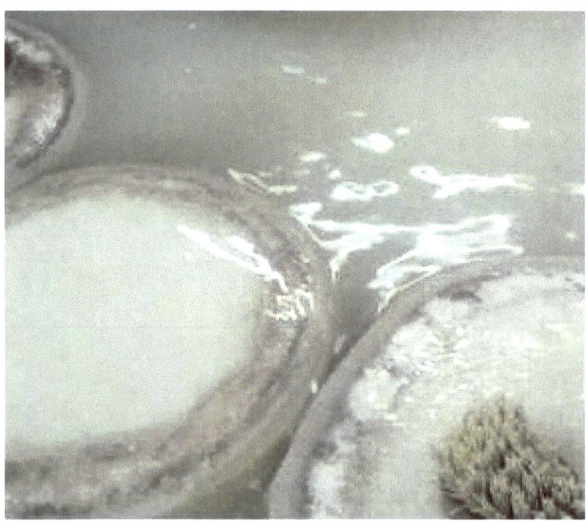

Fig. (22). Washing the cheese.

Fig. (23). Expose the cheese to sun.

Fig. (24). Cheese ripening.

Physical and Chemical Properties of the Traditional Kashkaval

The cheese quality depends on its composition, especially the moisture and dry matter, salt, pH, and the percentage of fat in dry matter (Fox, 2000). The dynamics of the physicochemical characteristics of the traditional kashkaval from Galicnik during the ripening were reported by Santa & Srbinovska (2017). In Tables **4** and **5**, the average results for both kashkaval cheeses are presented, for the regions of Galichnik and Lazaropole. The ripening of cheese resulted in the decrease of the pH value and the increase of protein, fat, ash and NaCl. The difference in the fat content of both cheeses occurs due to the difference in the composition of the raw material and the processing of the cheese itself. In the ripening process, there is a significant reduction in moisture content. The results of the dynamics of moisture during the cheese ripening and storage show that on the 180th and 360th day, there was a significant decrease in the moisture content. Similar results were reported by Niro and *et al.* (2014) in examining the cheese Caciocavallo, where the moisture content on the day of production was 45.4%, while on the 60th day of ripening it decreased to 39.8%. Hattem *et al.* (2012), in their study on the hard cheese Ras made from unpasteurized cow milk, reported lower results on the moisture content which ranged from 40.28% on the first day to 31.26% after the 6th month of ripening. The difference in results is due to the different origin of milk.

Table 4. Physicochemical characteristics of the kashkaval from Galichnik (A) and Lazaropole (B).

Time (days)	Moisture (%)		Fat (%)		Protein (%)	
	A	B	A	B	A	B
1	46,48e[1]	45,81b	20,75a[1*]	25,38a*	11.88a[1*]	18,78a*
30	43.8de	41,13b	21,75ab*	27,25ab*	12.06a*	23,77b*
60	39,21cd*	32,68a*	24,38bc*	32,25c*	21.03b*	25,83bc*
100	36,64bc	32,63a	27,88cd	30,00b	23,05bc*	26,65cd*
180	32,92ab	31,20a	29,00d*	33,00c*	25,50cd	27,62cd
360	27,54a	27,98a	31,40d	32,50c	28.41 d	28,39d

[1]Means in the same column with different letters are significantly different ($p < 0.05$)
*Means in the same period for different variants are significantly different ($p < 0.05$)

Table 5. Physicochemical characteristics of kashkaval from Galichnik (A) and Lazaropole (B).

Time (days)	NaCl (%)		Ash (%)		Acidity (pH)	
	A	B	A	B	A	B
1	1,20a[1]	0,79a	3,54a[1]	3,57a	5,47b[1]	5,61b
30	1.29a	1,45ab	4,01ab	4,50ab	5,42ab	5,26a
60	1,48ab	1,84abc	4,24abc	5,23bc	5,35ab*	5,14a*
100	2,02abc	2,17bc	4, 85bcd	5,27bc	5,34ab	5,16a
180	2,44 bc	2,84cd	5,43cd	6,49cd	5,28ab	5,15a
360	2,90c	3,94d	5,81d*	7,62d*	5,26a	5,29a

[1]Means in the same column with different letters are significantly different ($p < 0.05$)
*Means in the same period for different variants are significantly different ($p < 0.05$)

The results on the fat content correspond with previous studies on kashkaval from different regions. According to Kindstedt *et al.* (2004), the content of milk fat in Bulgarian cheese is 30%, in Greek cheese it is 33.8%, and in Turkish cheese it is from 25.5% to 31.4%, in Egyptian cheese it is from 21.1% to 26.3%, while the content of fat in the countries of former Yugoslavia ranged from 27 to 32%. Also,

Mijačević *et al.* (2005) in their study on Pirotski kashkaval got similar results for the fat content which ranged from 28.8% to 31.1%.

Furthermore, Gobbeti *et al.* (2002), examining the dynamics of the chemical composition during ripening of the Caciocavallo Pugliese produced from cow milk, reported that the fat content was 24%, which is an expected lower value than the one in sheep cheese. Similar results for the fat content in cheese were detected in the Pecorino cheese (Coda *et al.*, 2006) and traditional Italian cheeses (Di Cagno *et al.*, 2007). Dry salting, which is a part of the technology process of kashkaval, leads to a higher loss of water and the cheese where dry salting is used is usually characterized by greater firmness (Jovanović, 2001). The degree and the method of salting is different for different cheeses and it depends on the technology process, the salting technique, the consumer's preference, etc. Similar findings for salt content were reported by Kindstedt *et al.* (2004) for the kashkaval from the former Yugoslavian area. It is the cheese-maker's professional skills that ensure the various phases in the cheese making process are carried out correctly. He relies on his years of experience and training when making these decisions (De Roest, 2000).

Free Fatty Acid Profile of Traditional Kashkaval

Lipids in foods may undergo hydrolytic or oxidative degradation. Although some lipolysis occurs in most or all cheeses, it is most extensive in some hard Italian varieties and in blue cheese (McSweeney, 2004). While short-chain fatty acids contribute directly to the cheese flavour, free fatty acids (FFAs) also contribute indirectly to the cheese flavour by acting as precursors for the production of volatile flavour compounds through a series of reactions collectively known as the metabolism of fatty acids. Lipolysis in kashkaval cheese is not very intense. According to Kozev (2006), the kashkaval made from cow milk has low fat hydrolysis and thus does not have a very intensive taste and aroma. More evident and intensive hydrolysis in cheese is noticed in the kashkaval from sheep milk. Cheeses made from raw milk or heat-treated milk in sub-pasteurisation conditions tend to contain higher amounts of FFAs, because the indigenous milk lipase remains intact or is not entirely inactivated, respectively. The major FFA acids present in the kashkaval from the Bistra mountain are palmitic (C16:0), oleic (C18:1) and myristic acid (C14:0) (Table **6**).

The content of butyric acid (C4:0) corresponds to the results obtained by Gobbetti *et al.* (2002) in the Caciocavallo Pugliese cheese. The results of our research are similar to the results of the research by Woo *et al.* (1984) on the concentration of butyric acid (C4:0) in Parmesan and Provolone. A similar fatty acid profile was reported by Ergönül *et al.* (2011) on Haloumi cheese.

Table 6. Free fatty acid profile of kashkaval (n=4), g•100g^{-1} of the cheese (kashkaval) from Galichnik (A) and Lazaropole (B).

FFA g•100g^{-1}	A	B	FFA g•100g^{-1}	A	B
C4:0	2.09±1.98	3.32±0.06	**C17:0**	0.51±0.04	0,61±0,1
C6:0	/	2.66±0.29	**C17:1**	0.22±0.01	0,67±0,06
C8:0	0.83±0.18	1.45±0.29	**C18:0**	7.68±2.43	8,79±0,67
C10:0	6.46±0.60	5.08±0.46	**C18:1 c**	24.89±1.05	23,60±0,32
C11:0	/	0.28±0.05	**C18:1 t**	/	/
C12:0	5.20±1.28	4.87±0.59	**C18:2 t**	0.47±0.05	0,58±0,05
C13:0	/	0.13±0.03	**C18:2 c**	2.12±0.99	2,33±0,07
C14:0	11.01±0.69	10.82±0.46	**C20:0**	0.44±0.10	0,05±0,11
C14:1	0.26±0.10	0.30±0,08	**C18:3 n-6**	0.92±0.29	0,62±0,05
C15:0	1.35±0.13	0.40±0,07	**CLA**	4.13±0.74	3,28±0,34
C15:1	0.28	/	**C18:3 n-3**	0.85±0.22	1,11±0.11
C16:0	30.57±4.17	26.71±1,36	**C21:0**	1.66±0.08	/
C16:1	/	2.27±0,12	**C20:4 n-6**	0.38±0.01	0,34±0.57

PROTEOLYSIS AND VOLATILES OF MACEDONIAN TRADITIONAL CHEESES

Kashkaval is a Pasta-filata type of cheese of large popularity after the white-brined cheese (Belo sirenje) in Macedonia (Sulejmani *et al.*, 2014a). Its production dates back to the 11th and 12th century, however, historical references suggest that it has an even older tradition (Santa & Srbinovska, 2014). Sulejmani *et al.* (2016) evaluated the differences in the heat treatment effect (60, 70 and 90 °C for 5 min) in several biochemical and technological properties of the Kashkaval cheese. The chemical composition of 1 day old Kashkaval cheeses ranged between 0.2–0.3% for titratable acidity (as g of lactic acid per 100 g of cheese), 58.2–60.8% for total solids, 23.9–25.2% for protein, 29.7–31.3% for fat, and 0.7%–0.9% for salt. The use of a different stretching temperature significantly affects the pH, the soluble nitrogen fractions and the colour parameters of the cheeses ($P < 0.05$) (Table 7). Urea-PAGE patterns of the water–insoluble fractions of the cheeses after 30, 60, 90 and 120 days of ripening are presented in Table 7. During the ripening of the Kashkaval cheese, as1-casein (f24-199) and c-caseins were produced early, indicating chymosin and fairly high plasmin activity (Kindstedt 1993). However, it was seen that the as1-casein hydrolysis pattern was higher than that of b-casein during all stages of ripening. As a consequence of as1-casein hydrolysis, the bands corresponding to as1-I-casein (as1–casein f102–199) were observed after 60 days of ripening in K60 and K70 cheeses. Other bands characterized by slower mobilities than as1-I-casein (as1-CN f24–199) are shown in 120 days-old K60 and K70 cheeses (Fig. **25**). These may be produced by the actions of the residual coagulant and the native milk proteinases.

Table 7. The levels of pH, water-soluble nitrogen (WSN), 12% Trichloroacetic acid-soluble nitrogen (TCA-SN), total free amino acids (TFAA) of Kashkaval cheeses stretched at different temperatures (60, 70 and 90 °C).

	Cheeses	1	60	90	120
pH	K60	$5.21{\pm}0.01^{a}$	$5.09{\pm}0.02^{aA}$	$5.14{\pm}0.02^{A}$	$5.18{\pm}0.01^{A}$
	K70	$5.29{\pm}0.04^{ab}$	$5.08{\pm}0.01^{aA}$	$5.24{\pm}0.02^{B}$	$5.15{\pm}0.03^{AB}$
	K90	$5.43{\pm}0.03^{b}$	$5.29{\pm}0.10^{aB}$	$5.17{\pm}0.09^{AB}$	$5.21{\pm}0.07^{A}$
WSN	K60	$3.95{\pm}0.10^{ab}$	$7.79{\pm}0.32^{aA}$	$9.54{\pm}0.05^{cB}$	$10.69{\pm}0.02^{bC}$
(% of TN)	K70	$3.99{\pm}0.22^{b}$	$7.55{\pm}1.04^{aAB}$	$9.11{\pm}0.12^{bAB}$	$10.47{\pm}0.04^{bB}$
	K90	$2.73{\pm}0.56^{a}$	$5.38{\pm}1.34^{aA}$	$6.39{\pm}0.06^{aA}$	$9.00{\pm}0.08^{aB}$
TCA-SN	K60	$1.73{\pm}0.17^{a}$	$4.46{\pm}0.21^{bB}$	$5.41{\pm}0.03^{cBC}$	$6.55{\pm}0.56^{bC}$
(% of TN)	K70	$1.48{\pm}0.02^{a}$	$4.11{\pm}0.67^{abAB}$	$4.94{\pm}0.06^{bB}$	$6.51{\pm}0.02^{bC}$
	K90	$1.39{\pm}0.09^{a}$	$2.38{\pm}0.13^{aB}$	$3.00{\pm}0.03^{aC}$	$4.43{\pm}0.04^{aD}$
TFAA	K60	NA	$4.37{\pm}1.51^{bAB}$	$4.61{\pm}1.10^{bB}$	$3.53{\pm}0.04^{cAB}$
mg Leu/g	K70	NA	$2.69{\pm}0.24^{abB}$	$3.45{\pm}0.20^{abC}$	$2.50{\pm}0.01^{bB}$
	K90	NA	$0.96{\pm}0.05^{aAB}$	$1.11{\pm}0.13^{aBC}$	$1.35{\pm}0.05^{aC}$

$^{a\text{-}c}$Means ± SD with different letters were significantly different among cheese samples within the column of the similar ripening period ($P < 0.05$). $^{A\text{–}D}$ Means ± SD with different letters were significantly different among storage periods within the rows of each cell ($P < 0.05$). NA: Not analyzed. (*Adapted from Sulejmani et al. 2016*)

After 30 days of ripening, K60 and K70 cheeses exhibited pronounced proteolysis which resulted as 1-I-casein peptide (as1-CN f110-199) (Sulejmani *et al.,* 2016). From c-caseins (the polypeptides generated by the activity of plasmin), the c3- (f108–209) and c1- (f29–209) caseins were present in the highest amounts, attended by c2-casein b-(f106–209). Also, Sulejmani *et al.* (2016) studied the residual coagulant activity in the Kashkaval cheese, ranged between 7.1 and 71.6%. During ripening, K60 and K90 cheeses showed the highest and lowest residual coagulant activity, respectively. The mean residual coagulant activity of Kashkaval cheeses were higher than other types of cheeses reported in the literature (Roa, Lopez, & Mendiola 1999). Similarly, Boudjellad *et al.* (1994) observed a higher amount of residual coagulant activity (86%) in a Swiss type of cheese.

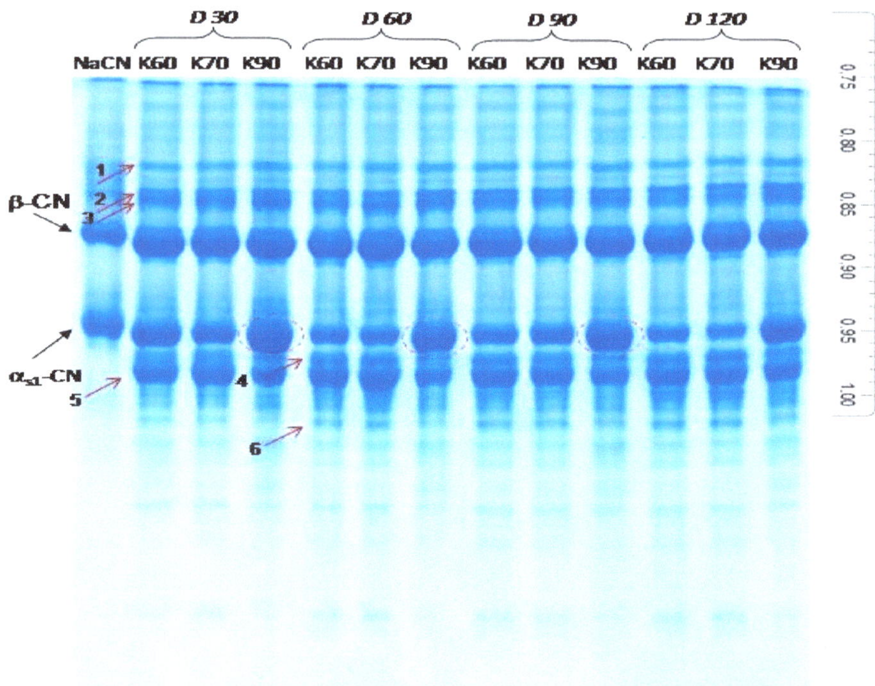

Fig. (25). Urea-polyacrylamide gel electrophoretograms (left) of the water-insoluble fractions of Kashkaval cheeses made using different stretching temperatures (60, 70 and 90 _C) during 120 days of ripening. Casein bands, 1: c2-CN (b f 106–209), 2: c1-CN (b f 29–209), 3: c3-CN (b f108–209), 4: as1-CN (f24-199), 5: as1–CN (f 102–199), 6: as1-CN (f 110–199) (*Adapted from Sulejmani et al. 2016).*

Sulejmani *et al.* (2016) emphasized that the volatile compounds present in the volatile fraction of Kashkaval cheeses consisted of 8 acids, 13 esters, 11 alcohols, 24 hydrocarbons, 7 ketones, and 3 other groups. Acids and esters composed the main chemical classes (mean concentration of 57.1% and 26.8% w/w of total volatile compounds, respectively). Statistically significant differences for thirty-eight and thirty-two volatile compounds were determined in the terms of storage and type of production, respectively. As most of the volatile compounds increased, a few decreased and new compounds were formed during maturation. Compared with the 1st day, an important increase in the total amount of volatile compounds was found after 120 days of ripening. Carboxylic acids are one of the principal volatile groups in Kashkaval as all sampling times and stretching temperatures influenced the volatile concentration (Sulejmani *et al.*, 2016). Butanoic acid was the most important acid found in the headspace of Kashkaval cheese. The most noticeable differences between the Kashkaval cheeses were in the concentration of butanoic acid, which was identified at higher levels in K60 cheese samples. However, the content of acetic acid significantly increased ($P < 0.05$) after 60 days of ripening and it was the second main compound isolated in the headspace of Kashkaval cheeses scalded at 90 °C. Esters were the second most abundant volatile group isolated in the headspace of Kashkaval, with a percentage of 26–28% of total volatiles. Among the esters, 7 ethyl esters, 5 methyl esters and 1 methyl propyl ester were detected. Lower levels of heat treatment of the curd on the first day in K60 cheeses increased the levels of total FAAs, which are one of the precursors of ethanol by Strecker degradation (Sulejmani *et al.*, 2016). Hydrocarbons were the largest group of volatiles (24 compounds) identified in the headspace of Kashkaval cheeses. These compounds do not directly make a considerable contribution to the aroma of Kashkaval cheese, as these compounds have high threshold values.

PROTEOLYSIS AND VOLATILES OF BEATEN CHEESE

Beaten cheese ("bieno sirenje") is a traditional autochthonous type of cheese produced in Macedonia and other Balkan countries with a relatively high nutritional value and performance features (Sulejmani *et al.*, 2014a). The traditional cheese-making procedure is still used in farms and villages. The scalding procedure is a crucial step in the production of Beaten cheese and it substantially affects the physical, biochemical and sensory properties of the cheese (Dubrova Mateva *et al.*, 2008).

Raw sheep milk is used in the production of Beaten cheese by the traditional method, however, milk is pasteurized in large-scale dairies. The raw/pasteurized sheep milk used for the manufacture of Beaten cheese was coagulated by using animal rennet (Chr. Hansen) at a level of 20 mL/100 L. The milk was allowed to coagulate at 34°C for 45 min. Following the coagulation, cheese curd was mechanically stirred and beaten 40 to 50 times with a special metal tackle (horizontally and vertically knitted metal wires) for about 5 min, and then left for 30 min to facilitate the drainage of whey. The traditional cheese curd is usually stirred and beaten manually with a wooden stick until a homogenous compact structure is obtained. After the drainage, the curd was cooked in hot water between 70°C to 90°C for 20 min. Following the cooking, the curd was again beaten for 2 to 5 min. and then left to rest for 15 to 20 min. Following the rest period, the curd was molded into balls, which were then each transferred to a cotton cloth to hang by their own weight or by a weight pressure of 3 kg/kg for about 24 h. Then, the curd was pre-ripened for 3 days at 35°C or in sunlight for the traditional way of production. During this period, the curd gained a specific yellowish colour due to the proportional increase of β-carotene among casein micelles (Fig. **26**). Then, the cheese loaf was cut in pieces 30 to 40 cm long and 4 to 5 cm wide. The cheese pieces were salted by the dry-salting method for 24 h and then followed by the brine salting. The pieces were then collected and packed in plastic barrels containing brine with a concentration of 18 to 20% (wt/vol) NaCl for long-term storage. Sulejmani *et al.* (2014a) reported that the total FAA contents were the highest in the industrial Beaten cheeses, probably due to the lower scalding temperature used in industrial production compared with the traditional method. The level of total FAA reached 1.48 mg of Leu/g of cheese on day 180 in industrial cheeses. However, it was 1.18 mg of Leu/g of cheese for traditional cheeses during the same sampling time. The level of total FAA in Beaten cheeses showed a similar trend to the soluble nitrogen fractions, including WSN-SN, during ripening, as reported by other researchers (Andic *et al.*, 2010). Some differences were observed during the ripening for the fractions of peptides, which were eluted from industrial cheeses at higher levels than the ones eluted from traditional cheeses (Sulejmani *et al.*, 2014a). Increases were observed in the concentrations of peptides that were mainly eluted between min 20 and 40 from the industrial cheeses (Fig. **27**). In the chromatogram, between min 60 and 66, the peak heights in traditional cheeses were generally much higher than in industrial cheeses. In the case of traditional cheeses, the particular peaks with retention time between min 22 to min 26 appeared and varied strongly, depending on the type of cheese and ripening process. However, the pattern and extent of proteolysis varied due to differences in the manufacturing practices (industrial and traditional methods), which caused differences in the moisture content and residual coagulant activity. Peaks that appeared at a retention time of min 60 to min 69 corresponded to large hydrophobic peptides, which are strongly correlated with the bitterness of cheeses (Bumberger & Belitz, 1993), and were higher mostly in industrial cheeses. Sulejmani *et al.* (2014b) extensively investigated the volatile compounds in beaten cheeses obtained from different local retailers in Macedonia, in 2012 (Table **8**). These compounds belonged to different chemical families including acids, ketones, esters, and alcohol. A total of 11 acid compounds were detected, and the main acidic compounds in Beaten cheese were butanoic, hexanoic, octanoic and decanoic acids. The principal acid within this group was butyric and its concentration was about 59.5 mg/kg per cheese. Butyric acid is produced principally by the catabolism of lactic acid due to butyric acid fermentation. The reduction of salt in dry matter (SDM) in Beaten cheese is claimed to significantly increase the susceptibility to butyric acid fermentation, which was also reported for Gouda cheese (Van den Berg *et al.* 1986).

Ketones were the second most abundant compounds isolated in beaten cheeses. In total, 11 ketones were identified; 2-butanone, 2-pentanone, and acetoin were the most abundant ketones in Beaten cheese samples. Acetoin originates from the metabolism of citrate as a reduction of diacetyl by the action of lactic acid bacteria, and they are very important compounds for dairy products. Also, they

Fig. (26). Beaten sheep milk cheese with visible holes (*Adapted from Sulejmani et al., 2014b).*

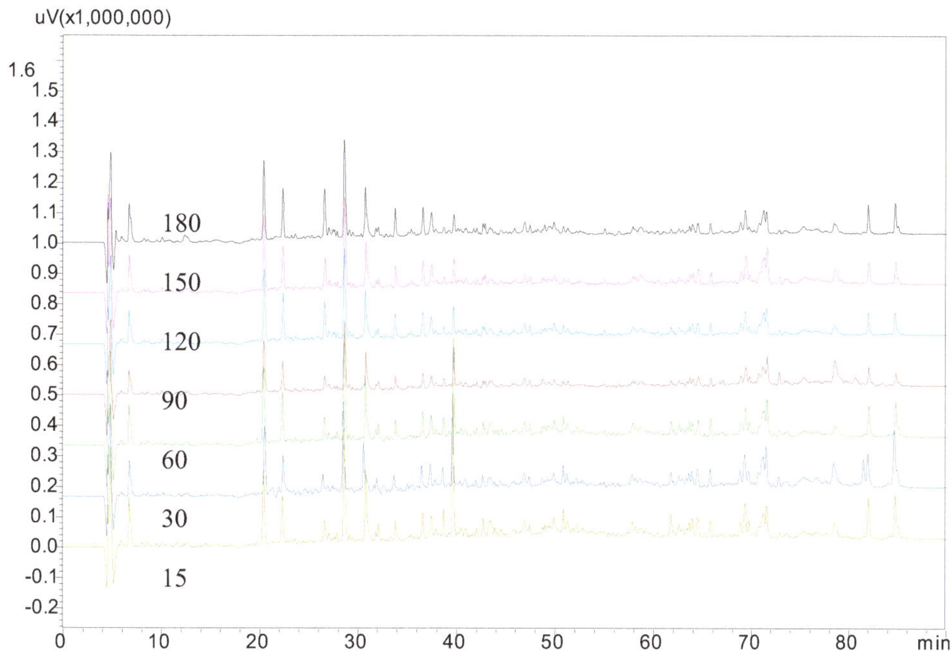

Fig. (27). Reversed-phase HPLC profiles of the water-soluble fraction of industrial beaten sheep milk cheese during ripening. Detection was at 214 nm. mAU = milliabsorbance units. *(Adapted from Sulejmani et al. 2014a).*

have very particular odours and low perception thresholds (McSweeney & Fox, 2004). Concentration of acetoin detected in Beaten cheeses was about 650 μg/kg cheese. A total of 16 ester compounds with a total concentration of 2731 µg/kg were identified in Beaten cheese. Ethyl butyrate (1505 µg/kg), ethyl caproate (806 µg/kg) and ethyl octanoate (121 µg/kg) were the most abundant esters in Beaten cheeses. Sulejmani & Hayaloglu (2018) extensively investigated the effect of heat treatment of cheese milk on the biochemical and volatile properties of white brined cheese made using goat milk, during its maturation. The raw goat milk (GR) cheeses were characterised by a higher amount (78%) of total volatile compounds than the pasteurized goat milk (GP) cheeses, during maturation. Compared to day 1, an important decrease in the total amount of volatile compounds (except ketones and alcohols) was found after 120 days of maturation.

Table 8. Volatile compounds (µg/kg cheese) in beaten sheep milk cheeses (*n=12*).

	Aroma Characteristics	RI	Mean± SE
Butyric acid	Cheesy	838	59484±17165
3-Methyl butanoic acid	Cheesy, sour	891	3641±983
2-Methyl butanoic acid	Acidic, dirty,	923	410±48
Hexanoic acid	Sweaty, sour	1016	27618±7453
Octanoic acid	Fatty, waxy	1192	13704±3097
Nonanoic acid	Fatty, waxy	1267	11±2
Decanoic acid	Rancid, sour	1369	6619±1426
Dodecanoic acid	Fatty	1558	256±74
Myristic acid	Waxy, soapy	1757	4±1
Oleic acid	Fatty, vegetable oil	1939	46±12
Palmitic acid	Waxy, soapy	1958	206±31
Total acids			***111999***
Ketones			
Acetone	Solvent	<600	371±135
2-Butanone	Etheral, fruity	<600	1419±695
2-Pentanone	Sweet, banana-like	679	1227±304
Acetoin	Buttery, creamy	708	650±177
2-Hexanone	Etheral	791	52±12
2-Heptanone	Fruity, spicy	893	563±179
8-Nonen-2-one	Fruity baked	1084	41±1
2-Nonanone	Dirty, earthy	1092	36±18
Delta octalactone	Coconut, creamy	1288	3±1

(Table 8) cont.....

	Aroma Characteristics	RI	Mean± SE
2-Undecanone	Fruity, creamy cheese	1292	2±1
Deltadecalactone	Sweet, coconut	1501	7±1
Total ketones			***4371***
Esters			
Ethyl acetate	Fruity, grape	613	30±8
Ethyl butyrate	Fruity, apple	802	1505±383
Isoamyl acetate	Banana	884	45±10
Methyl caproate	Fruity, pineapple	928	51±5
Ethyl caproate	Fruity, green banana	998	806±195
Ethyl heptanoate	Pineapple, fruity	1096	10±5
Methyl octanoate	Fruity	1123	15±5
Ethyl octanoate	Fruity	1194	121±51
Phenylethyl acetate	Rosey	1260	22±0
Propyl octanoate	Coconut	1285	5±2
Isoamylbutyrate	Fruity, melon	1058	69±3
Butanoic acid butyl ester	Sweet, fatty	1373	6±1
Ethyl decanoate	Waxy, fruity	1392	36±9
Ethyl palmitate	Faty, waxy, balsam	1590	6±0
Ethyl laurate	Waxy, soapy	1612	3±1
Ethyl myristate	Sweet, waxy	1791	1±0
Total esters			2731
Alcohols			
1-Butanol	Fusel oil	653	784±276
Isoamylalcohol	Whiskey, fruity	738	599±98
2,3-Butanediol	Buttery, creamy	787	104±50
2-Heptenol	Green grass	896	96±37
2,6,-Dimethyl-4-heptanol	Fermented, yeasty	953	175±58
1-Octanol	Waxy, green	1073	12±4
Phenylethylalcohol	Rosey	1116	46±12
Alpha terpineol	Floral	1424	7±0
Total alcohols			***1823***

(Table 8) cont.....

	Aroma Characteristics	RI	Mean± SE
Terpenes			
Alpha pinene	Turpentine	936	56±17
3,7-Dimethyl-octa-1,6-diene	-	948	83±18
(Z)-3,7 Ddimethyl-2-octene	-	971	61±23
Carene	Citrus	996	2043±420
Phellandrene	Citrus, terpenic	1003	332±102
Carvomenthene	-	1023	80±24
p-Cymene	Terpy, woody	1026	60±6
D-Limonene	Limon, citrus	1031	1±0
Gamma terpinene	Terpy, lime-like	1062	1±0
Copaene	Woody	1385	7±1
Trans caryophyllene	Spicy	1432	4±1
Delta cadinene	Thyme herbal	1530	1±0
Total terpenes			***2729***

MS, mass spectral identification using NIST 07 and Wiley 7 library; RI, identification using alkane series (C10–C26) by calculation of retention indices at the same chromatographic conditions. S.E: Standard error. ; n, number samples occurred the corresponding compounds. RI: Retention index based on P-5 MS column. (Adapted from Sulejmani *et al.* 2014b).

CONSENT FOR PUBLICATION

Not applicable.

CONFLICT OF INTEREST

The authors confirm that this chapter contents have no conflict of interest.

ACKNOWLEDGEMENTS

Declared none.

REFERENCES

Aarestrup, FM, Butaye, P & Witte, W (2002) Non-human reservoirs of enterococci.*The Enterococci: Pathogenesis, Molecular Biology and antibiotic resistance* AMS Press, Washington, District of Columbia 55-100.
[http://dx.doi.org/10.1128/9781555817923.ch2]

Abdalla, MOM & Ahmed, OI (2010) Effect of heat treatment, level of sodium chloride, calcium chloride on the chemical composition of white cheese. *Research Journal of Animal and Veterinary Sciences,* 5, 69-72.

Abdalla, MOM & Omer, HEA (2017) Microbiological characteristics of white cheese (Gibna bayda) manufactured under traditional conditions. *Journal of Advances in Microbiology,* 2, 1-7. [http://dx.doi.org/10.9734/JAMB/2017/33152]

Abdel-Razig, KA & Al Gamry, AS (2009) Effect of natural acidifying agenst and storage Temperature on quality of unripend whey cheese. *J Sci Technol,* 10, 119-29.

Alichanidis, E & Polychroniadou, A (2008) Characteristics of major traditional regional cheese varieties of East-Mediterranean countries: a review. *Dairy Sci Technol,* 88, 495-510. [http://dx.doi.org/10.1051/dst:2008023]

Alonso-Calleja, C, Carballo, J, Capita, R, Bernardo, A & Garsía-López, ML (2002) Changes in the microflora of Valdeteja raw goat's milk cheese throughout manufacturing and ripening. *LWT - Food Science and Technology,* 35, 222-32. [http://dx.doi.org/10.1006/fstl.2001.0842]

Andiç, S, Gençcelep, H, Tunçtürk, Y & Köse, S (2010) The effect of storage temperatures and packaging methods on properties of Motal cheese. *J Dairy Sci,* 93, 849-59. [http://dx.doi.org/10.3168/jds.2009-2413] [PMID: 20172204]

Anifantakis, EM (1998) *Greek Cheeses, a Tradition of Centuries.* National Dairy Committee of Greece, Athens, Anonimous, 2001.

Anifantakis, EM (1986) Bulletin of IDF N;202.

Arenas, R, González, L, Bernardo, A, Fresno, JM & Tornadijo, ME (2004) Microbiological and physico-chemical changes in Genestoso cheese, a Spanish acid curd variety, throughout ripening. *Food Control,* 15, 271-9. [http://dx.doi.org/10.1016/S0956-7135(03)00067-7]

Bertozzi, L & Panari, G (1993) Cheeses with Appellation d'Origine Controlée (AOC). Factors that affect quality. *Int Dairy J,* 3, 297-312. [http://dx.doi.org/10.1016/0958-6946(93)90019-V]

Bintsis, T & Papademas, P (2002) Microbiological quality of white-brined cheese: a review. *Int J Dairy Technol,* 55, 113-20. [http://dx.doi.org/10.1046/j.1471-0307.2002.00054.x]

Bonetta, S, Coïsson, JD, Barile, D, Bonetta, S, Travaglia, F, Piana, G, Carraro, E & Arlorio, M (2008) Microbiological and chemical characterization of a typical Italian cheese: Robiola di Roccaverano. *J Agric Food Chem,* 56, 7223-30. [http://dx.doi.org/10.1021/jf8000586] [PMID: 18646764]

Boudjellad, N, Rolet-Repecaud, O & Collin, JC (1994) Detection of residual chymosin in cheese by an enzyme-linked immunosorbent assay. *J Dairy Res,* 61, 101-9. [http://dx.doi.org/10.1017/S0022029900028090]

Bouton, Y, Guyot, P & Grappin, R (1998) Preliminary characterization of microflora of Comté cheese. *J Appl Microbiol,* 85, 123-31. [http://dx.doi.org/10.1046/j.1365-2672.1998.00476.x] [PMID: 9721662]

Bumberger, E & Belitz, HD (1993) Bitter taste of enzymic hydrolysates of casein. I. Isolation, structural and sensorial analysis of peptides from tryptic hydrolysates of β-casein. *Z Lebensm Unters Forsch,* 197, 14-9. [http://dx.doi.org/10.1007/BF01202693] [PMID: 8356850]

Caridi, A, Micari, P, Foti, F, Ramondino, D & Sarullo, V (2003) Ripening and seasonal changes in microbiological and chemical parameters of the artisanal cheese Caprino d'Aspromonte produced from raw or termized goat's milk. *Food Microbiol,* 20, 201-9. [http://dx.doi.org/10.1016/S0740-0020(02)00116-8]

Coda, R, Brechany, E, De Angelis, M, De Candia, S, Di Cagno, R & Gobbetti, M (2006) Comparison of the compositional, microbiological, biochemical, and volatile profile characteristics of nine Italian ewes' milk cheeses. *J Dairy Sci,* 89, 4126-43.

[http://dx.doi.org/10.3168/jds.S0022-0302(06)72458-4] [PMID: 17032999]

Codex Alimentarius (2000) *Codex group standard for cheese in brine-codex stan 208.*FAO/WHO of UN, Rome.

Dahl, S, Tavaria, FK & Malcata, M (2002) Relationship between flavor and microbiological profiles in Serra da Estrela cheese through ripening. *Int Dairy J,* 10, 255-62.
[http://dx.doi.org/10.1016/S0958-6946(00)00042-X]

Davis, JG (1976) Cheese, Manufacturing methods. Churchill Livingston, Edinburg.Vol. III

De Roest, K (2000) *The production of Parmigiano- Reggiano Cheese Doctoral thesis Wageningen University*

Demarigny, Y, Beuvier, E, Buchin, S, Pochet, S & Grappin, R (1997) Influence of raw milk microflora on the characteristics of Swiss-type cheeses. II. Biochemical and sensory characteristics. *Lait,* 77, 151-67.
[http://dx.doi.org/10.1051/lait:1997110]

Di Cagno, R, Buchin, S, de Candia, S, De Angelis, M, Fox, PF & Gobbetti, M (2007) Characterization of Italian cheeses ripened under nonconventional conditions. *J Dairy Sci,* 90, 2689-704.
[http://dx.doi.org/10.3168/jds.2006-654] [PMID: 17517708]

Dilanyan, ZH (1967) Sirodelie, Picsevaja promislenost

Dimitrovska, G, Srbinovska, S, Joshevska, E & Jovanovska, V (2017) Quality and technology of indigenous traditional "Bieno" cheese in the region of Mariovo, Macedonia. *AGROFOR International Journal,* 2, 113-23.

Dimov, N (1963) *Mlekarstvo.*Zemizdat, Sofia.

Domingos-Lopes, MFP, Stanton, C, Ross, PR, Dapkevicius, MLE & Silva, CCG (2017) Genetic diversity, safety and technological characterization of lactic acid bacteria isolated from artisanal Pico cheese. *Food Microbiol,* 63, 178-90.
[http://dx.doi.org/10.1016/j.fm.2016.11.014] [PMID: 28040167]

Dubrova Mateva, N, Eftimova, E, Srbinovska, S, Karovska, E, Menkovska, M & Pacinovski, N (2016) Preserving and marketing farmhouse artisanal cheese. *J Hyg Eng Des,* 14, 54-9.

Enríquez, L, Díaz, JJ, Pinto, A, Mendoza, L & Fontecha, GA (2016) Characterization of microbiota isolated from traditional Honduran cheese. *Journal of Food & Industrial Microbiology,* 2, 1-7.

Ergönül, B, Ergönül, PG & Seçkin, AK (2011) Chemical and textural attributes of Hellim (Halloumi) cheese marketed in Turkey. *Mljekarstvo,* 61, 168-74.

Fitzsimons, NA, Cogan, TM, Condon, S & Beresford, T (2001) Spatial and temporal distribution of non-starter lactic acid bacteria in Cheddar cheese. *J Appl Microbiol,* 90, 600-8.
[http://dx.doi.org/10.1046/j.1365-2672.2001.01285.x] [PMID: 11309072]

Foulquié Moreno, MR, Sarantinopoulos, P, Tsakalidou, E & De Vuyst, L (2006) The role and application of enterococci in food and health. *Int J Food Microbiol,* 106, 1-24.
[http://dx.doi.org/10.1016/j.ijfoodmicro.2005.06.026] [PMID: 16216368]

Franciosi, E, Settanni, L, Cavazza, A & Poznanski, E (2009) Biodiversity and technological potential of wild lactic acid bacteria from raw cow's milk. *Int Dairy J,* 19, 3-11.
[http://dx.doi.org/10.1016/j.idairyj.2008.07.008]

Franz, CMAP, Holzapfel, WH & Stiles, ME (1999) Enterococci at the crossroads of food safety? *Int J Food Microbiol,* 47, 1-24.
[http://dx.doi.org/10.1016/S0168-1605(99)00007-0] [PMID: 10357269]

Talevski, G, Čobanova-Vasilevska, R, Srbinovska, S & Sireta, Z (2009) Quality of the sheep milk as a raw material in dairy industry of Macedonia. *Biotechnol Anim Husb,* 25, 971-7.

Gaglio, R, Francesca, N, Di Gerlando, R, Cruciata, M, Guarcello, R, Portolano, B, Moschetti, G & Settanni, L (2014) Identification, typing and investigation of the dairy characteristics of lactic acid bacteria isolated from "Vastedda della valle del Belìce" cheeses. *Dairy Sci Technol,* 9, 157-80.

[http://dx.doi.org/10.1007/s13594-013-0150-5]

Garde, S, Tomillo, J, Gaya, P, Medina, M & Nuñez, M (2002) Proteolysis in Hispánico cheese manufactured using a mesophilic starter, a thermophilic starter, and bacteriocin-producing *Lactococcus lactis* subsp. lactis INIA 415 adjunct culture. *J Agric Food Chem,* 50, 3479-85.
[http://dx.doi.org/10.1021/jf011291d] [PMID: 12033814]

Gardini, F, Tofalo, R, Belletti, N, Iucci, L, Suzzi, G, Torriani, S, Guerzoni, ME & Lanciotti, R (2006) Characterization of yeasts involved in the ripening of Pecorino Crotonese cheese. *Food Microbiol,* 23, 641-8.
[http://dx.doi.org/10.1016/j.fm.2005.12.005] [PMID: 16943063]

Gelsomino, R, Vancanneyt, M, Condon, S, Swings, J & Cogan, TM (2001) Enterococcal diversity in the environment of an Irish Cheddar-type cheesemaking factory. *Int J Food Microbiol,* 71, 177-88.
[http://dx.doi.org/10.1016/S0168-1605(01)00620-1] [PMID: 11789936]

Giraffa, G (2002) Enterococci from foods. *FEMS Microbiol Rev,* 26, 163-71.
[http://dx.doi.org/10.1111/j.1574-6976.2002.tb00608.x] [PMID: 12069881]

Gobbetti, M, Morea, M, Baruzzi, F, Corboc, AMR, Mataranteb, T, Considined, RDC, Guinee, T & Fox, PF (2002) Microbiological, compositional, biochemical and textural characterisation of Caciocavallo Pugliese cheese during ripening. *Int Dairy J,* 12, 511-23.
[http://dx.doi.org/10.1016/S0958-6946(02)00042-0]

Golić, N, Cadež, N, Terzić-Vidojević, A, Suranská, H, Beganović, J, Lozo, J, Kos, B, Sušković, J, Raspor, P & Topisirović, L (2013) Evaluation of lactic acid bacteria and yeast diversity in traditional white pickled and fresh soft cheeses from the mountain regions of Serbia and lowland regions of Croatia. *Int J Food Microbiol,* 166, 294-300.
[http://dx.doi.org/10.1016/j.ijfoodmicro.2013.05.032] [PMID: 23973841]

Hattem, HE, Taleb, AT, Manal, AN & Hanaa, SS (2012) Effect of pasteurization and season on milk composition and ripening of Ras cheese. *Journal of Brewing and Distilling,* 3, 15-22.
[http://dx.doi.org/10.5897/JBD11.010]

Hayaloglu, AA, Guven, M & Fox, PF (2002) Microbiological, biochemical and technological properties of Turkish white cheese "Beyaz Peynir". *Int Dairy J,* 12, 635-48.
[http://dx.doi.org/10.1016/S0958-6946(02)00055-9]

Jovanović, S (2001) *Uticaj obrazovanja koagregata proteina mleka na veće iskorišćenje polutvrdih sireva.*Doktorska disertacija. Univerzitet u beogradu. Poljoprivredni fakultet..

Kalantzopoulos, GC (1993) Cheese from Ewes and Goats Milk*Fox, PF (Eds) Cheese Chemistry, Physics and Microbiology* Major Cheese Groups, Chapman and Hall, Ltd London. 507-542.
[http://dx.doi.org/10.1007/978-1-4615-2648-3_16]

Kamber, U & Çelik, TH (2007) Some microbiological and chemical characteristics of Gorcola cheese. *YYÜ Vet Fak Derg,* 18, 87-92.

Kapac-Parkačeva, N, Čižbanovski, T & Lazarevska, D (1974) Hemiski sastav, osobine i reološka svojstva bijenog sira sa područja SR Makedonije. *Mljekarstvo,* 24, 74-82.

Kapac-Parkaceva, N (1988) *Tehnologija na prerabotki od mleko*

Kindstedt, P, Caric, M & Milanovic, S (2004) Pasta-filata cheeses. *Cheese: Chemistry, Physics and Microbiology* Major Cheese Groups 251-77.

Kindstedt, PS (1993) Mozzarella and Pizza cheese. *Cheese: Chemistry, physics and microbiology* Chapman & Hall, London 337-62.
[http://dx.doi.org/10.1007/978-1-4615-2648-3_12]

Kirdar, SS & Kursun, O (2011) Microflora and pathogen bacteria in traditional Anatolian Canak (Pan) cheese. *J Anim Vet Adv,* 10, 2511-5.
[http://dx.doi.org/10.3923/javaa.2011.2511.2515]

Kongo, JM, Gomes, AP & Malcata, FX (2008) Monitoring and identification of bacteria associated with

safety concerns in the manufacture of São Jorge, a Portuguese traditional cheese from raw cow's milk. *J Food Prot,* 71, 986-92.
[http://dx.doi.org/10.4315/0362-028X-71.5.986] [PMID: 18522034]

Kosikowski, F (1982) *Cheese and Fermented Milk Foods.*Brooktondale, New York.

Kostova-Mojsova, S (2013) *Microbiological biodiversity of traditional white brined sheep milk cheese.*

Kozev, A (2006) *Kashkaval Pareni sirena.*

Levkov, V & Kakurinov, V (2012) Microbiological properties of artisanal cheese (Bieno Sirenje). *J Hyg Eng Des,* 1, 325-9.

Levkov, V, Mojsova, S, Nastova, R, Srbinovska, S & Gjorgovska, N (2017) Identification and phenotypic characteristics of lactic acid bacteria isolated from some traditional cheeses produced in R. Macedonia. *Maced J Anim Sci,* 7, 79-87.

Levkov, V, Srbinovska, S & Gjorgovska, N (2014) Microbiological and chemical characteristics of traditional ewe's milk cheese from Mariovo region. *Mljekarstvo,* 64, 195-206.
[http://dx.doi.org/10.15567/mljekarstvo.2014.0307]

Levkov, V (2013) Composition and dynamics of technological significant microorganisms throughout traditional manufacturing of Beaten cheese, PhD thesis, Ss Cyril and Methodius University, Faculty of Agricultural Science and Food, Skopje

Li, J, Zheng, Y, Xu, H, Xi, X, Hou, Q, Feng, S, Wuri, L, Bian, Y, Yu, Z, Kwok, LY, Sun, Z & Sun, T (2017) Bacterial microbiota of Kazakhstan cheese revealed by single molecule real time (SMRT) sequencing and its comparison with Belgian, Kalmykian and Italian artisanal cheeses. *BMC Microbiol,* 17, 13.
[http://dx.doi.org/10.1186/s12866-016-0911-4] [PMID: 28068902]

Litopolou-Tzanetaki, E (1997) Staphylococci and micrococci in Kefalotyri cheese. *Milchwissenshaft,* 32, 211-4.

Litopolou-Tzantaki, E (1990) Changes in numbers and kinds of lactic acid bacteria during ripening of Kefalotyri cheese. *J Food Sci,* 55, 111-3.
[http://dx.doi.org/10.1111/j.1365-2621.1990.tb06029.x]

Macedo, CA, Malcata, FX & Hogg, TA (1995) Microbiological profile in Serra ewe's cheese during ripening. *Journal of Applied Biology,* 791-811.

Maćej, O (1989) *Proučavanje mogčnosti izrade mekih sireva na bazi koagregata belančevina mleka, Doktorska disertacija.*(In Serbian), Poljoprivredni fakultet, Zemun-Beograd..

Manolopoulou, E, Sarantinopoulos, P, Zoidou, E, Aktypis, A, Moschopoulou, E, Kandarakis, IG & Anifantakis, EM (2003) Evolution of microbial populations during traditional Feta cheese manufacture and ripening. *Int J Food Microbiol,* 82, 153-61.
[http://dx.doi.org/10.1016/S0168-1605(02)00258-1] [PMID: 12568755]

Marino, M, Maifreni, M & Rondinini, G (2003) Microbiological characterization of artisanal Montasio cheese: analysis of its indigenous lactic acid bacteria. *FEMS Microbiol Lett,* 229, 133-40.
[http://dx.doi.org/10.1016/S0378-1097(03)00816-4] [PMID: 14659553]

Mateva, N (2004) The influence of proteolitic enzymes on the course of the Beaten cheese ripeninig, Master thesis, Ss Cyril and Methodius University, Faculty of Agricultural Sciences and Food, Skopje.

Mateva, N (2012) Participation of whey proteins in production of white brine cheese by ultrafiltration, PhD thesis, University Ss Cyril and Methodius-Skopje, Faculty of Agricultural Sciences and Food-Skopje.

Mc Sweeney, PLH (2004) Biochemistry of cheese ripening. *Int J Dairy Technol,* 57, 127-44.
[http://dx.doi.org/10.1111/j.1471-0307.2004.00147.x]

McSweeney, PLH, Fox, PF, Cotter, PD & Everett, DW (2017) *Cheese, Chemistry, Physics & Microbiology.*Elsevier Ltd..

McSweeney, PHL & Fox, PF (2004) Metabolism of residual lactose and of lactate and citrate.*Cheese:*

Chemistry, physics and microbiology Elsevier Academic Press, London 361-71.
[http://dx.doi.org/10.1016/S1874-558X(04)80074-5]

Micari, P, Sarullo, V, Sidari, R & Caridi, A (2007) Physico-chemical and higienic characteristics of the calabrian raw milk cheese, Caprino d'Aspromonte. *Turk J Vet Anim Sci,* 31, 55-60.

Micev, N (1969) O nekim svojstvima makedonskog bijenog sira. *Mljekarstvo,* 12, 274-7 (In Croatian).

Micevski, K (1994) *Bistra III, Visokoplaninska vegetacija na planinata Bistra.* MANU (In Macedonian).

Mijačević, Z & Bulajić, S (2008) Sensory evaluation and microbiological characterization of autochthonous Sombor cheese. *Acta Vet (Beogr),* 58, 531-41.

Mijačević, Z, Petrović, MP & Bulajić, S (2005) The specific characteristics of Pirot kachkaval. *Biotechnol Anim Husb,* 21, 375-9.
[http://dx.doi.org/10.2298/BAH0506375M]

Minervini, F, Montagna, MT, Spilotros, G, Monaci, L, Santacroce, MP & Visconti, A (2001) Survey on mycoflora of cow and buffalo dairy products from Southern Italy. *Int J Food Microbiol,* 69, 141-6.
[http://dx.doi.org/10.1016/S0168-1605(01)00583-9] [PMID: 11589553]

Mojsova, S (2016) Enterococci from Macedonian cheeses: population structure, antibiotic susceptibility, biogenic amines and presence of bacteriocins and virulence determinants, PhD Thesis, University Ss Cyril and Methodius-Skopje, Faculty of veterinary medicine-Skopje.

Mojsova, S, Krstevski, K, Dzadzovski, I, Popova, Z & Sekulovski, P (2015) Phenotypic and genotyupic characteristics of enterocin producing enterococci against pathogenic bacteria. *Maced Vet Rev,* 38, i-viii.
[http://dx.doi.org/10.14432/j.macvetrev.2015.08.052]

Mojsova, S, Jankuloski, D, Sekulovski, P, Angelovski, Lj, Ratkova, M & Prodanov, M (2013) Microbiological properties and chemical composition of Macedonian traditional white brined cheese. *Maced Vet Rev,* 36, 13-8.

Montel, MC, Buchin, S, Mallet, A, Delbes-Paus, C, Vuitton, DA, Desmasures, N & Berthier, F (2014) Traditional cheeses: rich and diverse microbiota with associated benefits. *Int J Food Microbiol,* 177, 136-54.
[http://dx.doi.org/10.1016/j.ijfoodmicro.2014.02.019] [PMID: 24642348]

Mrkonjic Fuka, M, Engel, M, Skelin, A, Bogovic Matijasic, B, Redzepovic, S & Schloter, M (2013) Culture-independent quantitative approach to monitoring the dynamics of bacterial population during Istrian cheese ripening. *Food Technol Biotechnol,* 51, 414-21.

Niketić, M, Tomović, G, Melovski, Lj, Stevanović, V & Matevski, V (2014) New species for the vascular flora of Republic of Macedonia and their distribution in the Balkan Peninsula. *Bot Serb,* 38, 57-67.

Niro, S, Fratianni, A, Tremonte, P, Sorrentino, E, Tipaldi, L, Panfili, G & Coppola, R (2014) Innovative Caciocavallo cheeses made from a mixture of cow milk with ewe or goat milk. *J Dairy Sci,* 97, 1296-304.
[http://dx.doi.org/10.3168/jds.2013-7550] [PMID: 24440270]

Özdemir, S, Yangilar, F & Ozdemir, C (2010) Determination of microbiological characteristics of Turkish Karin kaymagi cheeses packaged indifferent materials. *Afr J Microbiol Res,* 4, 716-21.

Özkalp, B, Özden, B, Tuncer, Y, Şanlibaba, P & Ançelik, M (2007) Technological characterization of wild-type *Lactococcus lactis* starins isolated from raw milk and traditional fermented milk products in Turkey. *Lait,* 87, 521-34.
[http://dx.doi.org/10.1051/lait:2007033]

Papageorgiou, DK & Marth, EH (1989) Behavior of Listeria monocytogenes at 4 and 22° C in whey and skim milk containing 6 or 12% sodium chloride. *J Food Prot,* 52, 625-30.
[http://dx.doi.org/10.4315/0362-028X-52.9.625] [PMID: 31003293]

Parente, E & Cogan, TM (2004) Starter culture: General aspects *(Eds) Fox, PF, McSweeney, PH, Cogan, TM&Guinee TP, Cheese* Chemistry, Physics and Microbiology., General Aspects, Academic press. 123–148.
[http://dx.doi.org/10.1016/S1874-558X(04)80065-4]

Pereira-Dias, S, Potes, ME, Marinho, A, Malfeito-Ferreira, M & Loureiro, V (2000) Characterisation of yeast flora isolated from an artisanal Portuguese ewes' cheese. *Int J Food Microbiol,* 60, 55-63. [http://dx.doi.org/10.1016/S0168-1605(00)00323-8] [PMID: 11014522]

Pešić-Mikulec, D & Jovanović, L (2005) Microbiological study of fresh white cheese (A Serbian craft variety). *Appl Ecol Environ Res,* 4, 129-34. [http://dx.doi.org/10.15666/aeer/0401_129134]

Phelan, JA, Renaud, J & Fox, PF (1993) Some non European Cheese Varities *(Eds) Fox, PE, Cheese: chemistry, phisics and microbiology.* Chapman & Hall, London..

Picon, A, Garde, S, Ávila, M & Nuñnez, M (2016) Microbiota dynamics and lactic acid bacteria biodiversity in raw goat milk cheeses. *Int Dairy J,* 58, 14-22. [http://dx.doi.org/10.1016/j.idairyj.2015.09.010]

Psoni, L, Tzanetakis, N & Litopoulou-Tzanetaki, E (2003) Microbiological characteristics of Batzos, a traditional Greek cheese from raw goat's milk. *Food Microbiol,* 20, 575-82. [http://dx.doi.org/10.1016/S0740-0020(02)00153-3]

Roa, I, Lopez, MB & Mendiola, FJ (1999) Residual clotting activity and ripening properties of vegetable rennet from Cynara cardunculus in La Serena cheese. *Food Res Int,* 32, 413-9. [http://dx.doi.org/10.1016/S0963-9969(99)00098-8]

Ruzic-Muslic, D, Petrovic, MM, Bijelic, Z, Pantelic, V, Perisic, P & Bogdanovic, V (2011) Traditional production and characteristics of Sjenica cheese and Pirot Kachkaval. *Bulg J Agric Sci,* 17, 664-72.

Santa, D (2015) Technology and Dynamic of ripening of traditional sheep Kashkaval, PhD thesis, University Ss Cyril and Methodius – Skopje, Faculty for Agricultural Sciences and Food - Skopje.

Santa, D & Srbinovska, S (2014) Traditional production of Galichki kashkaval. *Mljekarstvo,* 64, 119-26.

Santa, D & Srbinovska, S (2017) *Comparative study of sheep milk, whey and traditional kashkaval from Galichnik and Lazaropole, Symposium proceedings of 3rd International Symposium for agriculture and food, ISAF 2017, Ohrid, Macedonia.*

Şengül, M (2006) Microbiological characterization of Civil cheese, a traditional Turkish cheese: microbiological quality, isolation and identification of its indigenous Lactobacilli. *World J Microbiol Biotechnol,* 22, 613-8. [http://dx.doi.org/10.1007/s11274-005-9079-z]

Simonovska, T (2018) Dynamics of acidity (pH) during the production of traditional galichki kachkaval, MSc thesis, University Ss Cyril and Methodius – Skopje, Faculty for Agricultural Sciences and Food – Skopje.

Simos, EN, Nikolaou, EM & Zoiopoulos, PE (1996) Yield, composition and certain physicochemical characteristics of milk of the Epirus mountain sheep breed. *Small Rumin Res,* 20, 67-74. [http://dx.doi.org/10.1016/0921-4488(95)00780-6]

Srbinovska, S (2013) *Zastita na geografska oznaka (ZGO) BISTRA Ovco belo salamureno sirenje, elaborat.*MZSV.

Srbinovska, S & Cizbanovski, T (1994) *Godisen zbornik na Zemjodelski fakultet 39* Skopje 187-195.

Steele, JL (1995) Contribution of lactic acid bacteria to cheese ripening. *Adv Exp Med Biol,* 367, 209-20. [http://dx.doi.org/10.1007/978-1-4615-1913-3_12] [PMID: 7572362]

Stević, B (1962) Tehnološka mikrobiologija stočnih proizvoda i ishrane stoke., Naučna knjiga, Beograd.

Sulejmani, E (2010) *Quality of White brined cheese produced by different temperature, rate of processing and starter-culture MSc, Ss.*Cyril and Methodius University, Skopje, Macedonia.

Sulejmani, E, Pollozhani, H & Idrizi, Xh (2011) Sensory profiling and rheological properties of white brined cheese produced by different starter cultures. *J Hyg Eng Des,* 1, 309-11.

Sulejmani, E, Hayaloglu, AA & Rafajlovska, V (2014) Study of the chemical composition, proteolysis,

volatile compounds, and textural properties of industrial and traditional Beaten (Bieno sirenje) ewe milk cheese. *J Dairy Sci,* 97, 1210-24. a
[http://dx.doi.org/10.3168/jds.2013-7092] [PMID: 24377800]

Sulejmani, E, Rafajlovska, V, Güneşer, O, Karagül-Yüceer, Y & Hayaloglu, AA (2014) Volatile compounds and proteolysis in traditional Beaten (Bieno sirenje) ewe's milk cheese. *Int J Dairy Technol,* 67, 1-10. b
[http://dx.doi.org/10.1111/1471-0307.12159]

Sulejmani, E & Hayaloglu, AA (2016) Influence of curd heating on proteolysis, residual coagulant activity and volatiles of Kashkaval cheese during ripening. *Food Chem,* 211, 160-70.
[http://dx.doi.org/10.1016/j.foodchem.2016.05.054] [PMID: 27283619]

Sulejmani, EI & Hayaloglu, AA (2018) Characterisation of Macedonian white-brined cheese: Effect of raw or heat-treated caprine milk. *International Journal of Dairy Technology,,* 70, 584-593.

Talevski, G (2012) Traditional production of Beaten cheese, Third International Scientific Symposium "Agrosym Jahorina 2012".

Tarakci, Z, Sagun, E, Sancak, H & Durmaz, H (2004) The effect of salt concentration on some characteristics in Herby cheese. *Pak J Nutr,* 3, 232-6.
[http://dx.doi.org/10.3923/pjn.2004.232.236]

Terzić-Vidojević, A, Tonković, K, Leboš Pavunc, A, Beganović, A, Strahinić, I, Kojić, M, Veljović, K, Golić, N, Kos, B, Čadež, N, Gregurek, Lj, Šušković, J, Raspor, P & Topisirović, Lj (2015) Evaluation of autochthonous lactic acid bacteria as starter cultures for production of white pickled and fresh soft cheeses. *Lebensm Wiss Technol,* 63, 298-306.
[http://dx.doi.org/10.1016/j.lwt.2015.03.050]

Turantaş, F, Unlütürk, A & Göktan, D (1989) Microbiological and compositional status of Turkish white cheese. *Int J Food Microbiol,* 8, 19-24.
[http://dx.doi.org/10.1016/0168-1605(89)90076-7] [PMID: 2701567]

Turkoglu, H, Ceylan, ZG & Dayisoylu, KS (2003) The microbiological and chemical quality of Orgu cheese produced in Turkey. *Pak J Nutr,* 2, 92-4.
[http://dx.doi.org/10.3923/pjn.2003.92.94]

Vafopoulou-Mastrogianaki, A, Litopoulou-Tzanetaki, E & Tzanetakis, N (1990) Effect of Pdiococcus pentosaceus on ripening changes of Feta cheese. *Microbiologie-Alimentis Nutrition,* 8, 53-62.

Van den Berg, G, de Vries, AE & Stadhouders, J (1986) The salt content of Gouda cheese. *Voedingsmiddelentechnologie,* 19, 37-9.

Woo, AH, Kollodge, S & Lindsay, RC (1984) Quantification of major free fatty acids in several cheese varieties. *J Dairy Sci,* 67, 874-8.
[http://dx.doi.org/10.3168/jds.S0022-0302(84)81380-6]

Zeppa, G, Fortina, MG, Dolci, P, Acquati, A, Gandini, A & Manachini, PL (2004) Characterization of autochthonous lactic acid bacteria from an artisanal Italian cheese. *Acta Agric Slov,* 84, 3-9.

CHAPTER 2

Montenegrin Traditional Cheeses

Slavko Mirecki[1,*] and **Kirdar Seval Sevgi**[2]

[1] *Biotechnical Faculty, University of Montenegro, Podgorica, Montenegro*

[2] *University Burdur, Food Agriculture and Livestock Vocational Higher Education School, Burdur Mehmet Akif Ersoy, Burdur, Turkey*

Abstract: Milk production and processing have been the main activities of most of the Montenegrin agricultural population, for centuries. That's why, there is a wide range of traditional dairy products that take primary place in Montenegrin meal, both during history and today. From a wide range of traditional dairy products, the most famous, the highest quality and, according to the quantity of production and consumption, the most common cheeses are Pljevaljski, Njeguški, Lisnati, and Prljo. They are all characterized by high nutritional value, piquant, unique and specific aromas. Pljevaljski cheese is full fat, soft cheese that ripens in the wooden vat for at least 4 weeks. During ripening the cheese acquires a pleasant milky-sour taste and odour, white to yellowish colour and a compact texture suitable for slicing. Lisnati cheese belongs to pasta filata cheeses. The name Lisnati was given because of its very unusual form of slices that are extremely thin as a leaf ("lisnati" = "leafy"). It is partially skimmed, semi-hard to soft cheese that can be consumed as fresh or as ripened cheese. The most popular semi-hard cheese in Montenegro is Njeguški cheese. It is typical semi-hard, full-fat cheese. The specialty of this cheese is that, after ripening on shelves, cheese can be put into the olive oil where it continues to ripen. Mature cheese, especially cheese from olive oil, has an extremely spicy taste. The low-fat, semi-hard cheese Prljo is produced in the northwestern, mountainous region of Montenegro. The purpose of this chapter is to discuss the main phases in production technology and the chemical composition of the most important traditional Montenegrin dairy products.

Keywords: Cheese Technology, Cheese Quality, Montenegro, Traditional Cheese.

INTRODUCTION

Montenegro is a state located in Southeastern Europe who lies on Adriatic Sea coast. It is bordered in the east and northeast with Serbia, in the west and northwest with Croatia and Bosnia and Herzegovina, and in the south-east with

* **Corresponding author Slavko Mrecki:** Biotechnical Faculty University of Montenegro Podgorica Montenegro; Tel: +38269507394; E-mail: slami@ac.me

Celile Aylin Oluk & Oya Berkay Karaca (Eds.)

Albania. In Montenegro, as in the rest of Balkan countries, farm-based traditional production is still very active due to a very high percentage of rural households.

Traditional dairy products in Montenegro can be divided into three main groups: cheeses, skorup (kaymak) and fermented kinds of milk. Montenegrin traditional cheeses consist of white brine cheeses, pasta filata cheeses, hard/semi-hard cheeses, and low-fat cheeses (Fig. **1**). (Dozet *et al.*. 1996, Mirecki, 2012).

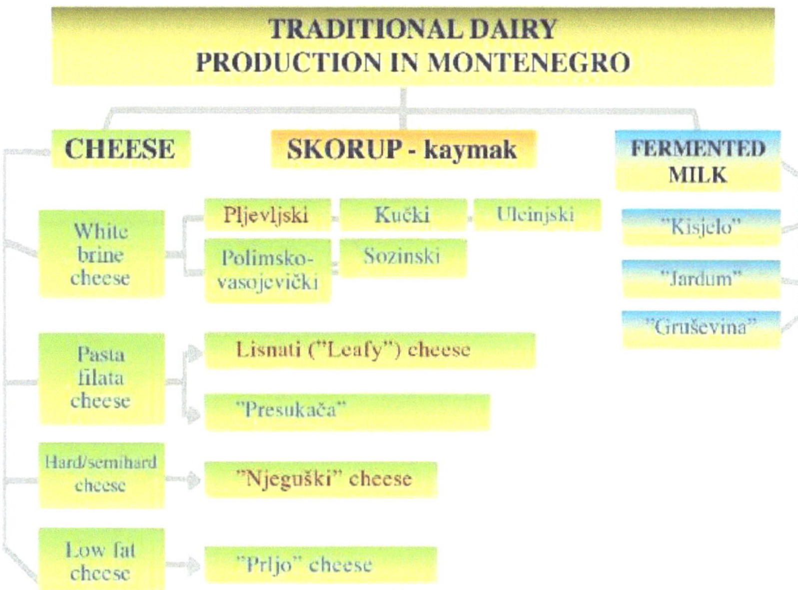

Fig. (1). Traditional Dairy Products in Montenegro (Mirecki, 2012).

Milk production and processing were the main activities of the agricultural population of Montenegro. The centuries-long tradition in the processing of milk is the reason why Montenegrin households developed a large number of different dairy products. A wide range of cheeses includes white brine, semi-hard, stretched-curd (pasta filata) and low-fat cheeses. In Montenegro, the most commonly produced and consumed are white brine cheeses. Those cheeses often bear the name of the geographical area, town or village where they are mostly produced (Bojanić-Rašović *et al.*. 2010). The main areas of production are northern and northeastern parts, as well as in the far south of Montenegro (Fig. **2**). Pljevaljski cheese is not only the most famous white cheese but also the most famous Montenegrin traditional dairy product (Mirecki and Konatar, 2014). Stretched-curd or pasta filata cheese is produced in the central part of Montenegro under the two name: Lisnati cheese, thanks to its recognizable and very unusual form of slices, or Kolašinski cheese, according to the name of the city in which it is most produced (Mirecki *et al.*, 2013). The coastal area of Montenegro is known by the production of semi-hard, full-fat Njeguški cheese (Mirecki *et al.*, 2015). The only low-fat cheese that is traditionally produced in Montenegro is Prljo cheese. A common to all these cheeses is the fact that none of them is standardized, and their technologies vary from manufacturer to manufacturer. Besides that, based on the quality, produced quantities, the market demand and popularity among the consumers, Pljevaljski, Lisnati, Njeguški, and Prljo cheese meet most requirements and should be proposed for the Protected Designation of Origin or Protected Geographical Indication, on the national and after that international level. Specification of the traditional cheese is given in Table **1**.

Fig. (2). Traditional cheeses in montenegro.

PLJEVALJSKI CHEESE

Pljevaljski cheese is one of the most popular and widely produced Montenegrin traditional cheese. It is a white, brined, soft to semi-hard, fatty cheese. Traditionally, cheese is made from ewe's milk. A significant reduction in the number of sheep in Montenegro caused that this cheese is more frequently produced from cow's milk and mixed milk (cow + sheep). The production area is in the north-west part of Montenegro, in the municipality of Pljevlja.

Table 1. Specification of cheeses in montenegro.

Name	Njeguški Cheese	Pljevaljski Cheese	Lisnati Cheese	Prljo cheese
Type of cheese	Full-fat, Semi-hard to hard	White brined, Soft to semi-hard, Fat to full fat	Pasta filata, semi-hard, medium fat	Low-fat, semi-hard to hard,
Area of production	Peripheral areas of the Lovćen Mountain and the Adriatic Sea coastline, Njeguši village, South of Montenegro	Geographical area of the Pljevlja municipality, North of Montenegro	Central and northern part of Montenegro	North-west part of Montenegro, Area of mountain Durmitor
Raw material	Ewe's, cow's and goat's milk The mix of these milks in different combinations	Ewe's milk Cow's milk The mix of cow's and ewe's milk	Cow's milk	Milk remaining after the production of Skorup- Kaymak Ewe's, cow's and mix of ewe's and cow's milk
Allowed additive	Rennet, salt	Rennet, salt	Rennet, salt	Rennet, salt
Characteristic of cheese Shape	Height 3-5 cm Weight 1-3 kg Diameter 15-25 cm	Height 1.5-3.0 cm Length 7-12 cm Width 6-10 cm Weight 100-200 g	Thin slice Height up to 5mm	The diameter of lumps 0.5–1 cm
Appearance	The shape of round, low cylinder	Irregular triangle or rectangle Without crust	İrregular quadrangle, elastic, thin like leaf or paper sheet	Small chunks or lumps
Color	Curst golden yellow Curd white to pale yellow	White, White to yellowish	Porcelain white	Pale yellow to greenish
Consistency	compact, well connected, easy to cut, not brittle	Gentile, easily breakable, brittle	Elastic, closed, plastic	Nubbly
Composition of cheese	Fat in dry matter 45 – 55% Moisture 35 – 45%	Fat in dry matter 45-55% Moisture 40-55%	Fat in dry matter 35-45% Moisture 40-55%	Fat in dry matter 5-15% Moisture 35-55%

(Table 1) cont.....

Name	Njeguški Cheese	Pljevaljski Cheese	Lisnati Cheese	Prljo cheese
Taste and odour	Pleasant smell, slightly sour-milky, moderately salty aroma	Pleasant, milky-sour aroma, moderately salted	Pleasurable milky-sour taste	very intensive milky-sour smell, moderately smell of sheepskin, strong milky-sour taste, salty, moderately to boiled milk taste
Structure	Closed, with a few cheese eyes,	Closed, soft curd, without or with a few cheese holes	Closed, elastic, smooth	Closed
Ripening	Min. 4 week Additionally could be smoked or submerged in olive oil	Min. 30 days	No ripening or not more than 7 days	Ripening in sheepskin for 60-90 days
Packing	No packing (own curst) or Vacuum-pack (whole cylinder of cheese)	Wooden vats, 30-40 kg Buckets 10-12 kg Confectioning in packs 0.25-1.0 kg	Buckets 5-10 kg Confectioning in packs 0.25-1.0 kg	Sheepskin Confectioning from sheepskin to packs 1-2 kg

Most of the cheese is produced in households by technology, and even "equipment" for cheese production (wooden plank, stone and wooden vat), that is the same as 100 years ago. A long tradition of production, a clearly defined production area, excellent quality, and a specific taste and smell have enabled Pljevaljski cheese to be the first Montenegrin dairy product to which the designation of protection of origin was assigned.

Physical Structure and Appearance

Pljevaljski cheese is white, brined, semi-hard, and fat to full-fat cheese. The cheese is produced and sold on the market in the form of slices. The shape of slices is triangle or rectangle, whose dimensions are 7-12 cm long side and 6-10 cm short side. The height of slices is 1.5-3.0 cm and weight is 100-200 g (Fig. **3**). The slices are without a crust and white colour. The consistency of the cheese is gentle with closed curd, very slightly porous, brittle and slices break during cutting. The odour of the cheese is pleasant with clearly expressed milky-sour flavour. The aroma has, also, milky-sour taste and it is moderately salty. The yield of the Pljevaljski cheese depends on the type of milk that is used for its production. Depending on the lactation period, 4-6 kg of ewe's milk, 7-9 kg of mixed milk, and 8-10 kg cow's milk are necessary for 1 kg of ripened cheese (Mirecki *et al.*. 2013).

Traditional Method of Production

Pljevaljski cheese is made from raw milk. The cheese is made from whole ewe's, cow's or mixed milk (ewe's and cow's). Immediately after milking, the milk is filtrated through cheesecloth, and rennet is added at the temperature of 28-32^{0}C. The optimal duration of coagulation is 60 minutes. After the coagulation, the resulting curd is cut into cubes with the size of 2x2 to 3x3 cm and allowed to settle to the bottom of the vessel. Afterwards, the curd is stirred slowly and, if necessary, additionally cut. The whole process takes 10-25 minutes and then the curd is collected into cheesecloth and squeezed to drain off the whey (Fig. **3a**). Separation of whey from the curd is done in several stages. First, the cheesecloth with curd is hanged above the vessel and separation of whey is carried out due to the self-pressing (Fig. **3b**). After 25-30 minutes of self-pressure, the curd is placed on the table for whey separation and is pressed with a wooden plank that has the shape of a circle. The final whey separation is carried out by additional loading of a wooden plank with stones of different sizes or a bucket of water where water is added. After, approximately, 6 hours of pressing, the curd becomes flat and has the shape of a circle. Cheese is cut into slices (Fig. **3c**). Slices are stacked into wooden tubs and salted during stacking. This phase lasts 5 - 15 minutes (Fig. **3d**). It is desirable for slices to have the form which best fits the form of tubs. With such ways of cutting and stacking it is possible to avoid empty space between the slices, which is important for obtaining anaerobic conditions for ripening of Pljevaljski cheese. The cheese ripens for 4 weeks in brine prepared from salted water or whey. When the wooden tub is full, the cheese is pressed with a wooden circle and stone (Fig. **3e**), since the entire cheese has to be beneath the surface of the brine (Fig. **3f**). Periodically, the surface of brine should be cleaned, or, if necessary, completely replaced with fresh brine (Mirecki & Adžić, 2006; Mirecki & Konatar, 2014). Table **2** shows the parameters of the technology of Pljevaljski cheese.

Table 2. Parameters of technology of Pljevaljski cheese (Mirecki *et al.*. 2006).

Technological Step	Parameter
Rennet adding	28-32^{0}C
Synerezis of whey	30 min
Curd forming	60 min
Curd stirring	10-25 min
Curd pressure	Wooden plank and stone 6 hours
Salting	Brine (salted water or whey)
Ripening	4 week

Fig. (3). Pljevaljski cheese production steps: **a**) curd is collected into a cheesecloth; **b**) drain off the whey; **c**) cut into slices; **d**) slices are stacked into wooden tubs; **e**) ripening; **f**) Pljevaljski cheese.

Chemical Properties

As it is mentioned before, during the last several decades, the significant reduction in the number of sheep in Montenegro is recorded. That is why, nowadays, most of the Pljevaljski cheese is made from cow's milk, rarely from mixed (cow's + ewe's), and it is very hard to find cheese made from ewe's milk. The quality of the milk used for Pljevaljski cheese production is presented in Table **3**. The optimum ripening period of Pljevaljski cheese is 4 weeks. But some manufacturers leave the cheese to mature from two to 8 weeks. The ripening period depends on the demands of customers, *i.e.* whether they prefer mild or piquant cheese taste. Also, depending on the period of ripening, the content of dry matter, fat, proteins, salts and other parameters of the chemical quality of Pljevaljski cheese that are placed on the market, varies considerably. Not only different customer requirements are the reason for this, but the lack of standardized technology is also a major cause. The technology of production of Pljevaljski cheese differs from producer to producer, and proof of this are researches, *i.e.* the results of the analysis of the chemical quality of Pljevaljski cheese presented in Table **4**. If we compare the results of the authors presented in the table, with Codex Stan 208-1999, it can be concluded that the Pljevaljski cheese belongs to a group of semi-hard cheeses, and by fat content in dry matter to fat and full-fat cheeses (Mirecki & Konatar, 2014).

Table 3. The quality of the milk for pljevaljski cheese production.

Parameters %	Mirecki & Adžić (2006)	Vesković-Moračanin et.al. (2012)
Fat, %	3.96	3.82
Proteins, %	3.30	3.22
Lactose, %	4.49	4.52
SNF*, %	8.57	8.47
FPD, -^{0}C**	0.528	0.528

*) solids non-fat; **) freezing point depression

Table 4. Chemical properties of ripened pljevaljski cheese.

Parameters %	Dozet *et al.* (1996)	Mirecki and Adžić (2006)	Vesković-Moračanin *et al.* (2012)	Mirecki *et al.* (2013)	Mirecki & Konatar (2014)
Dry matter	49.17	50.02	50.41	47.28	46.21
Fat	25.44	23.80	26.88	24.10	24.79
Fat in DM	51.55	47.58	53.11	50.92	53.01
Proteins	19.73	19.77	21.78	18.23	-
Salt	2.43	3.15	2.99	2.71	-

The traditional production of Pljevaljski cheese is the most common method of processing milk in area of Pljevlja municipality. It is one of the most important traditional dairy products in Montenegro. Traditional technology is not changing (technology, equipment, wooden board, stone and wooden vessel), for more than 100 years (Mirecki & Konatar, 2014). Pljevaljski cheese has good nutrition value. According to a dry matter content cheese is soft and semi-hard cheeses, and fat content in dry matter with fat and full-fat cheese.

NJEGUŠKI CHEESE

Njeguški cheese is a Montenegrin autochthonous full-fat, semi-hard to hard cheese. The original production area is the peripheral areas of the Lovćen Mountain and the Adriatic Sea coastline (Fig. **2**). The center of production, historically and today, is the village Njeguši, where the mountain and Mediterranean climates collide. This collision has an effect on the herbal composition of pastures, milk quality, and ripening conditions, which gives the cheese its piquancy. According to some literature data, during the Roman Empire, Rome was supplied with the cheese *Caseus doclestes*. Since Doclea was an old state that existed in the area of present-day Montenegro, it is assumed that this cheese is an ancestor of Njeguški cheese (Dozet et.al. 1996). Heinrich Stieglitz, in his travelogue *"Ein Besuch auf Montenegro"* (1841), commented that he was served with extremely tasty cheese during his visit to Njeguši (Mirecki *et al.* 2015).

Physical Structure and Appearance

Njeguški cheese belongs to semi-hard to hard, full-fat cheeses. It has the shape of a round, low cylinder whose diameter is 15-25 cm and height 3-5 cm. The weight of the cheese is approximately 1-1.5 kg. The yield of the Njeguški cheese depends on the type of milk and period of lactation. For 1 kg of cheese, it is necessary to use 5 kg of ewe's milk and 10-12 kg of cow's milk. If mixed milk is used, the yield depends on the ratio of used cow and ewe's milk. Njeguški cheese has a compact surface, crust without cracks, and on curst it could be visible the structure of cheesecloth that is "imprinted" on the surface. The colour of the crust is golden-yellow. The cut is white to pale yellow, mostly closed but it is allowed to has a few "cheese eyes" evenly distributed, round, shiny, which diameter is up to 0.5 cm. The texture of the cheeses is compact, well connected, easy to cut, and it is not brittle. The smell is pleasant, and the aroma has a slightly sour-milky and moderately salty taste. Njeguški cheese ripened minimum 4 weeks, and after that period it can, also, be exposed to smoke or submerged in olive oil for additional ripening. The cheese exposed to smoke, or ripened in olive oil has a "sharper" and more piquant taste.

Traditional Method of Production

The original Njeguški cheese is produced from ewe's milk, but nowadays it is producing from cow's, goat's and mixed milk. Rennet is added to the fresh milk after milking at a temperature of 32–35°C. Coagulation is forty-five to sixty minutes. After forming, the curd is cut to the size of walnuts, and after five minutes of resting, to the size of corn grains (Fig. **4a**). Cheese curd is formed in the whey gradually heated to the temperature "until the hand can withstand" (approx. 45 ^{0}C). The curd has to be formed by hand in the whey until it gains a globular shape. This phase lasts fifteen to twenty minutes. The formed curd is placed in cheese making cloth (Fig. **4b**), then in a cheese mold and pressed (Fig. **4c**). Pressing is carried out with a wooden circular plank and stone added on a plank. After twenty-four hours of pressing, the cheese is removed from the mold, salted with dry salt, and placed in a wooden box (Fig. **4e**). The cheese stays in this box for two days and is salted two to three times, and then it goes to ripening on dry wooden shelves (Fig. **4f**). Ripening on the shelves lasts at least four weeks, and after that, the cheese is ready for consumption. Longer ripening gives cheese more piquant taste.

Fig. (4). Njegusi cheese production steps: **a**) The curd is cut to size of walnuts; **b**) The formed curd is placed in cheesecloth; **c**)Cheese mold and pressed; **d**) pressing with a wooden cicular plank and put into stone; **e**) salted and placed in wooden box and ripening; **f**) Njegusi cheese.

Ripened Njeguški cheese may be exposed to smoke for a few days or dipped into olive oil where it matures from 90 to 150 days (Table **5**). Longer ripening and ripening in olive oil gives the cheese a more piquant taste. The finished product has a low cylindrical shape, a weight of 1 to 3kg compact surface, golden-yellow crust, closed structure with a few "cheese eyes," pleasant smell, slightly sour-milky and moderately salty aroma.

Table 5. Parameters of technology of njegusi cheese (Mirecki *et al.* 2012).

Technological Step	Parameter
Rennet adding	32-35^{0}C
Curd forming	45-60 min
Synerezis of whey	15-20 min
Curd pressure	Wooden plank and stone 12 hours
Salting	Dry salt
Ripening	90-150 days

Chemical Properties

Traditionally, Njeguški cheese is made from ewe's milk, but it is also producing from cow's, goat's and mixture made from all three mentioned type of milk (Table **6**). The quality of the milk used for Njeguški cheese production is presented in Table **7**. In general, the quality of milk for Njeguški cheese production is very good. The milk fat content depends on the type of milk.

Table 6. The quality of the milk for njeguški cheese production.

		Parameters			
References	**Type of Milk**	**Fat %**	**Proteins %**	**Lactose %**	**SNF* %**
Adžić *et al.* (1984)	Cow Cow + goat sheep cow + sheep	3.70 3.97 6.80 4.75	3.18 3.32 5.18 3.64	4.16 4.23 3.96 4.27	8.49 8.37 10.72 9.00
Dozet *et al.* (1996)	cow + sheep	4.76	3.92	4.20	-
Mirecki (1998)	goat	4.84	3.29	4.48	8.54
Popović (2014)	goat cow goat + sheep	4.27 4.08 3.52	3.68 3.44 4.59	3.47 4.20 4.14	7.91 8.38 9.46
Mirecki *et al.* (2015)	sheep	4.92	4.59	4.14	9.46

*) solids non-fat

Table 7. The chemical properties of njeguški cheese.

References		Parameters				
	Type of Milk	**Fat %**	**Proteins %**	**FDM* %**	**DM** %**	**Salt %**
Adžić *et al.* (1984)	sheep + goat cow goat	38.12 33.00 31.24	26.28 33.92 22.31	50.99 45.14 49.78	74.65 73.10 62.56	2.52 2.16 3.00
Dozet *et al.* (1996)	sheep + goat	37.96	30.84	50.54	74.97	2.83
Mirecki (1998)	goat	23.12	19.23	47.61	62.56	1.15
Popović (2014)	goat sheep + goat cow	27.81 29.97 26.04	28.95 21.89 24.99	43.37 51.73 45.15	64.12 57.93 57.67	2.36 1.94 1.24
Mirecki *et al.* (2015)	sheep	29.97	21.89	51.73	57.93	1.94

*) dry matter; **) fat in dry matter

Ewe's milk and a mixture of other types of milk with ewe's milk had a high percentage of milk fat. Same is with protein content. Good quality of milk, above all favourable fat and protein content, is a good basis for the production of quality cheese. This is confirmed by the results from Table **4** in which the chemical quality of Njeguški cheese was presented by various authors. Although the content of fat, fat content in dry matter, the content of dry matter significantly varies due to non-standardized technology. However, these parameters indicate that the quality of Njeguški cheese is very good, because, according to Codex Standard 283 (1978) all cheeses whose quality is presented in Table **4** belongs to full fat, hard type of cheeses.

LISNATI CHEESE

Lisnati cheese is "pasta filata" cheese produced in the central and northern part of Montenegro (Fig. **2**). Rare historical data indicate that nomadic tribe Vlachs brought the technology to this area. Considering the high level of depopulation of rural areas in Montenegro, there are fewer producers of the original autochthonous "Lisnati cheese". Lisnati cheese got the name due to its unique foliate, leafy shape of slices, but it is also known as Kolašinski cheese, by name of the city where such cheese is mainly being produced. The cheese has a distinctive, soft, elastic and smooth structure, mild sour aroma and porcelain white colour. The most recognizable feature is an unusual form of slices that are extremely thin, like a sheet of paper or leaf. Hence the name "lisnati" = "foliate, leafy" (Mirecki *et al.* 2012).

Physical Structure and Appearance

Lisnati cheese belongs to "pate fillee" or "pasta filata" cheese. It is semi-hard, medium-fat cheese whose the most recognizable feature is an unusual form of slices that are extremely thin, up to 5mm. The shape of the slices is an irregular quadrangle, but dimensions of slices (width and length) are not defined precisely, also as the weight of slices (Fig. **5d**). Only cow's milk is used for Lisnati cheese production, and the yield is 9-11 kg of milk for 1 kg of cheese and it depends on the lactation period. The cheese doesn't have a crust, and the colour of the curd is white to pale yellow. The curd is elastic, layered, semi-hard, and leafy, and texture is closed, plastic and layered. Lisnati cheese has intensive fresh milky-sour smell and pleasurable milky-sour taste. There is no ripening period or that period is no longer than 7 days. Texture and taste are similar to mozzarella, lightly salted and slightly acid. The aroma is of fresh milk (Mirecki *et al.*, 2012).

Traditional Method of Production

The technology of this cheese is specific and unique and does not match the technology of any other cheese in Montenegro. Table **8** shows parameters of the technology of Lisnati cheese. The unique technology is the result of a special way of preparing raw milk for cheese production and specific way of coagulation and curd processing (Fig. **5a**). Only cow's milk is used for the production of Lisnati cheese. Cow's raw milk is heated up to 45^{0}C and left for 12 hours for acidification by the activity of natural microorganisms from milk. Thereafter, milk is skimmed and mixed with fresh raw milk and 10-15% of water. Rennet is added at 35^{0}C. After 30-45 minutes, coagulation was completed and the coagulum is cut and collected in the curd. Collecting the curd is done in whey (Fig. **5a**). Then, the curd is removed from the whey and pressed. Pressing is done by a wooden plank and putting the heavy stone on the plank (Fig. **5c**).

During pressing, the curd should be topped by warm whey (up to 55^{0}C) and very often folded. First, it is folded in half of the total amount of curd and later to ¼. This is repeated several times in order to gain as much and as thinner slices. It can be consumed immediately or left for ripening (Fig. **5d**).

Table 8. Parameters of technology of lisnati cheese (Mirecki *et al.*, 2012).

Technological Step	Parameter
Heating of raw milk	45^{0}C
Acidification of Raw milk	12h
Adding water to mixture	10-15%
Rennet adding	38-42^{0}C
Synerezis of whey	30 min
Curd forming	15-30min
Curd pressure	Wooden plank and stone
Curd folding	First on ½, than on ¼
Salting	Dry salt

Chemical Properties

Lisnati cheese is made from milk that needs to be prepared in a special way. Mirecki *et al.* (2012) analyzed milk prepared for Lisnati cheese production. The milk contained 2.39% of fat, 2.98% proteins, 3.65 lactose and solid non-fat was 7.41%. The milk of the evening milking stays overnight and in the morning the cream is separated. Milk of morning milking is added to evening milk, but also a certain amount of water. Because of this, the milk for the production of Lisnati cheese has low-fat content. The content of other milk components is also relatively low. Unfortunately, few studies have been done so far, which included this specific cheese. Chemical properties of Lisnati cheese are presented in Table **9** (Mirecki *et al.* 2012).

Fig. (5). Lisnati cheese production steps: **a**) Coagulation of milk; **b**) Curd in whey in cutting form; **c**) Curd pressing in warm whey; **d**) Lisnati cheese.

Table 9. Chemical properties (n=10) (Mirecki *et al.* 2012).

Parameters (%)	**Composition**
Fat	20.38
Proteins	22.40
Dry matter	48.43
Salt	2.60

Processing of milk and cheese is widespread in the mountains of Dinara, as in other mountainous areas of the Balkan Peninsula. The Kolasin area is the mountain area with many pastures suitable for cattle breeding. The cow's milk is processed into the famous soft cheese characteristic for this locality and it is not produced elsewhere in Montenegro. Texture and taste are similar to mozzarella, lightly salted and slightly acid. The aroma is of fresh milk. Lisnati cheese is still present in the area of Kološin municipality.

PRLJO CHEESE

A growing interest of consumers for low-fat and traditional dairy products was recently observed. Prljo is Montenegrin traditional low-fat cheese that meets both of the consumer's requirements. Another particular feature of this cheese is the fact that it's ripening takes place in sheepskin. The cheese is produced in the north-west part of Montenegro, in the area of the mountain Durmitor (Fig. **2**). Its production is directly connected with the production of Montenegrin famous extra-fatty dairy product Skorup. Since raw material for the production of Prljo is milk from which, after several days of separation, the fat layer (Skorup) is picked up, it can be said that Prljo is a by-product of Skorup production. But, on the contrary, Prljo cheese is a quality product that is very popular, especially for the local population. The content of fat in a dry matter that is less than 10% and dry matter content approx. 50% has classified Prljo cheese as low-fat, semi-hard cheeses. Because of the specific technology and quality, Prljo cheese deserves to be labeled by Protected Designation of Origin (PDO).

Table 10. Parameters of technology of Prljo cheese (Mirecki *et al.*, 2012).

Technological Step	Parameter
Heating of raw milk	32^{0}C
Rennet adding	32^{0}C
Curd forming	30 min
Curd pressure	Wooden plank and stone
Synerezis of whey	24 hours
Salted	Dry salt
Ripening	60-90 days

Traditional Method of Production

The production of Prljo cheese starts by warming up the milk from which Skorup was previously removed (Table **10**). When the milk temperature reaches 32^{0}C, the rennet is added. After 30 minutes curd is formed (Fig. **6a**). During the next 10 minutes curd is cut, first on bigger and after on smaller cubes, and then warmed up together with whey (Fig. **6b**). The curd parts are collected by hand under the whey, and cheese is formed in the form of a ball. Then the cheese is put in cheesecloth, squeezed and pressed by wooden plank and stone to extract as much whey as possible (Fig. **6c**). After 24 hours, the cheese is crumbled or cut into slices (Fig. **6d**), salted and put in sheepskin (Fig. **6e**). The ripening of Prljo cheese is 60 to 90 days (Fig. **6f**). The ripened Prljo cheese has hard, nubbly consistency and strong specific taste.

Physical Structure and Appearance

Prljo cheese is low-fat, semi-hard cheese. The cheese is in shape of small slices or chunks of even size. The cheese doesn't have a crust, and the colour of the curd is pale to intensive yellow, sometimes become greenish (Fig. **6**). The consistency is nubbly and closed. The yield of the cheese depends on the type of milk. For 1 kg of not ripened cheesed it is necessary to use 5-6 kg of ewe's milk or 7-10 kg of mixed (cow's+ewe's) milk. Prljo cheese has a very intensive milky-sour smell, and very often there is moderately smell of sheepskin. The aroma is characterized by strong milky-sour taste and salty, sometimes taste of boiled milk.

Fig. (6). Prljo cheese production steps: **a**) The curd; **b**) Curd dripping; **c**) pressing of a wooden plank with stones; **d**) cut into slices; **e**) Put into sheepskin and ripening 60-90 days; **f**) Prljo cheese.

Chemical Properties

Prljo cheese is a dairy product that is highly appreciated in Montenegro, especially in the areas where it is originally produced, northwestern mountainous area (Fig. **2**). Unfortunately, in spite of its popularity and importance, from a scientific point of view, this cheese almost doesn't exist. There are just a few scientific researches on the technology and quality of cheese, and in so few papers, the quality of milk was not investigated at all. The available results on quality of Prljo cheese are presented in Table **11**.

Table 11. The chemical properties of Prljo cheese.

Parameters	Dozet *et al.* (1996) Experimental Production	Dozet *et al.* (1996) Autochthonous Production	Adžić (2011)
Fat	4.11	4.73	5.10
Proteins	34.25	32.69	32.55
DM*	46.52	43.95	45.48
FDM**	8.91	10.73	11.21
Salt	5.12	4.32	2.48

*) dry matter; **) fat in dry matter

CONCLUSIONS

Montenegro has a wide range of very good quality traditional (autochthonous) dairy products. Pljevaljski cheese already has the protection of origin on the national level. But the other cheeses that are the subject of this scientific article, as Njeguški, Lisnati and Prljo cheese also deserve to have a label of protected designation of origin (PDO) or protected geographical indication (PGI). Relevant experts and institutions of Montenegro are working hard to provide a PDO and PGI labels for the best, not only dairy but also other agricultural products.

CONSENT FOR PUBLICATION

Not applicable.

CONFLICT OF INTEREST

The authors confirm that this chapter contents have no conflict of interest.

ACKNOWLEDGEMENTS

Declared none.

REFERENCES

Adžić, N, Ljumović, M, Petrović, S, Tomović, M, Dozet, N, Stanišić, M & Perović, M (1984) Kvalitet mlijeka za proizvodnju njeguškog sira. *Mljekarstvo,* 34, 76-81.

Adžić, J (2011) *Tehnologija proizvodnje autohtonog sira Prljo Specijalistički rad Metalurško-tehnološki fakultet Podgorica* . Crna Gora.

Bojanić-Rašović, M, Mirecki, S, Nikolić, N, Vučinić, S, Ivanović, I & Rašović, R (2010) *Microbiological and chemical quality of autochthonous cheeses in Montenegro (in Montenegrin) Prehrambena industrija, Mleko i mlečni proizvodi* Serbia, Novi Sad 127-33.

Codex Standard 283 (1978) Codex general standard for cheese. Codex Stand 283-1978.

Dozet, N, Adžić, N, Stanišić, M & Živić, N (1996) *Autohtoni mliječni proizvodi Poljoprivredni institut Podgorica.*SILMIR, Beograd.

Mirecki, S (1998) Kvalitet kozijeg njeguškog sira, kozijeg mlijeka i surutke, III Jugoslovenski simpozijum prehrambene tehnologije, Zbornik radova. *Sveska,* IV, 85-9. (Beograd).

Mirecki, S & Adžić, N (2006) The chemical composition and hygienic quality of Pljevaljski cheese (in Montenegrin). *Prehrambena Industrija,* 17, 64-8.

Mirecki, S (2012) Protection of traditional dairy products in Montenegro. *Poljopr Sumar,* 57, 27-42. [Podgorica, Montenegro].

Mirecki, S, Ivanović, I & Nikolić, N (2012) Characteristics of Montenegrian autochtonous "Lisnati cheese". *J Hyg Eng Des,* 1, 320-4.

Mirecki, S, Konatar, Z, Marković, M, Vuković, N, Nedović, B, Ćorović, LJ & Tomović, T (2013) *Zaštita oznake porijekla Pljevaljskog sira Elaborat NVO"Pljevaljski sir.*Pljevlja.

Mirecki, S & Konatar, Z (2014) Technology and Quality of Pljevaljski cheese – Traditional Montenegrin Dairy Product. *J Hyg Eng Des,* 6, 208-14.

Mirecki, S, Popović, N, Antunac, N, Mikulec, N & Plavljanić, D (2015) Production technology and some quality parameters of Njeguški cheese. *Mljekarstvo,* 65, 280-6.

Popović, N (2014) Njeguški sir *Specijalistički rad Biotehnički fakultet Podgorica, Podgorica.*

Vesković-Moračanin, S, Mirecki, S, Trbović, KD, Turubatović, RL, Kurćubić, SV & Mašković, ZP (2012) Traditional manufacturing of white cheeses in brine in Serbai and Montenegro – Similarities and differences. *Acta Periodica Technologica, APTEFF,* 43, 107-13.
[http://dx.doi.org/10.2298/APT1243107V]

CHAPTER 3

Traditional Cheese Maturing in Lambskin Sacks from Dalmatian Region

Jadranka Frece[1,*], **Deni Kostelac**[1], **Marija Vrdoljak**[2], **Iva Čanak**[1], **Željko Jakopović**[1], **Marko Jelić**[2] and **Ksenija Markov**[1]

[1] *Faculty of Food Technology and Biotechnology, Department of Biochemical Engineering, University of Zagreb, Pierottijeva 6, HR 10000 Zagreb, Croatia*

[2] *University of Applied Sciences "Marko Marulić", Petra Krešimira IV 30, HR-22300 Knin, Croatia*

Abstract: Sack cheese is a highly regarded traditional cheese matured anaerobically in lambskin, still produced in local communities in the south of Croatia (Dalmatian region). Even though the production has been adapted to new technologies, the basic production parameters and cheese properties remain unchanged and unique. Lambskin cheese is prepared from unpasteurised sheep's, cow's and goat's milk and their mixtures, according to the technology of semi-hard cheese production, without the addition of starter cultures. Milk is acidified with its natural microflora. The specific characteristic of this cheese is an anaerobic maturation that takes about 2 months in lamb or goatling skin locally called a sack. Due to the intense processes of lipolysis and proteolysis during ripening, mature cheeses have a specific and highly appreciated spicy flavour and aroma. According to the moisture content in the non-fat dry substance, these cheeses may belong to a group of hard or semi-hard cheeses, and by the fat content in the total dry matter, they may be full or reduced fat cheeses. Production of sack cheese takes place in small batches on family farms and has a mostly local character. The best way to preserve specific quality of traditional cheese is to use an indigenous starter culture, however, lambskin cheese is traditionally produced without adding one. Biochemical analysis of milk, used for lambskin cheese, identified indigenous bacterial species, among which the most important are *Lactobacillus plantarum*, *Lactococcus lactis*, *Lactobacillus curvatus* and *Staphylococcus xylosus*.

Keywords: LAB, Lambskin, Ripening, Sack cheese, Starter culture.

INTRODUCTION

Different types of cheese that ripen in an animal skin sack are produced in various geographic areas and have different local names, such as "sir iz mišine" (Croatia),

* **Corresponding author Jadranka Frece:** Faculty of Food Technology and Biotechnology Department of Biochemical Engineering University of Zagreb Pierottijeva 6, HR 10000 Zagreb Croatia; Tel/Fax: +385 1 4836083; E-mail: jfrece@pbf.hr

Celile Aylin Oluk & Oya Berkay Karaca (Eds.)

"sir iz mijeha" (Bosnia and Herzegovina) and "Tulum" (Turkey) (Kivanc, 1989; Yilmaz *et al.*, 2005; Tudor Kalit *et al.*, 2010). On family farms in the Republic of Croatia, traditional regional dairy products are produced according to old recipes. In Croatia, about thirty different types of autochthonous cheeses are insufficiently investigated and are threatened to perish with time (Matutinović *et al.*, 2007). Traditional autochthonous cheeses represent a valuable heritage, and the recipes for them have been passed down for generations, as is the case with Croatian cheese maturing in lambskin sacks. Production technology has been adapted to new technical achievements, but despite all these modifications, the basic production parameters and characteristics of cheese in lambskin sack have remained unique and recognizable. Traditional production is usually carried out without the addition of starter cultures, so fermentation of these products depends on naturally present microflora that comes from milk or the environment and can result in difference in taste, consistency and microbiological quality of the final product. Depending on the type of microorganisms, microbial cultures in cheese production have different activity: acid production, flavour formation, gas formation, proteolysis, lipolysis, and inhibition of undesirable microorganisms (Markov *et al.*, 2010). In the cheese production, the dominant microbial population are lactic acid bacteria (LAB), which play a significant role in protection against degradation, since they produce lactic acid and bacteriocins, and thereby inhibit the growth of pathogenic bacteria and improve safety, stability and durability of the product. Recently, addition of starter cultures during production of cheese in lambskin sack was investigated (Vrdoljak, 2016; Frece *et al.*, 2016). Apart from the application of lactic acid bacteria as bioprotectors, their contribution to aromas, texture and nutritional value of the product is also important (Admassie, 2018).

GENERAL CHARACTERISTICS

Cheese in lambskin or sheepskin sacks (Sir iz mišine) is one of Croatian traditional farm products dating back to the times when Illyrians and Thracians herded their sheep on the pastures of Dinara Mountain (Frece *et al.*, 2016). They used sheepskin sacks to store and transport cheese from the mountain to the settlements in the valley (Kalit *et al.*, 2014a). Similar cheese matured in sack made from animal skin is also produced in Bosnia and Herzegovina, Montenegro, Turkey, and Lebanon (Kalit *et al.*, 2014b). In contrast to industrial food products, traditional cheese is of particular interest to the consumers who care about the nature, origin, and nutritional value of food (Lavasani *et al.*, 2012). They owe their reputation to the unique organoleptic properties and indigenous microorganisms living in raw milk or natural starter cultures (Serhan *et al.*, 2013). Cheese in lambskin sack is produced on the family farms of Dalmatian Zagora, Velebit and one part of Lika, where it is highly appreciated as a product by the local population (Hayaloglu *et al.*, 2013a). The cheese has a rectangular shape. The colour of the cheese depends on the type of milk used in production and varies from white to yellowish. The consistency of the cheese is firm, but not too hard or dry. Regardless of their geographical origin and production technology, cheeses that ripen in animal skin have a distinct smell and characteristic spicy flavour depending on the type of milk. They are produced mainly on family farms, with an annual production of about 5,000 to 10,000 kg. Recently, beneficial effects of naturally produced food have been emphasized. As a result, more and more consumers are interested in the potential of functional foods to improve their health. Probiotic bacteria are also among the functional supplements in the diet, and if used in sufficient amounts, they have a positive effect on the balance of intestinal microflora and can be used as biotherapeutics as well. Functional food is considered to be food containing ingredients that positively affect one or more targeted functions in the body. Raw milk can also contain pathogenic bacteria that have been raising public health concerns since the beginnings of dairy industry. The most common pathogenic bacteria in raw milk and milk products are *Salmonella*, *Listeria monocytogenes*, *Staphylococcus aureus* and *Escherichia coli* (Markov *et al.*, 2011). It is known that the microbiological quality of dairy products can be improved by adding starter cultures, such as LAB, which prevent the growth of pathogenic and spoilage microorganisms. The best way to preserve specific qualities of traditional cheese, such as flavour and aroma, is to use an indigenous starter culture (Frece *et al.*, 2016). However, lambskin cheese is traditionally produced without adding one, but newer research states that adding probiotic cultures improved the aromatic profile of the sack cheeses, as they showed better sensory properties than the control cheese (without added culture). Furthermore, they did not change the authentic taste of the traditional cheese produced without culture addition (unpublished data).

PRODUCTION OF CHEESE IN A SACK

The sack can be described as a "bag" made of skin of a whole lamb or goat (Kalit *et al.*, 2010). One of the assumptions is that the production of cheese in a lambskin sack started quite accidentally by spontaneous fermentation of milk in a sack which was used to store the milk. The cheese in a sack was originally produced from raw sheep's milk obtained from sheep extensively fed on the natural pastures of the sub-Mediterranean area. Today, this cheese is made from sheep, cow, goat's milk or their mixtures, and cheese production technology differs to a certain extent from the previously used one. In addition to the difference in the milk type, the breed of the sheep is significantly different. Earlier, the sheep breeds used were exclusively autochthonous (Dalmatian and Bosnian strains), but today more productive breeds such as Eastern Frisian sheep are used. The size of the flock is usually 20 to 30 sheep, although in recent years the production has intensified and flocks with 150 to 300 sheep or more are present. A significant feature that has survived from ancient times is the way of cattle-breeding, in which sheep stay in nearby mountain pastures from May to October. During that time, the production of sack cheese also takes place. The intensification of production in recent years led to the improvement of breeding conditions, and there is also a tendency for earlier separation of lamb from sheep in case of a more significant production of cheese in a sack. Furthermore, a significant change is the use of industrial rennets instead of natural ones, as well as the use of modern dairy and cheese production equipment. Two factors influenced the improvement of sheep's milk processing in Dalmatia: the first was the gradual introduction of factory brewing and the second the establishment of sheep-dairy cooperatives. Such improvements lead to better quality of sack cheese, which was often inadequate, especially because of the poor hygiene and technological conditions. In summer and autumn, sack cheese is ready for sale on a local market, although today almost everything is sold directly on the farm.

Preparation of the Sack

The preparation of the sack (in the vicinity of Knin) is carried out as follows: after cutting off the head, carpal and tarsal joints, the skin is removed from the body. Then the remains of muscle and fat tissue from the inside of the skin are removed and the skin is turned so that the skin hair is facing outwards and the hair is removed from the skin using a sharp razor. When shaving and cleaning, special care should be taken to prevent the skin from being damaged because during ripening of the cheese, any air entering would spoil the cheese production. The sack is then immersed in cold water and washed well. All openings of the sack are then closed and tied tightly with a clean rope. Around the neck opening, small holes (approximately 2 cm wide) are left, where wooden rods with a diameter of 0.5 cm and a length of 20 cm are placed, preventing the sliding of the rope when filling the sack with air. Air filling is carried out for as long as the sack does not reach the drying appropriate shape. The sack is placed in a shady place for drying for three to five days in the summer period, and for up to 20 days in the winter, followed by smoking. The dried sack is then vented by the openings on the neck and on the bottom (tail), and immersed in cold water or whey to soften for easier manipulation before the first use.

During ripening, the filled sack needs to be nurtured; in the first days, the sack is turned over each day two to three times, and in the later stages of ripening, every two to three days. Improper processing and a poorly dried sack are ideal for the growth and reproduction of spoilage microorganisms that can cause changes in organoleptic properties of this cheese. The most common mistakes that come up during the production of the sack are skin cutting during removal of fat or shaving, improper skin removal, and residual tissue parts on the skin preventing the proper drying of the sack.

Quality of Milk for the Preparation of Cheese in a Sack

Cheese in a sack is traditionally produced from full-fat or skimmed cow's, sheep's and goat's milk, or their mixtures. The common properties of most autochthonous cheeses, as well as cheese in a sack, are the use of raw, unpasteurized milk for ripening. Milk affects the specificity (autochthonousness) of cheeses, mainly due to lactic acid fermentation bacteria, which contribute to the characteristic and recognizable taste, smell and consistency of a given cheese. The quantity and composition of milk are influenced by numerous factors: breed, stage and lactation, conditions of keeping and breeding, nutrition, health, climatic conditions, age of animal, season, and others (Mioč *et al.*, 2007). Sheep's milk contains more dry matter, milk fat, protein and ash compared to cow's and goat's milk, and is therefore very suitable for processing of cheeses (Table **1**). Cheeses from sheep's milk are highly valued, and the production yield/income is higher than in other species. The amount of milk fat in this milk varies during lactation (from 3.5% at the beginning of lactation to 11.7% at the end); the characteristic fatty acid composition gives a specific taste and smell to both milk and cheese (Mioč *et al.*, 2007).

Table 1. The average chemical composition of sheep's milk for the production of cheese in a sack (Source: Tudor *et al.*, 2008).

Parameter	X ± SD
Dry matter (g / 100 g)	19.65 ± 1.84
Milk fat (g / 100 g)	8.22 ± 1.50
Proteins (g / 100 g)	6.37 ± 0.74
Lactose (%)	4.23 ± 0.29
Non-fat dry matter (%)	11.40 ± 0.46
pH	6.26 ± 0.33
freezing point (°C)	-0.597 ± 0.05

Hygiene and quality of milk are crucial for the production of quality cheese in a sack. Hayaloglu *et al.* (2007) suggest that the use of different types of milk (cow's, sheep's or goat's) is significantly influenced by Tulum cheese microbiology during 210 days of ripening. Research on the hygienic quality of sheep's milk for the production of cheese in a sack has shown that most samples do not meet the statutory regulations (Kaić *et al.*, 2008). In their research, which also included sheep's milk samples from Knin area, Frece *et al.* (2016) detected total aerobic bacteria *Enterobacteriaceae* and *Staphylococcus aureus* in the amount higher than the one required by the regulations. They also isolated naturally present dominant LAB cultures: *Leuconostoc mesenteroides and Lactococcus lactis* ssp. *lactis.*

During the research conducted by Kalit and Havranek (2001) on Tounj's cheese, it was proved that without the use of sanitary agents in milk production, the prescribed hygienic quality of milk and cheese cannot be met.

Cheese Production

Cheese in a sack is traditionally produced from raw full-fat sheep's, cow's or mixed sheep's and cow's milk without the use of dairy culture, and according to semi-hard cheese production technology. Since cheese in a sack production technology is not standardized, it is different from manufacturer to manufacturer. Seeing that in the production of cheese in a sack on certain farms natural rennet is still being used, its proper preparation is a critical factor in the production of this cheese. For its preparation, calve, lamb, buckling and swine stomachs washed in warm water, air-dried and smoked are used. After smoking, they are washed again in warm water, sliced and put in a jar, submerged in whey and left alone for a couple of days. One litre of rennet is added to 25 litres of milk (Tejkal *et al.*, 1913). Some authors claim that it is best to use calf stomachs of the age of 1 to 4 to be sure that the rennet will be of sufficient strength. The milk from evening and morning milking is used in production of this cheese. The milk is constantly kept in a chilled state. The renneting of the milk is carried out with natural or commercial rennet and the renneting time varies between 30 and 60 minutes, depending on the temperature ranging from 32°C to 37°C and the strength of the rennet. The end of renneting is checked by a "finger prick" method, where the curd must sharply break. The curd (Fig. **1**) is then sliced with a knife and left to rest until whey starts to separate and becomes green to yellowish in colour. The grain is heated and mixed manually until the temperature reaches the range from 35°C to 40°C, and after drying it usually takes on the size of hazelnut or walnut. Cheese clot is placed in cloths or moulds and wringed. After squeezing, cheese clot is cut to pieces (Fig. **2**), then salted with large grain salt and placed in a sack. Since it is an anaerobic ripening, it is especially important when filling the sack to remove all the air from the sack. According to some authors, one-time filling of the sack ensures better cheese quality. Ripening in a sack (Fig. **3**) takes place in a ripening chamber for two to three months, at a temperature of 12°C to 15 °C and relative humidity of about 85%. During this time, the cheese develops characteristic organoleptic properties and becomes ready for consumption. In a single sack, an average of 15 to 20 kg of cheese is produced, sometimes even up to 40 kg. Both sides of the skin can be used to store cheese, *i.e.* when the outer layer of skin (wool side) is in contact with cheese or when the cheese is in contact with the inside of the skin.

Fig. (1). Cheese curd (Source: M. Vrdoljak).

Fig. (2). Cheese clot cutting with knife (Source: M. Vrdoljak).

The authors state that 1 kg of cheese requires 7 to 9 litres of fresh milk, and from 100 litres of full-fat milk 12 to 15 kg of cheese can be produced (Bijeljac, 2004). The available data also state that for 1 kg of cheese, 7 to 9 litres of fresh sheep's or mixed sheep's and cow's milk is necessary. Mature cheese Fig. (**4**) is extracted from sack and kept in low temperature conditions, adequate air humidity and prescribed hygiene. According to the content of fat in the dry matter, cheese in a sack belongs to the group of full-fat cheeses, and by the content of water in non-fat matter, to hard cheeses (Tudor Kalit *et al.*, 2009).

Fig. (3). Cheese in a lambskin sack during ripening. (Source: M, Vrdoljak).

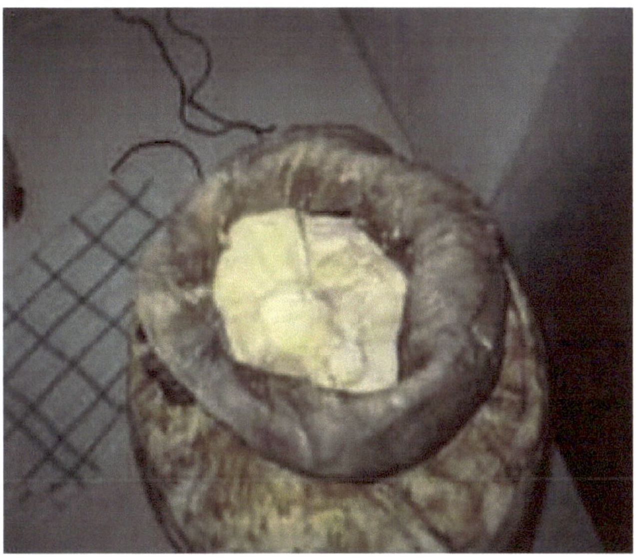

Fig. (4). Cheese in a lambskin sack. (Source: M. Vrdoljak).

MICROBIOLOGICAL AND PHYSICOCHEMICAL PROPERTIES OF CHEESES THAT RIPEN IN ANIMAL SKIN

Microbiological, chemical and physical properties of cheese ripening in animal skin depend on storage conditions, production conditions and procedures, milk quality, and the experience of the manufacturer. The natural population of lactic acid bacteria directly affects the sensory characteristics of traditional cheeses through biochemical reactions that occur as a result of the action of their enzymes (Casey *et al.*, 2006). Most bacteria come from raw milk but can also be added as starter cultures. Microorganisms involved in cheese-making and ripening can be divided into starter and adjunct cultures, and into non-starter LAB selected to improve the overall quality of the cheese (Ayad *et al.*, 2009). Unlike the starter bacteria, native bacteria are much more resistant, survive the solubility process, and therefore dominate during cheese ripening. The role of starter cultures is the formation of lactic acid, which lowers the pH value of milk. The resulting lactic acid, combined with the addition of rennet, by heating and mixing, results in the formation of clot and removal of whey, contributes to the enzymatic composition and affects the texture of the cheese, and also contributes to ripening by the produced enzymes (McSweney, 2004). The obtained cheese has lower water content (up to 35%) and a lower pH (4.6 to 5.2) than milk. Kaić *et al.* (2012) isolated and investigated the composition of the natural population of LAB present in the sack cheese. The presence of all three major genera of LAB: *Lactococcus*, *Enterococcus* and *Lactobacillus* was established. The most dominant genera were *Lactococcus* (6.02 to 8.10 log CFU/g), then *Lactobacillus* (5.65 to 6.91 log CFU/g) and *Enterococcus* (4.84 to 7.26 log CFU/g). Frece *et al.*, 2016 investigated cheese in lambskin sacks from local producers in the area of Knin, Croatia, and the obtained results showed that *Lactococcus lactis* and *Lactobacillus paracasei* prevailed in cow's cheese, whereas *Leuconostoc mesenteroides* and *Lactobacillus plantarum* prevailed in sheep's cheese. In the lambskin sacks, *Leuconostoc mesenteroides* and *Lactobacillus plantarum* were predominant. Beside LAB, proteolytic bacteria, lipolytic bacteria, Enterobacteriaceae, yeasts and moulds can also be present in traditional cheeses (Ceylan *et al.*, 2007; Frece *et al.*, 2016). Indigenous starter cultures, consisting of LAB isolated from traditional Mišina cow or sheep cheese, and their use might lead to the improvement of product quality and result in the preservation of characteristics that define such traditional cheese. This was proven by Vrdoljak, (2016) who demonstrated that the cheese produced with the *L. plantarum* B and *Lactococcus lactis* S1 cultures (previously isolated from cheese in lambskin sack) contained a higher proportion of milk fat, fat in dry matter and dry matter (except for 15 days of ripening) throughout all days of ripening, compared to cheese without added LAB cultures. Cheeses with added cultures showed sensory differences from the original cheese in aroma, texture, body, and taste. Adding cultures had a positive effect on the sensory properties and chemical profile of volatile aromatic compounds.

Available literature offers only a few publications on physical and chemical properties of cheese matured in an animal skin sack. The study by Tudor *et al.* (2009), conducted on a sheep sack Mišina cheese, determined that after 75 days of ripening, cheese in a sack turns into a full-fat cheese having a total fat content of 36.3% and protein content of 24.9%.

The results of other authors concerning cheeses in an animal skin sack produced in different geographical areas using different productions technologies and ripened for a different number of days, report their fat content to be 25.4 - 39.4% and their protein content to be 20.3 - 33.4% (Dozet *et al.*, 1980; Dozet *et al.*, 1996; Yilmaz *et al.*, 2005; Jovanović *et al.*, 2007). The ash content in such in-sack cheeses was reported to span from 4.44% to 7.08% (Dozet *et al.*, 1980; Dozet *et al.*, 1996).

At some time-point of ripening, free fatty acids (FFAs) are considered to be an indicator of the extent of lipolysis (Tudor Kalit *et al.*, 2014). FFAs play an important role in the cheese flavour development or as the precursors of volatile compounds, such as ketones, lactones, alcohols, esters and aldehydes (Collins *et al.*, 2003).

Lešić *et al.*, (2016) demonstrated the presence of 21 fatty acids in cow and sheep cheese ripened in skin sack. The analysis showed a significant difference in the content of 9 fatty acids depending on the cheese type (cow's or sheep's milk) and a significant difference between fatty acid groups present in 2 cheese types. This could be due to the fact that sheep milk cheeses have significantly more individual saturated (SFA) and polyunsaturated fatty acids (PUFA) in comparison to cow milk cheeses. On the other hand, Rodrigues *et al.*, (2012) obtained no statistically significant difference in SFA and monosaturated fatty acid (MUFA) content of probiotic cheeses during 60 ripening days, which could be the result of the reduced bacterial lipolytic activity under given conditions.

Enzymes originating from starter cultures (proteinases and peptidases) play a major role in the formation of small peptides and amino acids that serve as precursors of the compounds responsible for cheese flavour (Urbach, 1997). Vrdoljak, (2016) demonstrated that from the set of 66 investigated aroma components, 17 significantly affected sensory properties of cheeses in lambskin sack, produced with the addition of selected indigenous LAB strains (isolated from sack cheese), in comparison to the control cheese samples without the added LAB cultures.

CONCLUDING REMARKS

New production technologies that are based on research mentioned in this chapter strive for the future use of autochthonous LAB cultures in cheese production. The application of the aforementioned cultures in sack cheese production has many benefits, and undeniably improves the microbiological quality and authenticity of the product, while providing uniform taste and aroma. Consumers actively seek natural products that have beneficial effects on human health. The addition of autochthonous LAB cultures with probiotic potential makes cheese in a sack a valued functional food product that has great market potential because of consumer interest in traditional, natural and healthy products.

CONSENT FOR PUBLICATION

Not applicable.

CONFLICT OF INTEREST

The authors confirm that this chapter contents have no conflict of interest.

ACKNOWLEDGEMENTS

Declared none.

REFERENCES

Admassie, M (2018) A Review on food fermentation and the biotechnology of lactic acid bacteria. *World J of Food Sci and Techno,* 2, 19-24.
[http://dx.doi.org/10.11648/j.wjfst.20180201.13]

Ayad, EHE (2009) Starter culture development for improving safety and quality of Domiati cheese. *Food Microbiol,* 26, 533-41.
[http://dx.doi.org/10.1016/j.fm.2009.03.007] [PMID: 19465251]

Bijeljac, S (2004) *Sir iz mijeha Udruga" „Pramenka". Mostar.*

Casey, MG, Häni, JP, Gruskovnjak, J, Schaeren, W & Wechsler, D (2006) Characterisation of the non-starter lactic acid bacteria (NSLAB) of Gruyère PDO cheese. *Lait,* 86, 407-14.
[http://dx.doi.org/10.1051/lait:2006020]

Collins, YF, Mc Sweeney, PLH & Wilkinson, MG (2003) Lipolysis and free fatty acid catabolism in cheese: A review of current knowledge. *Int Dairy J,* 13, 841-66.
[http://dx.doi.org/10.1016/S0958-6946(03)00109-2]

Ceylan, ZG, Çağlar, A & Çakmakçi, S (2007) Some physicochemical, microbiological and sensory properties of tulum cheese produced from ewe's milk via a modified method. *Int J Dairy Technol,* 60, 191-6.
[http://dx.doi.org/10.1111/j.1471-0307.2007.00326.x]

Dozet, N, Stanišić, M, Bijeljac, S & Medan, N (1980) Mogućnost iskorištavanja mlijeka na brdsko-planinskom području Hercegovine. *Mljekarstvo,* 30, 67-72.

Dozet, N, Adžić, N, Stanišić, M & Živić, N (1996) *Autohtoni mliječni proizvodi Poljoprivredni institut.*Podgorica i Silmir, Beograd.

Frece, J, Vrdoljak, M, Filipčić, M, Jelić, M, Čanak, I, Jakopović, Ž, Pleadin, J, Gobin, I, Dragičević, TL & Markov, K (2016) Microbiological quality and variability of natural microbiota in Croatian cheese maturing in lambskin sacks. *Food Technol Biotechnol,* 54, 129-34. [http://dx.doi.org/10.17113/ftb.54.02.16.4418] [PMID: 27904402]

Hayaloglu, AA & Karabulut, I (2013a) SPME/GC-MS characterization and comparison of volatiles of eleven varieties of Turkish cheeses. *Int J Food Prop,* 16, 1630-53. [http://dx.doi.org/10.1080/10942912.2011.587625]

Hayaloglu, AA, Cakmakci, S, Brechany, EY, Deegan, KC & McSweeney, PLH (2007) Microbiology, biochemistry, and volatile composition of Tulum cheese ripened in goat's skin or plastic bags. *J Dairy Sci,* 90, 1102-21. [http://dx.doi.org/10.3168/jds.S0022-0302(07)71597-7] [PMID: 17297085]

Jovanović, S, Maćej, O, Kovačević, Z, Vučić, T & Marinković, S (2007) Kvalitet Ljubinjskog autohtonog sira iz mijeha *Proceedings of the 47th Croatian and 7th International Symposium on Agriculture,* Opatija 697-700.

Kalit, S & Lukač Havranek, J (2001) Primjena sanitacijskih sredstava u proizvodnji i preradi mlijeka. *Mljekarstvo,* 51, 197-204.

Kivanç, M (1989) A survey on the microbiological quality of various cheeses in Turkey. *Int J Food Microbiol,* 9, 73-7. [http://dx.doi.org/10.1016/0168-1605(89)90039-1] [PMID: 2641276]

Lešić, T, Pleadin, J, Krešić, G, Vahčić, N, Markov, K, Vrdoljak, M & Frece, J (2016) Chemical and fatty acid composition of cow and sheep cheeses in a lamb skin sack. *J Food Compos Anal,* 46, 70-7. [http://dx.doi.org/10.1016/j.jfca.2015.11.007]

Markov, K, Perši, N, Pleadin, J, Čvek, D, Radošević, V, Delaš, F, Duraković, L & Frece, J (2010) Characterization of natural microflora and chemical parameters in fresh domestic cheese. *Vet Stn,* 42, 211-8.

Matutinović, S, Rako, A, Kalit, S & Havranek, J (2007) Značaj tradicijskih sireva s posebnim osvrtom na Lećevački sir. *Mljekarstvo,* 57, 49-65.

McSweeney, PLH (2004) Biochemistry of cheese ripening. *Int J Dairy Technol,* 57, 73-190. [http://dx.doi.org/10.1111/j.1471-0307.2004.00147.x]

Mioč, B, Pavić, V & Sušić, V (2007) *"Ovčarstvo" Hrvatska mljekarska udruga, Zagreb.*

Serhan, M & Matar, J (2013) Characterization of four Lebanese artisanal goat milk cheeses: Darfiyeh, Aricheh, Shankleesh and Serdale by physico-chemical, microbiological and sensory analyses. *J Food Agric Environ,* 11, 97-101.

Shahab Lavasani, AR, Ehsani, MR, Mirdamadi, S & Mousavi, MAEZ (2012) Changes in physicochemical and organoleptic properties of traditional Iranian cheese Lighvan during ripening. *Int J Dairy Technol,* 65, 64-70. [http://dx.doi.org/10.1111/j.1471-0307.2011.00724.x]

Tejkal, LJ (1913) *"Sirarstvo u Dalmaciji" Nagragijena tiskara E.*Vitaliani, Zadar.

Tudor, M, Kalit, S, Havranek, J, Kaić, D & Vrdoljak, M (2009) Tehnologija proizvodnje i kvaliteta sira iz mišine. *Croatian and 4th International Symposium on Agronomy, Opatija* . Croatia

Tudor Kalit, M, Kalit, S & Havranek, J (2010) An overview of researches on cheeses ripening in animal skin. *Mljekarstvo,* 60, 149-55.

Tudor Kalit, M, Kalit, S, Delaš, I, Kelava, N, Karolyi, D & Kaić, D (2014a) Changes in the composition and sensory properties of Croatian cheese in a lamb skin sack (Sir iz mišine) during ripening. *Int J Dairy Technol,* 67, 255-64. [http://dx.doi.org/10.1111/1471-0307.12117]

Tudor Kalit, M, Marković, K, Kalit, S, Vahčić, N & Havranek, J (2014b) Application of electronic nose and electronic tongue in the dairy industry. *Mljekarstvo,* 64, 228-44. [http://dx.doi.org/10.15567/mljekarstvo.2014.0402]

Urbach, G (1997) The chemical and biochemical basis of cheese and milk aroma". *Microbiology and Biochemistry of Cheese and Fermented Milk:* 253-98. [http://dx.doi.org/10.1007/978-1-4613-1121-8_8]

Vrdoljak, M (2016) Probiotičke kulture Lactobacillus plantarum B i Lactococcus lactis subsp. lactis S1 u poboljšanju funkcionalnih svojstava sira iz mišine. *PhD Thesis,* Faculty of Agriculture, University of Zagreb.

Yilmaz, G, Ahmet, A & Akin, N (2005) The effect of microbial lipase on the lipolysis during the ripening of Tulum cheese. *J Food Eng,* 69, 269-74. [http://dx.doi.org/10.1016/j.jfoodeng.2004.08.017]

CHAPTER 4

Brazilian Traditional Cheeses

Maria C. E. Ribeiro[1,*], **Débora P. Baptista**[2], **Mayara S. Queirós**[2], **Karina S. Chaves**[3] and **Oya B. Karaca**[4]

[1] *Faculty of Technology of Capão Bonito, FATEC, Capão Bonito, SP, Brazil*

[2] *Food Technology Department, Faculty of Food Engineering, University of Campinas, SP, Brazil*

[3] *Federal University of Mato Grosso, UFMT, Campus Universitário do Araguaia, Barra do Garças, MT, Brazil*

[4] *Karatas School of Tourism and Hotel Management, Hotel Management Department Cukurova University 01330 Adana Turkey*

Abstract: Agriculture sector is considered one of the most important economic activities in Brazil contributing to the growth of the country's economy. According to the USDA (United States Department of Agriculture), Brazil is the fourth largest producer and consumer of cheeses and behind Russia, the United States, and the European Union. The cheese production in Brazil was consolidated with the arrival of European immigrants, especially Dutch, Danish, and Italians. Brazilian cheese market was diversified and currently presents a wide variety inspired in European cheeses like Prato cheese and typically Brazilian cheeses such as Requeijão, Coalho cheese, Manteiga cheese, and Minas cheese. The growing knowledge of biochemical and microbiological characteristics of the cheese allowed the establishment of manufacturing technologies for the production in the industrial scale of some cheese varieties that were previously made in a totally artisanal way. In order to strengthen the artisanal production and the regionality of some cheese varieties, including those produced from raw milk, the Brazilian legislation has been endeavoring to ensure cheese microbiological quality, especially regarding raw milk quality and minimum ripening time. This attempt for improving the Brazilian legislation aims to ensure the traditional production and the insertion of small producers into the market. This chapter will present an overview of the cheese industry in Brazil, the history, regional characteristics, manufacturing technology, physicochemical, microbiological and sensory aspects of the main cheeses, as well as the challenges encountered by the artisanal producers and the dairy industry.

Keywords: Endogenous microbiota, Geographical indication, Ripening, Traditional cheeses, Brazil.

* **Corresponding author Maria C. E. Ribeiro:** Faculty of Technology of Capão Bonito, FATEC, Rua Amantino de Oliveira Ramos, 60, Terras do Embiruçu, 18304-755, Capão Bonito, SP, Brazil; Tel/Fax: +55153542-2654; E-mail: maria.ribeiro10@fatec.sp.gov.br

Celile Aylin Oluk & Oya Berkay Karaca (Eds.)

INTRODUCTION

Brazil is the fourth largest producer and consumer of cheeses in the world, with production greater than 1 million tons in 2017, with a growth forecast of 2.5% for the year 2018 (Datamark, 2018; Food News, 2018). The cheese production was started in the colonization period by immigrants from European countries. They developed and adapted the technology of European cheeses manufacturing to each region of Brazil and favored the development of the great diversity of cheeses currently produced in the country. Fig. (**1**) shows the main cheeses produced in the different regions of the country.

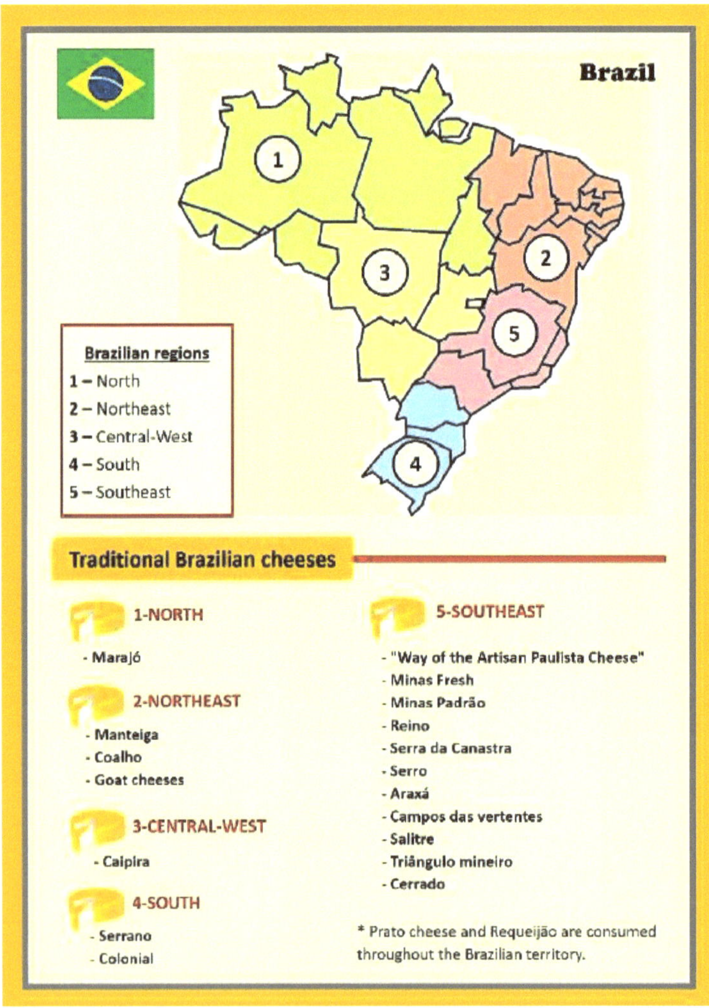

Fig. (1). Map of Brazil, with the main cheeses of the North, Northeast, Central-West, South and Southeast regions.

In the country, cheeses are produced both in artisanal and industrial scale. Artisanal cheeses are an important factor for the development of many small rural companies, with the appreciation of the culture and natural resources of each region of the country. The artisanal methods of production are quite rudimentary, passed through generations, do not present standardization and show variations among the producers, which promotes differences in the characteristics of the products. Due to the different types of artisanal cheeses in each region of the country, many states are seeking the protection of denomination of origin. In addition to the artisanal production, many kinds of cheese that are well-known throughout the national territory, present well-defined manufacturing technology on an industrial scale, and its manufacturing technology and characteristics are regulated by federal and state laws and resolutions.

In this context, the present chapter aims to present an overview of the history, origin, physicochemical and sensory characteristics, and manufacture protocol of the main types of Brazilian cheese produced both on artisanal and industrial scales.

PRATO CHEESE

Prato cheese is a typically Brazilian ripened cheese which is similar to the Dutch Gouda and Danish Danbo cheeses (Furtado & Lourenço Neto, 1994). The European cheeses Danbo and Gouda are characterized as cheeses obtained by low-scalding with hot water. It was firstly produced in the country by Danish immigrants in the southern region of the state of Minas Gerais in the 1920s (Furtado & Lourenço Neto, 1994; Spadoti *et al.*, 2011). Its manufacturing protocol was adapted to the local conditions, which resulted in the development of very particular sensory characteristics, different from Gouda and Danbo cheeses' (Furtado & Lourenço Neto, 1994). Today three different varieties of Prato cheese are found in the market: *Lanche*, *Cobocó,* and *Esférico*. The Prato cheese *Lanche*, widely used in sandwiches, was produced for the first time in the 1960s and is currently the most consumed Prato cheese variety in the country (Amarante, 2015).

Currently, Prato cheese is predominantly produced on an industrial scale. With an annual production of approximately 75 thousand tons in 2017, it is the third most produced cheese in Brazil and the main ripened cheese marketed in the country. Although Prato cheese is produced all over the country, the state of Minas Gerais is still the main producer, concentrating 27.30% of the national production (Datamark, 2018).

According to the Regulation of Industrial and Sanitary Inspection of Products of Animal Origin – RIISPOA, Prato cheese is obtained through the coagulation of pasteurized milk with rennet or other appropriate coagulating enzymes, supplemented by the action of specific bacteria. It is a semi-cooked, pressed, salted and ripened mass (Brazil, 2017). It is classified as a fatty (45.00 to 59.90% fat in dry matter), and medium moisture (36.00 to 45.90%) cheese (Brazil, 1996; Brazil, 1997a). This wide range of composition allowed by the legislation is indeed found as verified by several studies that characterized samples of Prato cheese marketed in Brazil. Moreno *et al.* (2002) found a mean moisture content of 40.68% (38.64 - 44.41%) and 54.07% of fat in dry matter (49.50 - 56.64%) in cheeses from four Brazilian states. Likewise, Baptista *et al.* (2017a), determined that the mean values of the composition of 10 commercial Prato cheeses were 42.92% (38.71 - 48.74%) moisture and 49.76% (44.46 - 55.75%) of fat in dry matter. Matera *et al.* (2018) analyzed 50 Prato cheese samples and reported a variation of 40.00% to 50.00% and 37.00% to 58.00% in moisture and fat in dry matter contents, respectively.

Prato cheese can be marketed in different forms, depending on its variety: *Lanche, Cobocó,* and *Bola (Esférico)*. Prato cheese *Lanche, Cobocó,* and *Bola* have a rectangular shape, a cylindrical shape, and a spherical shape, respectively (Brazil, 1997a). Table **1** presents the main sensory attributes of Prato cheese.

Table 1. Main sensory attributes of Prato Cheese (Brazil).

Sensory Characteristics of Prato Cheese	
Consistency	Semi-hard, elastic
Texture	Compact, smooth, with some evenly distributed round small eyes
Color	Yellow
Flavour	Characteristic
Aroma	Characteristic
Rind	Without, or smooth thin rind without cracks
Eyes	Some small eyes, evenly distributed, or without eye formation.

The manufacturing process of Prato cheese follows the general Dutch cheese-making protocol, such as Gouda cheese, with some adaptations Fig. (**2**). It is made from pasteurized cow's milk, and the first stages of production include the addition of starter culture, calcium chloride, urucum dye and coagulant (Silva, 2005). Prato cheese can be produced with the addition of homofermentative "O" type starter culture, consisting of *Lactococcus lactis* subsp. *lactis* and *Lactococcus lactis* subsp. *cremoris*, when the objective is to obtain a compact and closed texture. It can also be produced with the addition of heterofermentative "LD" type starter culture constituted of *Lactococcus lactis* subsp *lactis* biovar *diacetylactis* and *Leuconostoc*, which will lead to eye formation in cheeses (Furtado &

Lourenço Neto, 1994). Although "LD" type starter culture has been widely used for decades in the production of Prato cheese, with changes in consumption habits, "O" type starter culture has been widely used to favor the production of a compact cheese texture without eyes formation that is more suitable for slicing (Spadoti *et al.*, 2011).

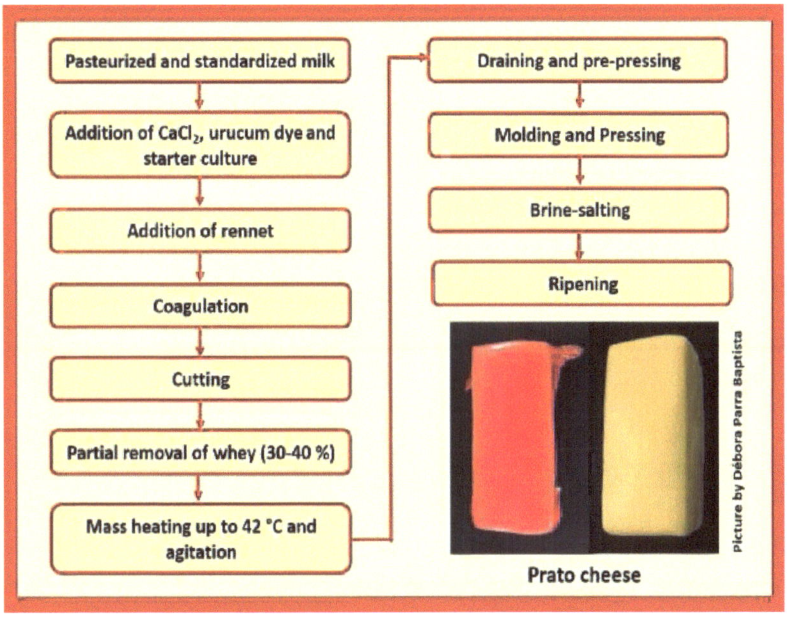

Fig. (2). Prato cheese manufacturing technology.

In the manufacture of Prato cheese, the milk is coagulated at a temperature of approximately 32 to 34 °C by the action of the coagulant enzyme. After coagulation, the curd is cut into 1 cm cubes and submitted to slow continuous stirring. Then, about 30.00 to 40.00% of the whey is drained off, and the curd is slowly heated until 42 °C (1 °C every 3 min) by the addition of hot water (70 to 80 °C) (Furtado & Lourenço Neto, 1994; Silva, 2005). Alternatively, the curd can be heated by indirect heating with the use of steam in the vat jacket. The traditional process of curd heating by the addition of hot water results in cheeses with a mild taste and not very high acidity, due to the dilution of the whey, while the indirect heating produces cheeses with higher acidity and a more intense flavor (Spadoti *et al.*, 2011). After cooking, the whey is drained off, and the curd is prepressed for 15-20 min (50-60 lbs/in^2) and placed into plastic molds. The cheeses are then pressed for about 2 hours with successive turns. After pressing, the cheeses made with the addition of Direct Vat Set (DVS) starter cultures can be kept at room temperature for fermentation to lower its pH to the range of 5.70 to 5.90 (Furtado & Lourenço Neto, 1994; Silva, 2005). Finally, cheeses blocks are salted in brine (20% NaCl) at a temperature of 10 to 12 °C for a variable period, according to the size of the cheese, to obtain a salt content between 1.60 and 1.70%. After salting, the cheeses are dried, vacuum-packed into heat-shrinkable plastic bags and ripened at a temperature of 12 to 14°C (Furtado & Lourenço Neto, 1994).

The development of the sensory characteristics of the cheese is directly related, among other factors, to the type of starter culture used, type of heating applied and the ripening time. Depending on the characteristics desired by the manufacturer, Prato cheese can be produced with the addition of different starter cultures. The use of "O" type starter cultures favors the manufacture of cheese without eye formation and a mild taste while the use of "LD" starter cultures favors eyes formation and flavor development due to citrate metabolism (Furtado & Lourenço Neto, 1994; Fox *et al.*, 2000).

The heat treatment of the curd during the cheese manufacturing process also has an important role in the formation of cheese sensory characteristics. The direct heating of the curd by the addition of hot water results in lower acidity and softer texture. Cheeses obtained by indirect heating of the curd show higher acidity and more intense taste (Spadoti *et al.*, 2011).

The ripening time is one of the main factors that affect the development of the sensory characteristics of cheeses. According to Brazilian legislation, Prato cheese must be ripened for at least 25 days (Brazil, 1997). However, it is known that a ripening period of at least 60 days is required to achieve the optimum flavor and consistency characteristics (Spadoti *et al.*, 2011).

The ripening of cheeses consists of a complex phenomenon characterized by several biochemical changes such as the glycolysis of residual lactose and catabolism of lactate, lipolysis, and metabolism of fatty acids, proteolysis, and catabolism of amino acids and citrate metabolism in some cheeses' varieties (Fox *et al.*, 2000). Prato cheese ripening had been widely described in the literature through studies that evaluated the effect of several factors such as the type of coagulant (Alves *et al.*, 2013), the addition of probiotic microorganisms (Chaves & Gigante, 2016), reduction in salt content (Baptista *et al.*, 2017b) and adjunct culture addition (Baptista *et al.*, 2018) on the biochemical changes that occur during the cheese ripening. These studies led to the characterization of Prato cheese ripening and, in particular, proteolysis, the main biochemical change that occurs in this cheese variety.

Proteolysis of Prato cheese is characterized by the initial hydrolysis of α_{s1}-casein by the action of the residual coagulant on the peptide bond Phe_{23}-Phe_{24} and slower degradation of β-casein during ripening time (Alves *et al.*, 2013; Chaves &

Gigante, 2016; Baptista *et al.*, 2017b). The hydrolysis of the Phe_{23}-Phe_{24} bond, which begins on the first days of ripening, is responsible for the softening of the cheese and, therefore, for the transition from a firm texture to the desirable semi-hard and elastic texture in the final product (Ardö *et al.*, 2017).

In a recent study, the peptide profile formed during Prato cheese ripening was investigated through the analysis of commercial cheeses. This study showed a very complex peptide profile with 44 peptides identified and a predominance of peptides derived from the α_{s1}-casein and β-casein (24 peptides derived from α_{s1}-casein, 14 from β-casein, 3 from κ-casein and 3 from α_{s2}-casein) (Baptista *et al.*, 2017a). The higher prevalence of peptides formed from the α_{s1}-casein hydrolysis reaffirms the importance of the hydrolysis of this casein fraction during Prato cheese ripening.

In Prato cheeses produced with heterofermentative starter culture constituted of *Lactococcus lactis* subsp *lactis* biovar *diacetylactis* and *Leuconostoc*, the citrate is metabolized leading to the release of diacetyl and CO_2, which leads to eye formation. Domingos (2015) analyzed Prato cheeses produced with starter culture composed of *Lactococcus lactis* subsp. *lactis*, *Lactococcus lactis* subsp. *cremoris*, *Lactococcus lactis* subsp. *lactis* biovar. *diacetylactis*, and *Leuconostoc* and identified the volatile compounds diacetyl and acetoin released by citrate metabolism and which were present in the cheeses since the beginning of ripening. The same study identified organic acids, esters, methyl ketones, and lactones, which are volatile compounds originated from lipolysis during Prato cheese ripening. Therefore, lipolysis is also an important biochemical change responsible for the release of volatile compounds and, consequently, for the development of the sensory characteristics of Prato cheese (Domingos, 2015).

REQUEIJÃO

Requeijão is a processed cheese very popular in Brazil. With an annual production of about 180 thousand tons, it is currently the second most produced cheese in the country. Although it is produced in 21 Brazilian states, *Requeijão* production is concentrated in the states of Minas Gerais and São Paulo, with 37.90 and 29.90% of the national production, respectively (Datamark, 2018).

Processed cheeses were firstly produced in the early 20th century to increase the shelf life of cheeses. In 1911, the first processed cheese, the Schachtelkäse, was produced in Switzerland. It was a stable melted cheese obtained by the heat treatment of Emmental cheese with the addition of sodium citrate. Subsequently, other types of cheeses began to be processed around the world to obtain stable products by the addition of other emulsifying salts (Guinee, 2017).

Requeijão is the main processed cheese consumed in Brazil and is different from the other processed cheeses produced around the world because it is produced from a fresh curd mass that can be obtained by acid or enzymatic coagulation (Furtado & Lourenço Neto, 1994). *Requeijão* was a homemade solution to use milk coagulated by the action of lactic acid bacteria. It was made from skimmed milk in semi-artisanal facilities. The skimmed milk was maintained undisturbed to favor the spontaneous coagulation and the curd obtained was used in the manufacture of *Requeijão* (Van Dender *et al.*, 2014).

Requeijão cremoso was produced for the first time in 1952, in Minas Gerais. It was initially marketed in the form of a tablet produced with the heat treatment of the curd obtained from skim milk with the addition of cream. Between 1955 and 1956, *Requeijão cremoso* began to be packaged in cups, which is the form it is currently marketed (Ligabue, 2008).

According to the Brazilian Legislation, *Requeijão* is the dairy product or dairy product compound obtained by melting the curd, cooked or uncooked, drained and washed, obtained by acid or enzymatic coagulation, or both, of milk, optionally with addition of cream, butter, anhydrous milk fat or butter oil, separated or in combination, with or without the addition of condiments, spices and other food substances (Brazil, 2017).

According to the Technical Regulation for Identity and Quality Fixation of this cheese, it can be classified as *Requeijão, Requeijão Cremoso*, or *Requeijão de Manteiga*. *Requeijão* must have fat in dry matter content between 45.00 and 54.90% and a maximum of 50.00% of moisture, while the *Requeijão Cremoso* must have at least 55.00% of fat in dry matter and a maximum of 65.00% of moisture and *Requeijão de Manteiga* must have a fat content in the dry matter between 25.00 and 59.90% and a maximum of 58% of moisture (Brazil, 1997b). The Brazilian legislation also establishes that the name *Requeijão* is reserved for the product in which the dairy base does not contain fat or protein of non-dairy origin (Brazil, 2017).

Requeijão Cremoso is characterized by a soft, unctuous and creamy consistency, smooth texture without granule. It has characteristic taste and aroma, slightly acid and salty, and homogeneous white-cream color (Rapacci *et al.*, 1999). Due to the characteristics consistency and flavor, it is widely used in the preparation of snacks, sauces, and pies (Van Dender *et al.*, 2012).

According to Minim *et al.* (2010), salty taste, consistency, and viscosity are sensory attributes that contribute to the acceptance of this product. The process of coagulation used to obtain the fresh curd used in *Requeijão* manufacturing is determinant for the development of the sensory characteristics of the cheese. The curd obtained from the milk fermentation by the action of lactic acid bacteria (LAB) presents a more intense fermented flavor (Rapacci *et al.,* 1999).

The manufacturing of *Requeijão*, like all other processed cheeses, consists of the molten of protein and fat with the aid of heat and mechanical shear. The addition of emulsifying salts is necessary to avoid separation of water and fat after heating. In general, three types of emulsifying salts are used in the manufacture of processed cheeses: citrates, monophosphates, and polyphosphates. The correct use of the emulsifying salt favors the development of the necessary characteristics: ion exchange, peptization, and dispersion of the protein, stabilization of the emulsion, pH regulation, buffering action, and cremification (Van Dender *et al.*, 2014).

The curd mass used for the production of *Requeijão* can be obtained from the pasteurized milk by enzymatic coagulation, lactic fermentation or direct acidification at high temperature, the latter two being the most used processes (Furtado & Lourenço Neto, 1994; Van Dender *et al.*, 2016a). In the case of coagulation by lactic fermentation, acidification of milk and destabilization of caseins occur due to the action of mesophilic lactic acid bacteria constituted of *Lactococcus lactis* ssp. *lactis* and *Lactococcus lactis* ssp. *cremoris* (Rapacci *et al.*, 1999). The resulting coagulum is quite weak and should have a pH of around 4.50. After gel formation, the clot must be broken, heated rapidly to a temperature of 50 °C and drained. After draining the whey off, the coagulum has a pH from 4.30 to 4.60. However, for the curd mass to be melted, the pH must be raised to the range of 5.20 to 5.50 by washing the coagulum. After washing the coagulum twice with mineral water, the final deacidification is performed with the addition of milk (25.00 to 30.00% of the initial milk volume) followed by stirring and heating up to 60 °C for milk coagulation. After that, the whey is drained off, and the coagulum is pressed for further use in the manufacture of the *Requeijão* (Van Dender *et al.*, 2016a).

Direct hot acidification coagulation is achieved by the addition of lactic acid to the preheated skimmed milk. In this process, 0.28% of 85% lactic acid diluted in water (1:10) is added to hot milk. When whole or semi-skimmed milk is used, it must be heated to 80-82 °C before acid addition and heating until 69-70 °C is enough when skimmed milk is used. The draining should be performed once precipitation is completed or after a rest period of 10 minutes. Due to the use of high temperature in the process of destabilization of the caseins, coagulation occurs in a pH range of 5.40 to 5.80, ideal for the subsequent melting process. Thus, in this process, there is no need to wash the mass for acidity correction (Van Dender *et al.*, 2016b).

The processing of *Requeijão Cremoso* is usually carried out in a homogenizer-grinder machine with direct and indirect steam injection (Rapacci *et al.*, 1999). *Requeijão de Corte* can be processed in the same equipment, although it is traditionally produced in an open tank with indirect heating with steam (Furtado & Lourenço Neto, 1994).

Processing of the *Requeijão Cremoso* starts by mixing the grinded curd mass and the emulsifying salt Fig. (**3**). After that, the mixture is heated with simultaneously stirring to approximately 80 °C. The water and the cream are then added, and the mixture is heated to 95 °C. The maintenance of this temperature for a few minutes favors the formation of the consistency of melted and creamy cheese and is also important to guarantee microbiological safety (Furtado & Lourenço Neto, 1994).

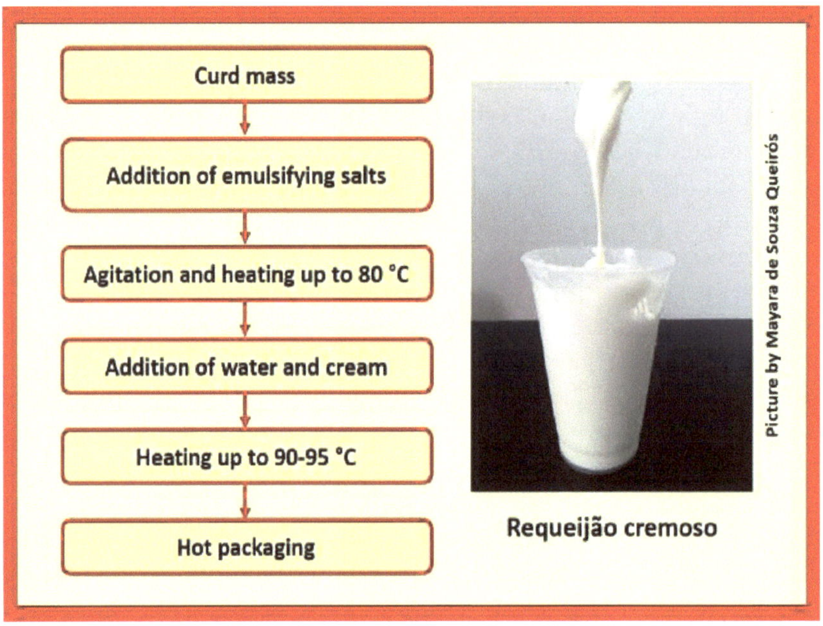

Fig. (3). Requeijão cremoso manufacturing technology.

In the production of *Requeijão de Corte*, the curd mass is heated under constant stirring in an indirectly heated tank. When it reaches 65 °C, the salt is added. If the melting does not occur, it is possible to add citrates and/or sodium phosphates salts. After the melting of the curd mass, the heating is continued until 86 - 90 °C with simultaneous addition of fresh cream. When reaching the desired point, the cheese should be packed while it is still hot (Furtado & Lourenço Neto, 1994).

After processing, *Requeijão* must be cooled. In general, processed cheeses which have the shape of blocks, as in the case of *Requeijão de Corte*, should have a relatively slow cooling for the product to become firm. However, slow cooling may intensify the Maillard reaction and the development of microorganisms (Van Dender *et al.*, 2014).

COALHO CHEESE

Coalho cheese is a typical product, and identity of the Northeast region of Brazil. It is originated from Serra da Estrela cheese from Portugal, which arrived in the country during the colonization period (Bastos *et al.*, 2013). This cheese has been produced and consumed for more than 150 years, with the cultural and regional characteristics of this region of the country, and is manufactured mainly in the states of Pernambuco, Ceará, Rio Grande do Norte, and Paraíba (Silva *et al.*, 2012; Almeida *et al.*, 2013; Medeiros *et al.*, 2017).

Currently, Coalho cheese is produced both on artisanal and industrial scales and has wide commercialization at the national level. The artisanal production in this region has significant importance in the economy and income of milk producers, especially those who do not have access to milk processing plants (Silva *et al.*, 2012; Medeiros *et al.*, 2017).

The artisanal technology of Coalho cheese manufacture has rather rudimentary methods which are passed through generations and do not present standardization. Therefore, variations among different producer's protocols are common, leading to differences in the quality and the characteristics of the product (Medeiros *et al.*, 2017; Cruz & Menasche, 2014). The industrial production technology is regulated by the Brazilian legislation (IN 30/2001, Anexo II), which describes the physicochemical, microbiological and sensory characteristics of the cheese (Brazil, 2001).

Coalho cheese is classified as high moisture (36.00 to 54.90%), semi-cooked or high-cooked cheese, with 35.00 to 60.00% of fat in dry matter (Brazil, 1996; Brazil, 2001). According to the Brazilian legislation, Coalho cheese is obtained through the coagulation of pasteurized milk with rennet or other appropriate coagulating enzymes, supplemented by the action of specific lactic acid bacteria, with the final production of a drained, semi-cooked or high-cooked curd, which is pressed, dried and may have variable shape and weight (Brazil, 2017). The cheese designation must be made according to the presence or absence of condiments. It is denominated as "*Queijo de Coalho*" (Coalho cheese), and when only one condiment is added, it must be declared at the product denomination. When more than one condiment is added, the product must be denominated "*Queijo de Coalho Condimentado*" (Condimented Coalho cheese), not requiring the citation of all condiments after the product denomination (Brazil, 2001). Table **2** shows the main sensory attributes of Coalho cheese.

Table 2. Main sensory attributes of Coalho Cheese (Brazil, 2001).

	Sensory Characteristics of Coalho Cheese
Consistency	Semi-hard, elastic
Texture	Compact, soft, with or without small eyes
Color	Yellowish white and uniform
Flavour	Mild, slightly acidic and might be salty
Aroma	Slightly acidic
Rind	Thin and without cracks

According to the Brazilian legislation (IN 30/ 2001), Coalho cheese is described as a product obtained by the coagulation of milk by the addition of rennet or other appropriate coagulating enzymes, supplemented or not by the action of selected lactic acid bacteria and normally commercialized with up to 10 (ten) days of storage after manufacture (Brazil, 2001).

Coalho cheese can be produced from either cow's milk or goat's milk. However, the availability of scientific information on the use of goat's milk is scarce. The use of goat's milk is associated to the fact that Northeast of Brazil is a major producer of goat's milk, representing 75% of the national production (with emphasis to Paraíba state that produces 18 milion liters a day). Due to the great goat's milk production, its use for dairy products manufacturing aims to improve its' value in order to create economic sustainability for the region, mainly for the rural areas (Queiroga *et al.*, 2013).

The artisanal and industrial Coalho cheese manufacturing technology is shown in Fig. (**4**), as described by Nassu *et al.* (2003).

Cow's or goat's milk can be used in the cheese manufacture, and both milk types can be used in whole or used together in different proportions. The use of both types of milk has been tested to obtain a product with acceptable sensory characteristics (Santos *et al.*, 2011; Queiroga *et al.*, 2013).

The milk reception is carried out either at the production plant (artisanal cheese) or at a reception platform, at larger processing units (industrial cheese). The raw milk received by the dairy industry must meet the physicochemical requirements established by Brazilian legislation (IN 62/2011a and IN 37/2000), which establishes the identity and quality standard of raw cow's and goat's milk, respectively (Nassu *et al.*, 2003; Brazil, 2000; 2011a). After the reception, the milk goes through the filtration process to remove dirt sediments. In the artisanal production, this process is optional, and when it is carried out, it is performed using a cotton cloth or a strainer made of stainless steel or plastic (Brazil, 2000; Nassu *et al.*, 2003; Silva *et al.*, 2010).

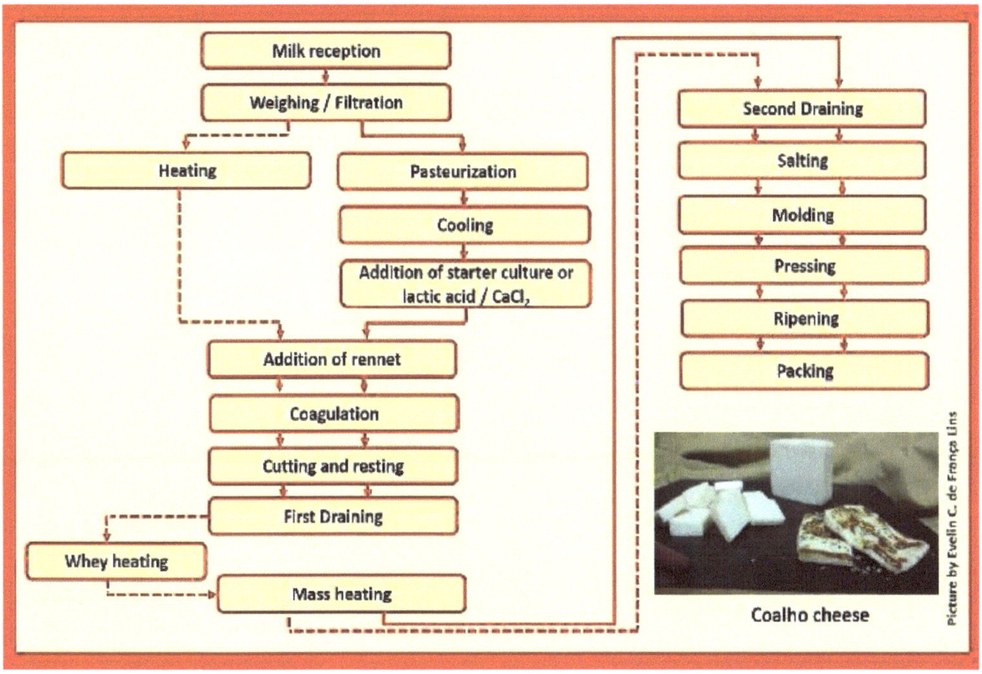

Fig. (4). Coalho cheese manufacturing technology. Artisanal production (--) and industrial production (-), described by Nassu *et al.* (2003).

Milk heat treatment is mandatory by Brazilian legislation, to guarantee the safety of the final product. The Brazilian legislation establishes that processing plants must use classical fast pasteurization (HTST) or equivalent heat treatment binomials according to the technological indication to produce dairy products. However, the use of this procedure in the artisanal production does not occur (Brazil, 2011a; Nassu *et al.*, 2003; Silva *et al.*, 2010). After the heat treatment calcium chloride, mesophilic starter culture (*Lactococcus lactis* subsp. *lactis* and *Lactococcus lactis* subsp. *cremoris*) and/or lactic acid and coagulant (liquid or powder) are added to the milk (35 °C), which is mixed and maintained undisturbed for approximately 40 minutes. After coagulation, the curd is cut and mixed, the whey is partially drained off, and the curd is heated by the addition of hot water or indirectly with steam until 45 °C (semi-cooked) or 45 to 55 °C (cooked). The heating procedure may also not occur for this cheese, which means that this cheese may also be made without scalding. After heating or not, the second drainage is carried out. Then, the salt (sodium chloride) is added to the curd, followed by the pressing, drying, packaging, and storage in the average temperature of 10 - 12 ºC usually up to ten days (Brazil, 2001; Nassu *et al.*, 2003).

The chemical and biochemical reactions in Coalho cheese play an important role in the development of sensory characteristics and, as well as other varieties of cheeses, occur simultaneously during the manufacturing process and refrigerated storage (Bezerra *et al.*, 2017a). Studies have shown particular final characteristics of this cheese because of the variations of the manufacturing technology such as milk type (cow or goat), heat treatment, starter culture, coagulant type, and curd cooking (Silva *et al.*, 2010; Silva *et al.*, 2012; Queiroga *et al.*, 2013; Bezerra *et al.*, 2017b; Merheb-Dini *et al.*, 2016; Fontenele *et al.*, 2017).

The flavor development in cheeses is strongly related to milk microbiota, which affects glycolysis, proteolysis, and lipolysis. The natural microbiota is composed of a large variety of microorganisms that directly or indirectly influence enzymatic reactions. However, data on the characterization of the natural microbiota of Coalho cheese are still limited in the scientific literature. Silva *et al.* (2012) verified the presence of heterogeneous microbiota (composed of *Lactobacillus* sp., *Enterococcus faecium*, *Enterococcus faecalis*, *Streptococcus thermophilus* and *Lactococcus lactis* subsp. *lactis*) at Coalho cheese in the Agreste area of the state of Pernambuco. Medeiros *et al.* (2016) reported that a variation in the geographical distribution of the genera of the LAB was observed in Coalho cheese samples in the state of Paraíba. The genera predominantly identified were *Lactococcus* (40.10%), *Enterococcus* (35.30%), *Streptococcus* (18.60%), *Lactobacillus* (5.50%), *Leuconostoc* (0.40%) and *Weissella* (0.20%). The data of these studies show that there is a similarity of LAB genera in artisanal cheeses. However, the predominance of these LAB has a diversity regarding the geographical distribution. This variability affects the biochemical events of the cheese and can lead to different sensory characteristics which can help in the protection of denomination of origin.

Proteolysis is the most complex and important biochemical process in the production and ripening of almost all types of cheeses, which involve the action of coagulant enzymes, natural milk proteases and microbial enzymes, which promote the hydrolysis of casein and the release of peptides and free amino acids (Bezerra *et al.*, 2017a; Costa *et al.*, 2018).

Merhed-Dini *et al.* (2016) evaluated the proteolytic profile of Coalho cheese and observed a greater increase in the pH 4.6 SN than TCA 12% SN, representing the hydrolysis of β-CN. The fraction α_{s1}-CN, which is hydrolyzed by chymosin at the peptide bond Phe_{23}-Phe_{24}, showed limited hydrolysis, resulting in low α_{s1}-I-CN formation. The authors associated the hydrolysis profile with the manufacturing process (high-cooking) that promoted the loss of chymosin activity and activation of plasminogen to plasmin, since no starter culture was used in the cheese production. The limited hydrolysis of α_{s1}-CN favored the melting resistance, which is an important functional characteristic of Coalho cheese since it is believed that the hydrolysis of this fraction of casein is responsible for the softening of cheese texture.

Complementary studies evaluated the peptide profile of Coalho cheese produced with cow's milk in different geographic areas of Brazilian Northeast (Silva *et al.*, 2012, Silva *et al.*, 2016, Fontenele *et al.*, 2017). Fontenele *et al.* (2017) identified 116 different peptides with a molecular weight between 702.3 Da and 3279.0 Da, which were identified by peptide sequencing. In the peptides sequencing, 74 fragments were identified as originated from β-CN, 4 fragments from βA2-CN, 4 fragments of βA3-CN, 25 fragments of α_{s1}-CN, 5 segments of α_{s2}-CN and 4 other fragments of κ-CN. The presence of 18 phosphorylated peptides, 17 from β-CN and 1 from α_{s1}-CN, corresponding to segment 110-119 was also identified. The peptide profile observed by the authors had a relationship with data described by Merheb-Dini *et al.* (2016) who observed greater hydrolysis of β-CN.

The production of organic acids, as a result from glycolysis, give important sensory characteristics of the flavor and aroma to the cheeses. The conversion of lactose to lactic acid is the first event of glycolysis, due to the action of starter cultures at the beginning of cheese production and the initial stages of ripening. Bezerra *et al.* (2017a) quantified the presence of lactic acid, oxalic acid, pyruvic acid, formic acid, citric acid, and propionic acid in Coalho cheese produced with starter culture and probiotic cultures, alone or in combination, and the results showed that lactic acid had higher concentration ($1.50g.100g^{-1}$) when compared to other organic acids. The citric acid in the cheeses showed a reduction throughout the storage, and it may be related to the metabolism of this acid by the LAB that produces volatile compounds, such as acetic acid, diacetyl, acetoin, and 2,3-butanediol, which confer butter aroma (Bezerra *et al.*, 2017a). The high concentration of organic acids, mainly lactic acid, in Coalho cheese influences the cheese's functional melting characteristic, since it reduces pH and alters salt balance, once pH value lower than 5.8 promote excessive melting (Silva *et al.*, 2010; Sousa *et al.*, 2014; Bezerra *et al.*, 2017a; Costa *et al.*, 2018).

Lipolysis, as well as glycolysis and proteolysis, contributes to the flavor development in Coalho cheese, especially due to the release of short chain fatty acids. In cheeses, the lipolysis can be associated with the action of lipase naturally present in milk, from the starter culture, NSLAB, and enzymes produced by contaminating microorganisms. Bezerra *et al.* (2017a) indicated that the effect of the bacteria of lactic acid on lipolysis is caused by the oxidation and hydrolysis of long-chain fatty acids and the formation of short-chain fatty acids, which enhance cheese aroma, especially goat flavor. Costa *et al.* (2018) attributed the lipolysis in the cheese to the natural milk lipases or contaminating microorganisms, since no starter culture was used in the production of Coalho cheeses. In the study, it was also observed that the sodium content affected the cheese fatty acid profile.

Differences in the fatty acids profile of Coalho cheese associated with milk type and animal metabolism were observed. Cow's and goat's milks have different fatty acid content, where goat's milk has significantly higher levels of short and medium chain length fatty acids in comparison with bovine milk (Sheehan *et al.*, 2009; Queiroga *et al.*, 2013). Queiroga *et al.* (2013) studied the fatty acid content of Coalho cheeses produced with cow's milk, goat's milk and the mixture of both. No significative difference was observed between cheeses produced with cow, goat or the mixture of them during 28 days of storage. However, cheeses had different individual C6, C8, C10, C12, C14, C16, C16:1 and C18:2 contents. The cheeses produced with the mixture of them and the goat's milk presented higher content of medium and short chain fatty acids than cheeses produced cow's milk, as well as C6, C8, C10 and higher amount of C12 was found in goat's milk cheese than the others after 28 days of storage.

MANTEIGA CHEESE

The Manteiga cheese or Sertão cheese, also known as *Requeijão do Sertão, Requeijão do Nordeste* and *Requeijão do Norte*, as well as the Coalho cheese, is a typical product of Northeast region of Brazil and presents great economic importance. This cheese also has a great acceptance in the North region of the country (Brazil, 2001; Guerra & Guerra, 2003; Cavalcante & Costa, 2005). The manufacturing technology of Manteiga cheese is still quite artisanal and empirical. There is lack of knowledge and technical support that affects the standardization and identity of the Manteiga cheese for its production, storage and distribution (Nassu *et al.*, 2003; Cavalcante & Costa, 2005). However, in order to standardize the production for industrial scale and the characteristics of the product, this cheese is regulated by Brazilian legislation (Normative Instruction 30/2001), which describes the physicochemical, microbiological and sensory characteristics of the cheese (Brazil, 2001).

The Manteiga cheese is classified as medium to high moisture (36.00 to 54.90%), semi-cooked or cooked cheese with fat content in dry matter ranging from 25.00% to 55.00%. The comercial designation must be "*Queijo de Manteiga*" or "*Queijo do Sertão*" (Brazil, 1996; Brazil, 2001). According to the Brazilian legislation, Manteiga cheese or Sertão cheese is the cheese obtained by coagulating the pasteurized milk with the use of organic acids, with the production of a drained, fused mass and the addition of *Manteiga de Garrafa* (bottle butter), which may have variable shape and weight (Brazil, 2017). The mandatory ingredients used in the cheese production are standardized whole, semi-skimmed or skimmed-milk,

Manteiga de Garrafa (*Manteiga da Terra*), food-grade organic acids such as lactic acid, citric acid and acetic acid (Brazil, 2001). Nassu *et al.* (2003) reported that in addition to the ingredients described by legislation, sodium bicarbonate and corn starch are also added and, in some cases, butter is replaced by soybean oil. However, these ingredients are not considered at the Brazilian legislation (Normative Instruction 30/2001 and Ordinance 146/1996) as optional ingredients for this product. Table **3** shows the main sensory attributes of Manteiga cheese.

Table 3. Main sensory attributes of Manteiga Cheese (Brazil, 2001).

Sensory Characteristics of Manteiga Cheese	
Consistency	Soft
Texture	Compact and semi-friable *
Color	Yellow
Flavour	Mild, buttery, slightly acidic and might be salty
Aroma	Little pronounced, buttery
Rind	Thin and without cracks

* Small mechanical holes containing liquid fat

The manufacturing technology of Manteiga cheese is shown in Fig. (**5**), as described by Nassu *et al.* (2003) and Cavalcante & Costa (2005). The milk reception, the first phase of processing, is carried out either at the production plant (artisanal cheese) or at a reception platform, at larger processing units (industrial cheese). The raw milk received by the industry must meet the physicochemical requirements established by Brazilian legislation (Normative Instruction 62/2011) (Nassu *et al.*, 2003; Brazil, 2011a). After the reception, the milk used in most of the processes is not subjected to heat treatment (pasteurization), except in industrial production, and the skimming when applied is carried out on different types of cream separators (Nassu *et al.*, 2003). Food grade organic acids are added to the heated milk (32 °C) for coagulation, or the milk is kept at rest for natural coagulation until acidity of 0.60 to 0.65% of lactic acid or pH 4.56 are reached (Brazil, 2001; Cavalcante & Costa, 2005).

Then, the curd is heated until 55 °C, with continuous mixing, in the tank with direct heating until the complete draining. The curd is drained off and washed once with skimmed milk and twice with cold water, and the volume of milk and water are about 50% of the whey volume. The washed curd is pressed and submitted to fusion by direct heating (85-90 °C) with the addition of *Manteiga de Garrafa* (*Manteiga da Terra*), sodium chloride and sodium bicarbonate. The mixture is than stirred until complete homogenization of the protein mass and butter is reached. After this step, the soft fused mass is transferred while still hot to polyethylene cheese molds, packed and stored under refrigeration (Nassu *et al.*, 2003; Cavalcante & Costa, 2005). The artisanal production of Manteiga Cheese is shown in Fig. (**6**).

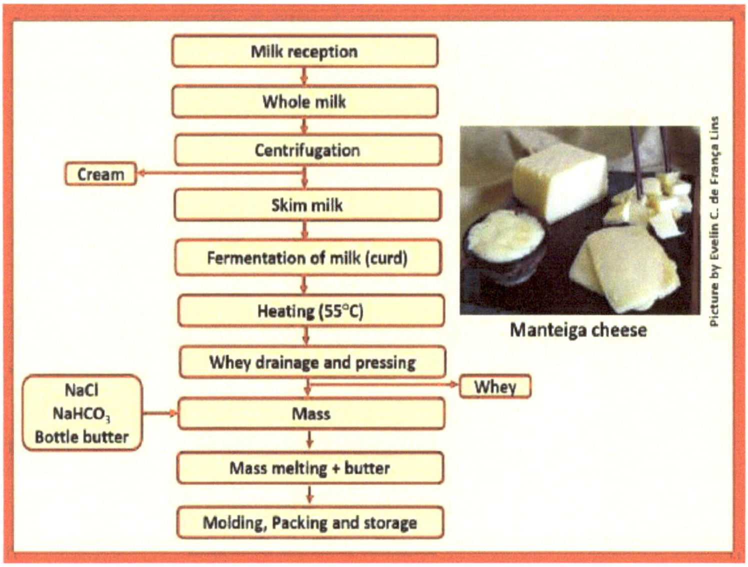

Fig. (5). Manteiga cheese manufacturing technology.

Fig. (6). Artisanal production of Manteiga cheese. (Pictures provided by Mônica Correia Gonçalves).

The literature about Manteiga cheese and its ripening biochemistry is rather scarce. Restricted information is available on the physicochemical and sensory characteristics of the cheese. As presented in the characteristics and production technology of Manteiga cheese, *Manteiga de Garrafa,* also known as *Manteiga da Terra,* is added during cheese processing. The *Manteiga de Garrafa* consists of anhydrous fat and is very susceptible to lipid oxidation (Ambrósio *et al.*, 2003). Therefore, due to the susceptibility of the *Manteiga de Garrafa* to lipid oxidation, one of the variables evaluated with regard to Manteiga cheese quality is the oxidative stability. Thus, oxidative stability is considered an important variable, since it reduces the nutritional and sensory (unpleasant odor) qualities and the shelf life of the product. Peroxide index values higher than 0.76 mEq/kg are considered a limiting factor for the product (Guerra & Guerra, 2003).

MINAS CHEESES

Minas Gerais State

In the southeastern region of Brazil, Minas Gerais has a privileged position, standing out as the main milk producing state in the country, with the production of 9 billion liters per year, equivalent to 26% of national production (IBGE, 2016). The geographical conditions of Minas Gerais, such as mountainous with altitudes above 700 meters, rich drainage, favorable soil, and climate, supports dairy farming and the production of milk and dairy products. Because of this extensive milk production, Brazil is the fourth largest producer and consumer of cheeses in the world (Datamark, 2018), especially, the state of Minas Gerais remains the leader in the national ranking, presenting a consolidated industrial and artisanal production (Perry, 2004).

The leadership of Minas Gerais in the dairy agribusiness is historical, and the first Brazilian and South American dairy industry was located in this state, which maintains the sector's representativeness also in the economy (Machado *et al.*, 2004). This industry has a history of strong European influence, both the influence of colonization with the Portuguese cheese tradition and the reception of European immigrants, especially Dutch and Danish, during the XIX and XX centuries. People from Minas Gerais learned with Europeans how to properly exploit milk and dairy products. Thus, in the early XX century, Minas Gerais had the first milk and dairy industry, making it essential to the state's economy, along with mining and coffee growing. The dairy industry in Minas Gerais has become an example of tradition and quality. Thus, it remains one of the most important economic sectors of the state. The cheese covers interdisciplinarity researches such as economic, geographical, historical, tourist, cultural, gastronomic studies related to food technology, with a focus mainly on the physicochemical and microbiological characterization of these cheeses (Netto, 2011).

Minas' cheeses are among the most popular cheeses consumed in Brazil. These cheeses include soft cheeses (*e.g.*, Frescal), semi-hard and hard (*e.g.*, Padrão,

Meia-cura, Serro, Canastra, etc.) (Campagnollo *et al.*, 2018a). In general, these cheese varieties are produced both artisanal by small farmers and on an industrial scale by large dairy industries. As the industrial process demands a larger production scale, in general, milk is bought from several small producers, and then cooled and stored in large cooling tanks until being transported to the dairy industry. Industrial trucks collect the milk that is stored in the collection points and transport it to the industry, where it will undergo quality control tests and then it is properly pasteurized. The pasteurized milk undergoes all stages of the cheese manufacturing process in mechanized installations. The artisanal production is performed manually at the farm following traditional manufacturing protocols.

In general, the state of Minas Gerais produces most of the cheeses marketed and consumed in Brazil, both on an industrial and artisanal scale. More than 500 cities in this state produce artisanal cheese, and almost all other cities produce dairy products (IPHAN, 2014). The general characteristics of the main industrialized and artisanal cheeses produced in the state of Minas Gerais, including the processing/manufacturing stages and their physicochemical, microbiological and sensory aspects, will be discussed in the following items.

INDUSTRIALLY PRODUCED MINAS CHEESES

Minas cheese is among the most consumed cheeses in Brazil. Within this category, the Minas Frescal and Minas Padrão cheeses have the highest production on an industrial scale.

Minas Frescal Cheese

Minas frescal cheese has this name because it was firstly produced in the state of Minas Gerais in XVIII century. This cheese does not require ripening since it is sold quite after its manufacturing. Therefore, it is characterized as a fresh cheese (Monteiro *et al.*, 2011).

According to the Brazilian legislation, Minas Frescal cheese "is the fresh cheese obtained through the enzymatic coagulation of pasteurized milk with rennet or other appropriate coagulating enzymes or both, which can be either supplemented by the action of specific bacteria or produced without the addition of lactic acid bacteria, with the production of a drained, not pressed and salted mass" (Brazil, 2017). According to the Brazilian legislation (Ordinance 352 from September 4th, 1997), which establishes the Technical Regulation for Identity and Quality Fixation of Minas Frescal Cheese, this cheese is classified as semi-fatty (between 25.00 and 44.90% fat), and has a very high moisture content, also known as soft cheese, with a moisture content exceeding 55.00% (Brazil, 1997c). Due to the high moisture content, this cheese has a very short shelf life from 9 to 20 days

(Sangaletti *et al.*, 2009; Souza *et al.*, 2008). Fig. (7) shows the Minas cheese manufacturing technology.

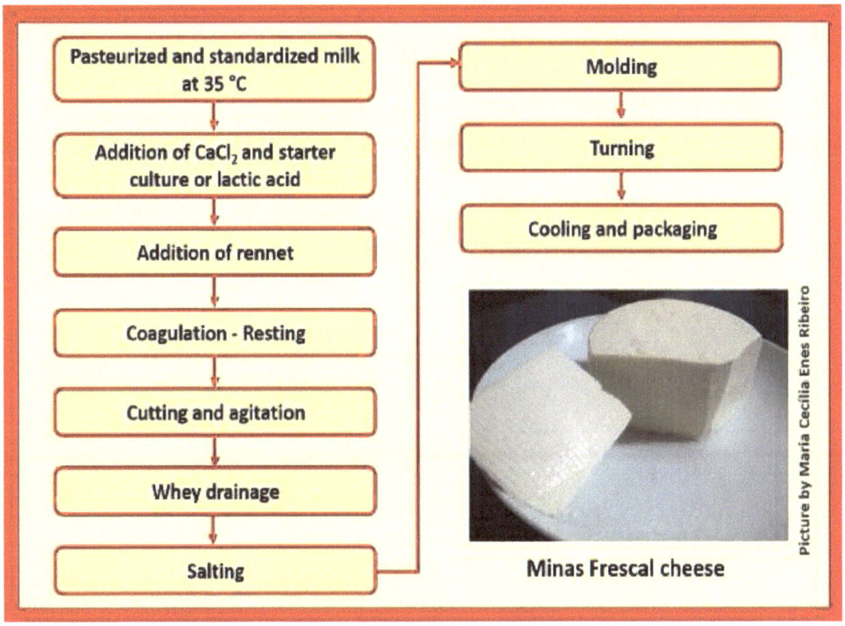

Fig. (7). Minas Frescal cheese manufacturing technology.

In general, milk is pasteurized (mandatory for this cheese type), the fat content is standardized to 3.00-3.20% and milk is kept at 32-35 °C. Calcium chloride and mesophilic starter culture constituted of *Lactococcus lactis* subsp. *lactis* and *Lactococcus lactis* subsp. *cremoris* are added.

One trend in the dairy area is to totally or partially replace the starter culture by direct acidification with lactic acid. This choice directly affects the flavor, yield and shelf life of the product, since the absence of lactic acid culture can contribute to obtaining a product that is not microbiologically safe. It happens because the addition of the culture leads to lactic acid production and consequently lower the pH continuously during the cheese storage, in addition to the production of antimicrobial compounds. The exclusive addition of lactic acid, however, results in lowering the pH only during the cheese making process (Buriti *et al.*, 2005).

After starter culture addition, the milk is coagulated by coagulant within 40-50 minutes. After this time, the curd is cut in a single step to obtain grains of approximately 1.50-2.00 cm. After cutting, the curd is maintained undisturbed for approximately 3 minutes, and then the stirring begins. The stirring is slow and careful, for 5 minutes following by 3 minutes of rest, successively with a total time of 25 minutes until obtaining firm and rounded grains, considered the ideal mass point. Once the step of stirring and checking the mass point, the whey is drained off. Then, 2.50% salt is added to the curd, and the cheese curd is placed in molds. The salting of this type of cheese can be carried out by the direct addition of salt to the curd after the whey is drained off or the salt can be added on the surface of the molded cheese. The cheeses are placed in cylindrical molds, which are proper for this cheese variety that does not has a mechanical pressing stage. After molding, two turns are performed, after 20 and 30 minutes. Four successive turns are recommended to achieve its characteristic. After the turns, the cheeses, still in the forms, are placed into the cold chamber (10-12 °C) until the following day for complete desorption. The cheeses are packaged and must be kept refrigerated throughout the storage, transport, and marketing (Monteiro *et al.*, 2011). Table **4** shows the main sensory attributes of Minas Frescal cheese.

Table 4. Main sensory attributes of Minas Frescal Cheese (Brazil, 1997c).

Sensory Characteristics of Minas Frescal Cheese	
Consistency	Soft
Texture	With or without mechanical holes
Color	Withish
Flavour	Mild or slightly acidic
Aroma	Mild, characteristic
Rind	Without, or thin rind

According to Brazilian legislation (1997), Minas Frescal cheese must have a cylindrical shape; it can be commercialized with a weight between 0.30 and 5.00 kg and temperature not exceeding 8 °C. The expected yield for the Minas Frescal cheese is 1.00 kg of cheese for each 6 to 7 liters of milk processed, according to the milk composition and cheese moisture. It presents simple processing and fast investment return (Monteiro *et al.*, 2011).

Minas Frescal cheese is one of the most consumed in Brazil, it has a high-water activity, pH above 5.00, low salt concentration, no preservatives added, high handling in the manufacturing process, which implies an excellent matrix for the growth of microorganisms (Chalita *et al.*, 2009; Souza & Saad, 2009). As already mentioned, this cheese must be produced with pasteurized milk. However, it is being produced from raw milk and the products obtained are marketed in the informal market without any inspection or regulation.

One of the main defects reported in the Minas Frescal cheese is the early blowing, due to the lack of hygiene in the process. This kind of cheese blowing is caused by bacteria from coliform group, which during their metabolic activity ferment lactose, releasing carbon dioxide, which results in the formation of small irregular eyes in the cheese and leads to the development of an undesirable flavor (Monteiro *et al.*, 2011; Ordóñez, 2005).

Minas Padrão Cheese

Minas Padrão is a traditional cheese, consumed extensively in Brazil and is considered one of the first cheeses produced on an industrial scale in the country. This cheese was firstly produced in Minas Gerais at a time in which the state faced legal problems to commercialize the cheeses produced there. Minas Padrão cheese was created through technological development with the objective to provide a standard to the many varieties of artisanal and regional cheeses in the state. It was produced by numerous small producers, with small variations in processing that gave different characteristics to cheeses called "Minas Cheese" (Melo *et al.*, 2009). According to the Brazilian legislation, Minas Padrão cheese "is the cheese obtained through the coagulation of pasteurized milk with rennet or other appropriate coagulating enzymes or both, which can be either supplemented by the action of specific bacteria or produced without the addition of lactic acid bacteria, with the production of a drained, mechanically pressed and salted mass, which can be fresh or semi-cooked" (Brazil, 2017). Fig. (**8**) shows the Minas Padrão cheese manufacturing technology.

In general, the milk fat content is standardized to 3.30-3.50%, and the mesophilic starter culture (*Lactococcus lactis* subsp. *lactis* and *Lactococcus lactis* subsp. *cremoris*) is added to milk to coagulate. In cheese without cooking procedure during manufacturing, the coagulation temperature is 32 °C and the coagulation time is 35 to 40 minutes. The curd cutting is performed in one single step, aiming to obtain small grains, of approximately 1.50 cm. After cutting, the curd is maintained undisturbed for 3-5 minutes before stirring (whey acidity: 10 to 11 °D). The stirring of the whey is slow and carried out for 40 to 50 minutes. The ideal mass point is when it is soft and slightly dry. After the stirring step, the whey is drained off, and the curd is pre-pressed with a pressure equivalent to 15 kg for 15 minutes. The curd is cut in regular blocks, and placed in cylindrical forms with cheesecloth, for later pressing. The cheeses are placed in the press and subjected to successive pressings. After pressing, the cheeses are salted in brine (20% NaCl) at a temperature of 10 to 12 °C for 18-24 h (Furtado, 2005; Monteiro *et al.*, 2011). Unlike the Minas Frescal cheese, Minas Padrão cheese is ripened for at least 20 days under refrigeration (10-12 °C) before being distributed to supermarkets throughout the country. It is a product that requires refrigeration during storage and commercialization since high temperatures can reduce its quality (Perry, 2004).

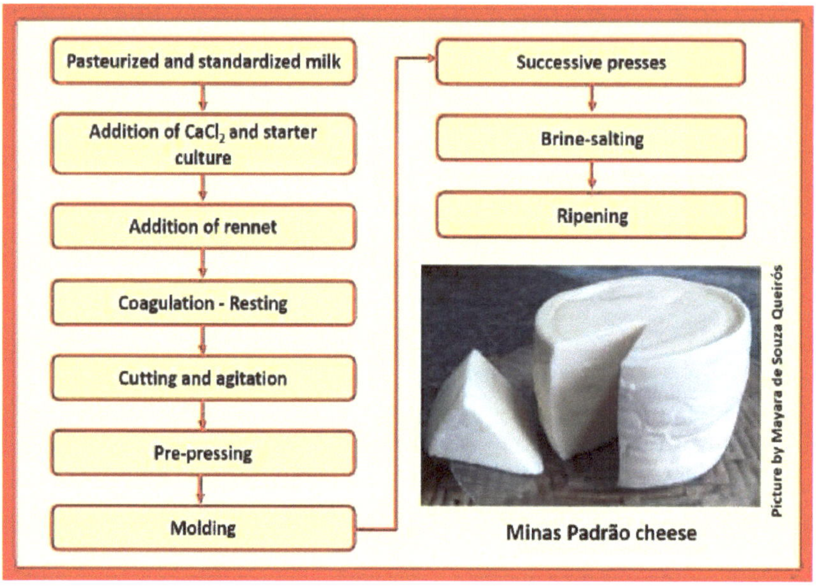

Fig. (8). Minas Padrão cheese manufacturing technology.

Minas Padrão cheese has a cylindrical shape, with 12 to 14 cm of diameter and weight between 0.80 and 1.20 kg. Its surface is smooth and yellowish, without the use of dyes, it has semi-hard consistency, tending to soft and brittle, with internal color cream and slightly acidic taste. The average composition of Minas Padrão cheese is 46.00-49.00% moisture, 51.00-54.00% total solids, 23.00-25.00% fat, 1.40-1.60% salt and pH between 5.00-5.10. The expected yield for Minas Padrão cheese is 1.00 Kg of cheese for each 7.50 to 8.50 liters of milk, according to the composition of the milk and the moisture content of the cheese (Furtado, 2005). Although it is one of the most produced cheeses in the state of Minas Gerais, it is possible to find variations in the physicochemical and sensory characteristics in cheeses marketed in the country, due to the absence of technical regulation of identity and quality of the product. Table **5** shows the main sensory attributes of Minas Padrão cheese.

Minas Artisanal Cheese

The artisanal cheese production plays an important social, economic and cultural role for the State of Minas Gerais and Brazil where this activity maintains thousands of families of small producers in the countryside. It is estimated that 220,000 tons of artisanal cheeses are produced in the seven main regions of Minas Gerais: Araxá, Canastra, Campo das Vertentes, Cerrado, Serro, Serra do Salitre and Triângulo Mineiro (Milkpoint, 2017; Campagnollo *et al.*, 2018a). Fig. **(9)**

shows a map of the state of Minas Gerais with the main producing regions of Minas Artisanal cheese.

Table 5. Main sensory attributes of Minas Padrão Cheese (Furtado, 2005).

Sensory Characteristics of Minas Padrão Cheese	
Consistency	Semi-hard, tending to soft
Texture	Mechanical holes
Color	Yellowish, homogenous
Flavour	Characteristic, slightly acidic
Aroma	Mild, characteristic
Rind	Smooth, thin and yellowish

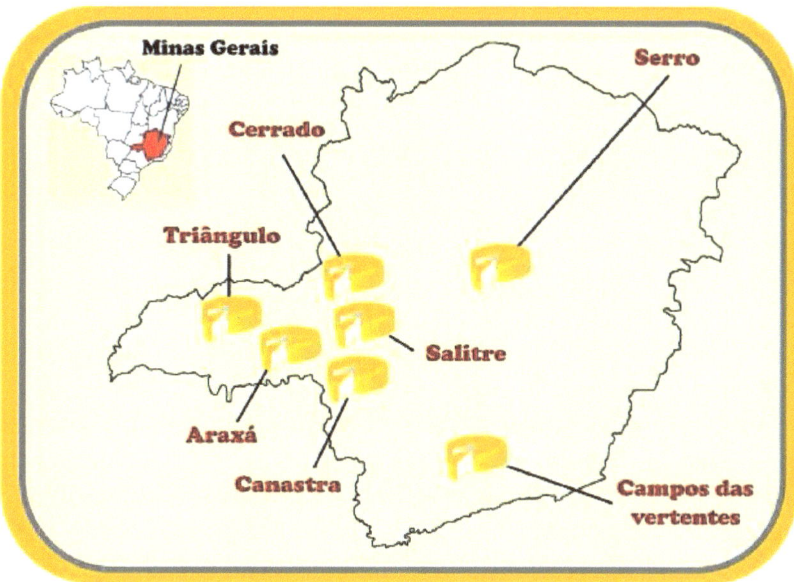

Fig. (9). Main producing regions of artisanal cheese in the state of Minas Gerais.

Besides the fact that it is the main income-generating activity of these regions, this activity characterizes the socio-cultural identity of the state (Furtado, 1980). The product that comes from each of these locations has an identity that varies according to altitude, temperature, soil type, pasture, and relative humidity. These aspects favor the development of certain microorganisms in the biological process of their production and ripening. The natural conditions and the know-how characteristic of each region give the Minas Artisanal cheese its own identity, according to the place where it is manufactured (Netto, 2011).

The last decades were of great importance for the Minas Artisanal cheese. The Law 20,549 (from December 18th, 2012) which is related to the production and marketing of artisanal cheeses from Minas Gerais, in order to regulate and establish criteria for the registration of producers, adaptation to the new parameters of good agricultural practices and its social organization. This law also includes the characterization of cheeses according to their regions of origin and sensory characteristics, with the aim to improve the production of Minas Artisanal cheese, without losing the tradition that qualifies it as a differentiated product (Minas Gerais, 2012). The Government of the State of Minas Gerais, through the Ministry of Agriculture and Livestock, establishes a series of Laws, Decrees and Normative Instructions to identify the registered regions, establish standards for sanitary protection of the herd, establish basic requirements for facilities and materials used in the manufacture of cheeses and regarding the health and hygiene of the food handlers (IMA, 2016).

Artisanal Minas cheese is manufactured on the farm from raw, whole and freshly milked milk, with the addition of rennet and natural starter culture. Also, the cheese does not go through any mechanical process, and the pressing is performed manually to complete its draining and molding. The Artisanal Minas cheese must have a firm consistency with or without mechanical eyes, cannot have the addition of dyes and preservatives, and is produced according to the historical and cultural tradition of the region where it is produced (Martins *et al.*, 2015). In the manufacture of Artisanal Minas cheese, processing should start up to ninety minutes after the milking is started. Besides the rennet and the salt, the artisanal producers use an endogenous microbiota, as a coadjuvant to the production of the cheese. The wheydrop (named "pingo") is a portion of fermented whey originated from the draining of cheeses produced the previous day and which is collected to be used as a starter culture. This endogenous microbiota contains several microbial groups that contribute to the fermentation and ripening of cheeses (Perin *et al.*, 2017; Campagnollo *et al.*, 2018b). This provides the product a diversified microbiota representative of the region in which gives the cheese unique sensory characteristics (Dores & Ferreira, 2012). The wheydrop is the main differential of the cheeses produced in the different regions recognized as producers of Minas artisanal cheese. The cheese production is carried out 364 days a year, except on Good Friday, due to the tradition (IPHAN, 2014).

The artisanal way of production was registered as a Brazilian intangible cultural heritage by the National Historical and Artistic Heritage Institute (IPHAN) in 2008, with its insertion in the Book of Knowledge (IPHAN, 2014).

Nowadays, consumers are looking for traditional/artisanal food products, and there is the understanding that such food can be a way to recover the economy, providing local resources and access to employment, which contributes to the improvement of the local economy, growth, entrepreneurship, and innovation (Perin *et al.*, 2017; Campagnollo *et al.*, 2018b).

Recently, the Law 13,680 (from June 14th, 2018) was established, and become effective with the addition of interstate marketing permission for food products produced in an artisanal way, with their own traditional or regional characteristics and methods, using good agricultural and manufacturing practices, subject to the supervision of public health organs of the states and of the Federal District. The artisanal product is identified, throughout the national territory, by a single stamp with the indication ART, according to the regulation (Brazil, 2018).

Although artisanal cheeses produced from raw milk have highly desirable organoleptic properties, there is a safety concern due to inadequate manufacturing and marketing practices, inefficient sanitary inspections and excessive handling, which may lead to microbiological contamination by pathogenic bacteria (Campagnollo *et al.* 2018a, 2018b). The *Listeria monocytogenes* is a pathogenic bacterium commonly transmitted by food, such as cheeses. This bacterium can cause serious problems for consumer health, such as intrauterine infection, meningitis, and septicemia. Several studies carried out in Brazil indicated that endogenous lactic acid bacteria (LAB) present in Minas Artisanal cheese can inhibit *Listeria monocytogenes* (Cavicchioli *et al.*, 2017; Ortolani *et al.*, 2010; Tulini *et al.*, 2013). The mechanisms of this antagonism are not fully elucidated; however, it is known that LABs can produce antimicrobial compounds, including lactic acid, hydrogen peroxide, diacetyl and bacteriocins (Guillier *et al.*, 2008). Currently, studies have been carried out with the aim to isolate and characterize strains of lactic acid bacteria with anti-pathogenic activity and modulate their effects on these microorganisms during the shelf-life of Minas Cheeses produced with raw milk (Campagnollo *et al.*, 2018a).

Consumers' concern to purchase microbiologically safe and high-quality Artisanal cheeses produced with raw milk is increasing, and endogenous microorganisms that play a key role in cheeses ripening are of utmost importance to ensure minimum ripening periods that guarantee consumers´ safety.

The legalization of the production and commercialization of artisanal cheeses at national level took place through the Normative Instruction n° 57 (Brazil, 2011b) which was subsequently replaced by the Normative Instruction n° 30 (Brazil, 2013). Since then, several studies have been carried out to verify the current situation of these cheeses so that they provide no public health risks.

In this context, the Government of Minas Gerais established in Ordinance n° 1,305 (IMA, 2013) the necessary ripening time for Artisanal Minas cheeses, indicated by scientific research, aiming to guarantee the quality and safety of the products. It is defined the ripening period of at least 17 days of Artisanal Minas cheeses produced at the Serro micro region and a minimum of 22 days for the cheeses produced at the Canastra, Cerrado, Araxá and Campo das Vertentes micro regions until further research is carried out ratifying or rectifying the defined ripening periods. The ripening should be carried out at room temperature, and the other regions traditionally recognized as producers of Minas Artisanal cheeses should submit to the IMA technical-scientific researches that attest the effectiveness of ripening period of less than 60 days, without prejudice to the quality and safety of the products.

In general, some aspects related to the quality of Artisanal Minas cheeses should be considered such as processing and hygienic-sanitary conditions during milk production, hygienic conditions of cheese processing, storage and transport and the quality of the water used in the milk and cheese production chain (Picoli *et al.*, 2006).

Minas artisanal cheeses are manufactured basically with the same technology, varying some processing steps, such as the pressing procedure. However, the final characteristics of the cheeses vary according to the region of the state where it is produced. Some characteristics of the main Minas artisanal cheeses, which are the Serro cheese and the Canastra cheese, will be presented as follows.

Serro Cheese

By the mid-1700s, the Serro region was rich in gold and diamonds, and for a century supplied the Portuguese court with these minerals. However, the depletion of the mineral deposits and the precarious conditions of the roads, have made the region economically stagnant. With the reduction in the exploration of gold and diamonds, the Serro region intensified its agricultural activity and, at the time of expansion, the cheese was the product that ensured the expansion of the business, by the quality and volume it represented for the market, however, with the distribution restricted to the region. The cheese technology, brought to Brazil by the Portuguese colonization, was adapted to the environment of cattle farms around the mines to support the exploration of gold. After the road connecting the Serro to other regions was built, the cheese was introduced to the market of the capital of the State of Minas Gerais and other states, guaranteeing a profit, as an autonomous element of the regional and state economy (Netto, 2011).

The Serro city expanded the production of cheese that takes its name to the neighboring regions. It is located in a mountainous region, in the eastern slope of the Serra do Espinhaço, in the Alto do Jequitinhonha region, in the State of Minas Gerais (EMATER, 2002). The beginning of the production of the typical cheese of the Serro region is related to the colonial period when the miners began to put into practice a Portuguese heritage. At the time of the mining (18th century), miners from Portugal in search of gold began to make the cheese with the same processing of the Serra da Estrela cheese, made from sheep's milk. In Minas Gerais, it began to be produced with cow's milk for their consumption. Due to the difficulty of access, the producers had to conserve the starter culture, thereby establishing the practice of using wheydrop (Santos, 2010). The production of Artisanal cheese from the Serro region received in 2002, in a pioneering way, a register, by the State Institute of Historic and Artistic Patrimony (IEPHA), as an intangible good of Minas Gerais culture.

The Serro micro region was identified by the Agriculture Institute of Minas Gerais state (Instituto Mineiro de Agropecuária - IMA) as a producer area of the Artisanal Minas cheese through Ordinance 546, dated October 29th, 2002 (IMA, 2002). However, only in 2011 (RPI n° 2136; December 13th, 2011) the Artisanal Minas cheese from Serro was officially recognized by an Indication of Origin (*Indicação de Procedência - IP*) which is a Geographical Indication provided by the National Institute of Industrial Property (INPI, 2012).

In the production of the Serro cheese, the quality attributes of the cheese involve the know-how, as well as the characteristics of cows, pasture, climate, relief, and it is also important to highlight factors such as milking, the use of wood in the production, techniques and ripening time, and especially the characteristics of the lactic acid bacteria and the famous wheydrop (Pinto *et al.*, 2011; Mata *et al.*, 2016). Serro cheese, which is considered a variety of Minas cheese, is manufactured from raw milk with the addition of wheydrop and rennet. Serro cheese production is totally artisanal without any mechanization and industrial techniques. The place where the cheeses are produced is called the "cheese house" or "cheese room" and is usually located next to the cow′s enclosure, connected to it by one window (Brant *et al.*, 2007; Mata *et al.,* 2016). Fig. (**10**) shows the Serro cheese manufacturing technology.

Immediately after milking, which can be manual or mechanical, the milk goes through the filtering process, with the aid of filter or synthetic fabric; then the milk is transferred to the manufacturing tank, and the wheydrop and rennet are added. The wheydrop, is defined as starter culture resulting from the draining of cheeses already salted, collected one day to be used in the next day, and gives the cheese specific physicochemical and organoleptic characteristics. The main genera found in this natural starter culture are *Lactobacillus*, *Lactococcus,* and *Streptococcus* (Cardoso *et al.*, 2015). After coagulation, the curd is kept undisturbed for about five minutes, followed by slow stirring for approximately half an hour until the firm curd is obtained. After that, part of the whey is removed, and the curd is manually pressed in the molds. The salting is done first on one side of the cheese and after 6 to 12 hours on the other side. The wheydrop is collected in a sanitized container to be used in the next manufacturing (Santos, 2010).

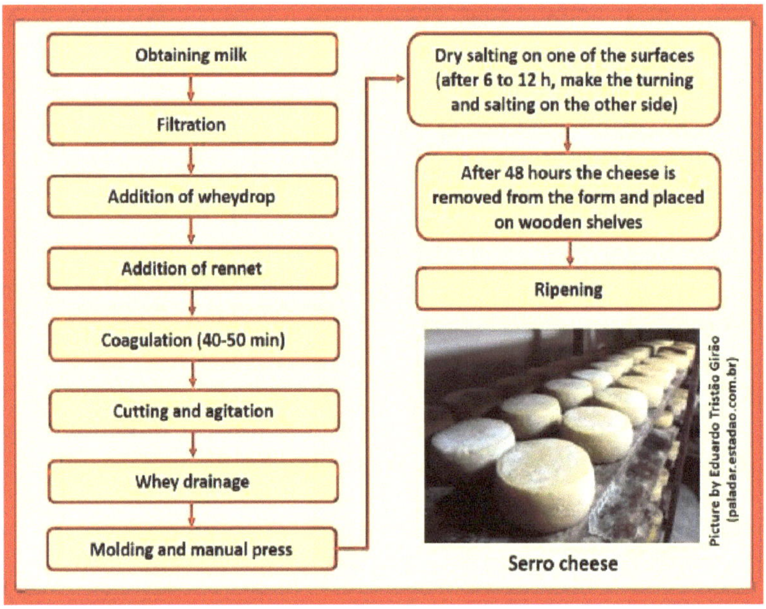

Fig. (10). Serro cheese manufacturing technology.

The cheeses are taken out from the mold on the next day or after two days, ripened at room temperature, for a period which varies from producer to producer, and finally commercialized. According to the Brazilian Legislation (Ordinance 1305; April 2013) (IMA, 2013), the minimum ripening period of the Minas cheese produced in Serro micro region is 17 days. However, most producers do not follow this requirement and sell the cheese after a short ripening period (3 to 15 days) (Cardoso *et al.,* 2015).

The Minas Artisanal cheese from Serro region has a cylindrical shape, with approximately 14 cm of diameter and height ranging from 4-6 cm. The rind is usually whitish, tending to turn into a thin yellowish crust when the cheese is ripened for a few days (Furtado, 1980). Table **6** shows the main sensory attributes of Minas Artisanal cheese produced in Serro region.

Table 6. Main characteristics of Serro Cheese (IMA, 2006).

Attributes of Serro Cheese	
Consistency	Semihard
Texture	Compact
Color	White-yellowish
Rind	Thin, without cracks
Aroma and Flavour	Mild, slightly acidic
Shape	Cylindrical
Height	4 to 6 cm
Diameter	13 to 15 cm
Weight	0.70 to 1.0 kg

Canastra Cheese

Among the Artisanal Minas cheeses, those produced at Serra da Canastra region are the best known and consumed. The Canastra micro region was identified by the Agriculture Institute of Minas Gerais state (Instituto Mineiro de Agropecuária - IMA) as a producer area of the Artisanal Minas cheese through Ordinance 694, dated November 17th, 2004 (IMA, 2004). However, only in 2012 (RPI 2149; March 13th, 2012), the Canastra Cheese was officially recognized through Indication of Origin (*Indicação de Procedência - IP*) which is a Geographical Indication provided by the National Institute of Industrial Property (INPI, 2012).

This Indication of Origin is currently represented by 7 cities in the state of Minas Gerais, São Roque de Minas, Vargem Bonita, Medeiros, Bambuí, Delfinópolis, Piumhi, and Tapiraí. The Serra da Canastra has more than 50 km of extension in its rocky formation, similar to a trunk, which takes the name of the region. The colonization of this region happened because families from other cities in the state went to the area due to diamonds and other minerals mining in the middle of century XIX. The village brought the tradition of the production of artisanal cheese elaborated from raw milk, which was produced in the rainy period when milk was abundant and was consumed only during the dry season. Since then, the quality behind this cheese has been discovered (IG Canastra, 2018).

Serra da Canastra National Park is located in this region, southwest of the state of Minas Gerais, where the spring of São Francisco River is located. The climate is considered as altitude tropical, typical of the Cerrado. The altitude varies from 637 to 1,485 m, and its relief is characterized by wavy (40%), mountainous (35%), and flat areas (25%). These conditions are favorable to the production of the cheese as it promotes the growth of typical bacteria that confer the characteristic and typical taste of Canastra cheese (IG Canastra, 2018). Fig. (**11**) shows the Canastra cheese manufacturing technology.

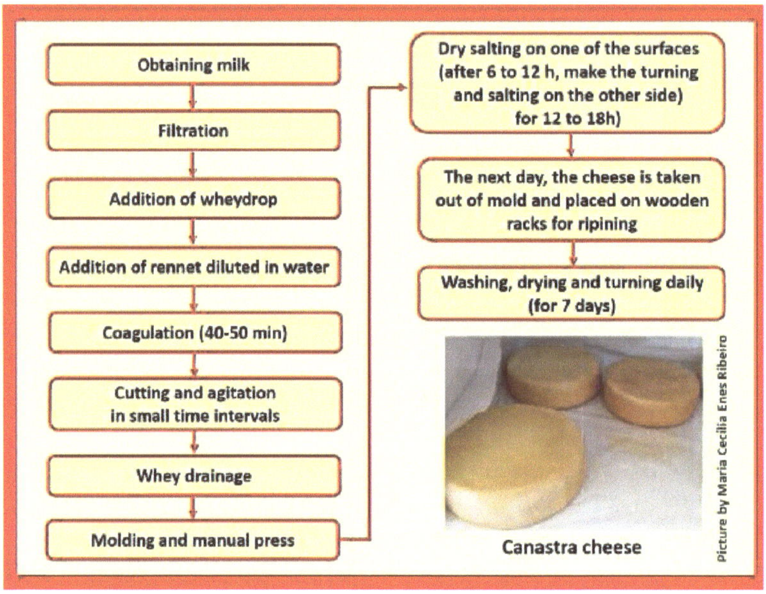

Fig. (11). Canastra cheese manufacturing technology.

The Canastra Minas Artisanal cheese has a cylindrical shape, approximately 16 cm of diameter and height ranging from 4 to 6 cm, smooth yellowish rind with a strong tone at the edges (IMA, 2006). The production of this cheese presents a strong economic and cultural milestone, since it is the only source of income of several farmers, besides owning the tradition passed from generation to generation (IG Canastra, 2018). In the region of Serra da Canastra and Salitre, women are responsible for the cheese production differently from the region of Serro where the men have this duty (IPHAN, 2014).

After milking, the raw milk goes through the filtering process, with the aid of filter or synthetic fabric; then the milk is transferred to the manufacturing tank, and the wheydrop and rennet sufficient to coagulate the milk within 40-50 minutes are added. After this time, the curd is cut and stirred in small intervals. After that, the draining is performed, and the cheese is placed into molds with 15 to 17 cm of diameter with cheesecloth. Afterwards, the manual pressing is carried out with the aid of the cheesecloth. Dry salting is carried out on one of the surfaces of the cheese, and the wheydrop is collected. After 6 to 12 hours, the cheese is turned and salted on the other surface. On the following day, the cheese is taken out of the mold and ripened on wooden shelves at room temperature. Canastra cheese must be ripened for at least 22 days. During the first 7 to 8 days, the cheeses are washed, dried and turned daily (IPHAN, 2014). Table **7** shows the main sensory attributes of Minas Artisanal cheese produced in Canastra region.

Table 7. Main characteristics of Canastra Cheese (IMA, 2006).

Attributes of Canastra Cheese	
Consistency	Semihard, tending to soft and buttery
Texture	Compact
Color	White-yellowish
Rind	Thin, yellowish, without cracks
Aroma and Flavour	Mild, not piquant
Shape	Cylindrical
Height	4 to 6 cm
Diameter	15 to 17 cm
Weight	1.0 to 1.2 kg

In a study conducted by Bemfeito *et al.* (2016) in which evaluated 10 samples of Canastra cheese aged between 22 and 25 days are evaluated, it was observed that the sensory characteristics are quite diverse. In this study, 60 cheese consumers gave scores between 6 (liked slightly) and 7 (liked moderately), with regard to the texture and flavor. In relation to the evaluated attributes through the temporal dominance of sensations, cheeses profile varied among the samples. However, creamy and soft or hard and firm were considered as positive attributes for cheese texture, and bitter, buttery, salty, and acid taste were mentioned as cheese acceptance.

Popularly, Canastra cheese is consumed with longer ripening time while Serro cheese is consumed with shorter ripening time. Nowadays, there is a great effort of the producers to adapt to the requirements of the legislation, however, one of the great challenges is the maintenance of the production tradition against the requirements imposed by law. In this context, the microbiological characterization of artisanal cheeses during ripening aims to ensure the quality and food safety of this product (IPHAN, 2014).

In the study conducted by Dores *et al.* (2013) while evaluating Canastra cheese samples from eight different producers, they observed the importance of complying with the minimum ripening time of 22 days at room temperature to guarantee the microbiological safety, especially concerning to the growth of

Staphylococcus aureus. The authors found that the ripening at refrigeration temperatures was not enough to reduce *Staphylococcus aureus* counts as required by legislation.

Nóbrega (2012) when characterizing the Canastra cheese, observed that mesophilic cocci and lactobacilli represented the dominant microbiota of this cheese in the rainy and dry season, and the main species were *Lactobacillus plantarum, Streptocossus gallolyticus, Streptococcus infantarius* subsp. *infantarius, Lactococcus lactis* subsp. *lactis, Enterococcus faecalis* and *Leuconostoc fallax.*

OTHER BRAZILIAN CHEESES

Marajó Cheese

The Marajó cheese is an artisanal cheese produced for more than 100 years in the Arari-Marajó micro region in the state of Pará, which has economic and social importance for the producers (Cruz *et al.*, 2015). The artisanal production of this cheese is regulated by the Law 7,565 (from October 25th, 2001), which allows its commercialization in the formal market, since this cheese is part of the food habits of the state and is present in the gastronomy of other regions (Figueiredo *et al.*, 2011; Pará, 2011; Cruz *et al.*, 2015).

Marajó cheese is classified as a cooked, not ripened and not pressed-cheese. The manufacturing technology uses skimmed raw cow or buffalo milk. The coagulation occurs through the natural fermentation of milk or by the addition of fermented whey (0.50 a 1.00% of the milk volume). The curd obtained is submitted to fusion, washings with hot water and milk, and the addition of butter or cream. Due to the addition of cream or butter, this cheese is denominated "*Queijo Marajó tipo creme"* (Marajó cheese type cream) and "*Queijo Marajó tipo manteiga*" (Marajó cheese type butter), respectively (Figueiredo *et al.*, 2011; Pará, 2013; Seixas *et al.*, 2015).

Goat Cheeses from Paraíba

The state of Paraíba, located in the northeastern region of the country, has gained prominence in the production of goat milk, with the production of 5,627,000 liters in 2017, the largest production in the country. Within the state, the cities with the highest milk production were Taperoá (543,000 liters), Sumé (507,000 liters), Parari (242,000 liters), Prata (209,000 liters), São Domingos do Cariri (207,000 liters), Cabaceiras (208,000 litros), São Sebastião do Umbuzeiro (182,000 liters), Zabelê (182,000) and Boqueirão (160,000 liters) (Capril Virtual, 2018). All these mentioned cities have industries and/or agroindustries that use goat's milk in the development of the most diverse types of cheeses, with different ripening periods.

In the city of Taperoá, the largest producer of goat's milk in 2017, there is the first dairy plant to produce goat cheese in the Northeast, the Laticínio Grupiara, from Fazenda Carnaúba, which started the goat cheese production in 1979. The cheeses produced are the Arupiara type (white), and Cariri type, seasoned with typical herbs of the region, such as aroeira, marmeleiro, alfazema and cumaru, and these cheeses have a ripening time of four to five months before commercialization (Sertão Bras, 2018). Besides Taperoá, other cities of Paraíba also use typical herbs of the region in the development of cheeses. Lima *et al.* (2017) worked on the standardization of the production of goat's cheese with cumaru (*Amburana cearensis*) developed by an agroindustry from the western Cariri (Santo André - PB), improving quality and strengthening the product identity of this region.

Due to the productivity of goat's milk in the state, especially in the semi-arid region, and the possibility of diversification of cheeses using goat's milk and herbs typical of the region, the process of creation of a Geographical Indication (GI) and Denomination of Origin (DO) for several products of animal origin is being discussed at the Northeast region (Filho & Silva, 2014; Fontenele *et al.*, 2017).

Reino Cheese

Reino cheese is considered one of the oldest cheeses in Brazil, inspired by the Dutch Edam cheese. It is typical in the state of Minas Gerais and is considered the first ripened cheese produced on an industrial scale in Brazil. It is a semi-hard, cooked cheese, and may present some oval, smooth and regularly distributed eyes. The Reino is produced in spherical molds, weights approximately 1.80 kg and its ripening may vary from 7 weeks to 3 months. Once the ripening process is concluded, the cheese is covered with fuchsine or magenta (2.00-3.00% in solution 50/50 water-alcohol) and packed in spherical cans made of tinplate. It has, on average, 36.00-40.00% moisture, 60.00-64.00% total solids, 25.00-30.00% fat and pH between 5.10 and 5.30. During the manufacturing process, urucum dye and *Lactobacillus helveticus*, which leads to the development of the characteristic taste of this cheese, can be added neat or in combination with "O" or "LD" type starter cultures (Furtado, 2005).

Although the production of this cheese is very old, it is not very consumed in Brazil, since it presents a high cost due to ripening and the packaging in spherical cans (Furtado, 2008). Even though the cheese is typical of the state of Minas Gerais, it is widely consumed in the Northeast region of the country, in the state of Pernambuco, especially at the *São João* party and at Christmas, since it symbolizes happiness and prosperity in these meetings (Shinohara *et al.*, 2015).

Serrano Cheese

Serrano cheese is produced in a region known as Campos de Cima, which covers an area between the states of Santa Catarina and the Rio Grande do Sul, in the south of Brazil. This region has specific characteristics, such as the presence of native pastures, high altitude, well defined climate throughout the year with severe winter, with the presence of frost and, occasionally snow. The techniques and knowledge required to produce the Serrano Cheese are maintained by rural families, with only minor changes in the process, which were passed from generation to generation. Their production uses milk from cows reared outdoor, predominantly beef cattle, raised on large ranches and fed mainly by native pastures, especially in spring and summer, when the grass is green and rich in nutrients (Cruz & Menasche, 2014).

According to Normative Instruction 7 (from December 9th, 2014), Serrano Cheese is defined as the ripened product that is obtained by enzymatic coagulation of raw milk through the use of industrial coagulants. It is a semi-fatty cheese, of medium moisture content, and its sales denomination can be Serrano Cheese or Artisanal cheese Serrano. It is characterized by elastic consistency tending to the unctuousness; compact and soft texture; yellowish or even straw yellow color; characteristic flavor. It may be slightly acidic, spicy or salty; with a characteristic, pleasant odor and has a uniform crust, of medium thickness, smooth and without cracks; with or without well distributed small mechanic eyes. The cheese shape and weight may vary.

The manufacturing of Serrano cheese includes the coagulation of whole raw milk for 45 to 60 minutes immediately after milking. The curd is broken into pieces larger than 2 cm, with or without the addition of hot water (70 to 80 °C). The curd is drained off and squeezed to facilitate syneresis, placed in molds and pressed mechanically for 24 hours. The salting can be done directly in the milk during the filtration, or in the curd mass, before molding. The ripening is carried out for at least 60 days at a temperature above 5 °C in a ventilated and dry place (Brazil, 2014). The ripening period can be reduced if it meets the criteria established by Normative Instruction 30 of August 7th, 2013, or another regulation that establishes specific conditions for the production of artisanal cheeses. The cheeses may be marketed without packaging or in microbiologically suitable and innocuous primary packaging to consumers, with labeling previously approved by the competent sanitary inspection service (Rio Grande do Sul, 2014).

Colonial Cheese

The colonial cheese is produced in the Rio Grande do Sul, and its way of production is a tradition that comes from the first colonies of Italians who settled in the South of Brazil. Thus, the term "colonial" refers to the culture and tradition brought by immigrants from non-Iberian Europe, their way of life and their agriculture. However, the system of production and the characteristics of the colonial cheese are unknown, since there is no standardization on its production. Due to the tradition that is passed from generation to generation, many producers produce the cheeses in the same way they learned with their relatives. Another common feature is the informality and the fact that most producers manufacture the cheese for their own consumption. The main difference between Serrano cheese and Colonial cheese is the production area. While the Colonial is found in several parts of southern Brazil, including the entire state, the Serrano cheese is restricted to the Campos de Cima da Serra and part of Santa Catarina. The colonial cheese produced in the Rio Grande do Sul does not have specific legislation. Its most common physical characteristics are the rounded shape, weighing around 1kg. It may not present rind when not ripened, thin and yellowish rind when ripened and hard and thicker rind when subjected to longer ripening.

Caipira Cheese

Caipira cheese is produced and consumed in the State of Mato Grosso do Sul, located in the Central-West region of Brazil. In general, the central-west region is characterized by cattle production, always focused on meat production, and low milk production (Nóbrega, 2012). However, the small milk production of the region is destined for the manufacture of Caipira cheese. No data on the production and characteristics of this cheese were found in the literature, but there is a current law that regulates the production process of Caipira Artisanal Cheese.

According to Law 2,820 (from May 4th, 2004), it is considered to be Caipira Artisanal Cheese, the cheese made according to the historical and cultural tradition of the region of the state where it is produced, from the fresh whole raw cows' milk, obtained and benefited in the property where it is produced. The processing starts until ninety minutes after the beginning of the milking, and natural lactic cultures such as wheydrop, fermented whey or whey starter culture, rennet, and salt can be used as ingredients. The manufacturing process is developed following the steps of milk filtration, the addition of natural starter culture and rennet, coagulation, curd cutting, agitation, draining, molding, manual pressing, dry salting, and ripening. The cheese must have a firm consistency, uniform mass, free of dyes and preservatives, with or without mechanical eyes (Mato Grosso do Sul, 2004).

For the commercialization of ripening and unpackaged cheese, it must have the printing of the state registration number of the producer in low relief. Moreover, for the commercialization of the packaged cheese, the registration of the packaging and the label in the State Agency of Animal and Plant Sanitary Protection (IAGRO) is required, using the same forms adopted for a product with state inspection.

Way of the Artisan Paulista Cheese

Currently, São Paulo state has a lot of artisanal cheeses. From a meeting of five women in 2016, the Way of the Artisan Paulista Cheese was created and consolidated in September 2017, becoming an example of the conservation status. In total, 12 people are part of this group that has already been recognized and awarded, once they have the support of the Secretary of Agriculture and Food Supply of the Government of the State of São Paulo. There are ten cheese factories that produce more than 100 different types of cheeses. The production of Artisan Paulista cheese is considered innovative since they have cheeses from cow, buffalo, goat and sheep milk, such as fresh or ripened varieties (Caminho, 2018).

CONSENT FOR PUBLICATION

Not applicable.

CONFLICT OF INTEREST

The authors confirm that this chapter contents have no conflict of interest.

ACKNOWLEDGEMENTS

Declared none.

REFERENCES

Almeida, SL, Júnior, FGP & Guerra, JRF (2013) Representação da Produção e Consumo do Queijo Coalho Artesanal *RIGS revista interdisciplinar de gestão social,* 2, 37-58. [http://dx.doi.org/10.9771/23172428rigs.v2i2.9870]

Alves, LS, Merheb-Dini, C, Gomes, E, da Silva, R & Gigante, ML (2013) Yield, changes in proteolysis, and sensory quality of Prato cheese produced with different coagulants. *J Dairy Sci,* 96, 7490-9. [http://dx.doi.org/10.3168/jds.2013-7119] [PMID: 24210486]

Amarante, JOA (2015) *Queijos do Brazil e do Mundo para iniciantes e apreciadores.*Summus Editorial, São Paulo.

Ambrósio, CLB, Guerra, NB & Mancini Filho, J (2003) Características de Identidade, Qualidade e Estabilidade da Manteiga de Garrafa. Parte II – Estabilidade. *Food Sci Technol (Campinas),* 23, 351-4. [http://dx.doi.org/10.1590/S0101-20612003000300009]

Ardö, Y, McSweeney, PLH, Magboul, AAA, Upadhyay, VK & Fox, PF (2017) Biochemistry of Cheese Ripening: Proteolysis.*Cheese: Chemistry, Physics and Microbiology.*Elsevier Academic Press, London. [http://dx.doi.org/10.1016/B978-0-12-417012-4.00018-1]

Baptista, DP, Araújo, FDS, Eberlin, MN & Gigante, ML (2017a) A survey of the peptide profile in Prato cheese as measured by MALDI-MS and capillary electrophoresis. *J Food Sci,* 82, 386-93. [http://dx.doi.org/10.1111/1750-3841.13618] [PMID: 28098943]

Baptista, DP, Araújo, FDS, Eberlin, MN & Gigante, ML (2017b) Reduction of 25% salt in Prato cheese does not affect proteolysis and sensory acceptance. *Int Dairy J,* 75, 101-10. [http://dx.doi.org/10.1016/j.idairyj.2017.08.001]

Baptista, DP, Galli, BD, Giacometti, FC, Negrão, F, Eberlin, MN & Gigante, ML (2018) *Lactobacillus helveticus* LH-B02 favours the release of bioactive peptide during Prato cheese ripening. *Int Dairy J,* 87, 75-83. [http://dx.doi.org/10.1016/j.idairyj.2018.08.001]

Bastos, MSR, Ulisses, IB & Fontenele, MA (2013) *Queijo de coalho do Jaguaribe: sabor perpetuado de uma tradição secular.*https://www.embrapa.br/busca-de-publicacoes/-/publicacao/976616/queijo-de- coalho-d--jaguaribe-sabor-perpetuado-de-uma-tradicao-secular

Bemfeito, RM, Rodrigues, JF, Silva, JGE & Abreu, LR (2016) Temporal dominance of sensations sensory profile and drivers of liking of artisanal Minas cheese produced in the region of Serra da Canastra, Brazil. *J Dairy Sci,* 99, 7886-97. [http://dx.doi.org/10.3168/jds.2016-11056] [PMID: 27497904]

Bezerra, TKA, Arcanjo, NMO, Garcia, EF, Gomes, AMP, Queiroga, RCRE, Souza, EL & Madruga, MS (2017a) 'Effect of supplementation with probiotic lactic acid bacteria, separately or combined, on acid and sugar production in goat 'coalho' cheese'. *Lebensm Wiss Technol,* 75, 710-8. [http://dx.doi.org/10.1016/j.lwt.2016.10.023]

Bezerra, TKA, Arcanjo, NMO, Araújo, ARR, Queiroz, ALM, Oliveira, MEG, Gomes, AMP & Madruga, MS (2017b) Volatile profile in goat coalho cheese supplemented with probiotic lactic acid bacteria. *Lebensm Wiss Technol,* 76, 209-15. [http://dx.doi.org/10.1016/j.lwt.2016.03.041]

Brant, LMF, Fonseca, LM & Silva, MCC (2007) Avaliação da qualidade microbiológica do queijo-de-minas artesanal do Serro-MG. *Arq Bras Med Vet Zootec,* 59, 1570-4. [http://dx.doi.org/10.1590/S0102-09352007000600033]

Brazil (1996), Ministério da Agricultura Pecuária e Abastecimento – MAPA, 'Portaria N°146, de 07 de março de 1996', Regulamento Técnico de Identidade e Qualidade de Queijos. Diário Oficial da União: Brasília, 11 de março de 1996. (Ministry of Agriculture, Livestock and Food Supply – MAPA, 'Ordinance No. 146, March 7, 1996', Technical Rules for Identity and Quality of Cheese. Official Gazette: Brazilia, March 11, 1996).

Brazil (1997a) Ministério da Agricultura Pecuária e Abastecimento – MAPA, 'Portaria N° 358, de 04 de setembro de 1997', Aprova o Regulamento Técnico para Fixação de Identidade e Qualidade do Queijo Prato. Diário Oficial da União: Brasília, 08 de setembro de 1997. (Ministry of Agriculture, Livestock and Food Supply – MAPA, 'Ordinance No. 358, September 4, 1997', Technical Rules for Identity and Quality of Prato Cheese. Official Gazette: Brazilia, September 08, 1997).

Brazil (1997b) Ministério da Agricultura Pecuária e Abastecimento – MAPA, 'Portaria N° 359, de 04 de setembro de 1997', Aprova o Regulamento Técnico para Fixação de Identidade e Qualidade do Requeijão ou Requesón. Diário Oficial da União: Brasília, 08 de setembro de 1997. (Ministry of Agriculture, Livestock and Food Supply – MAPA, 'Ordinance No. 358, September 4, 1997', Technical Rules for Identity and Quality of Prato Cheese. Official Gazette: Brazilia, September 08, 1997).

Brazil (1997c) Ministério da Agricultura, Pecuária e Abastecimento. MAPA, 'Portaria N° 352, de 4 de setembro de 1997', Regulamento Técnico para Fixação de Identidade e Qualidade de Queijo Minas Frescal. Diário Oficial da União: Brasília, 08 de setembro de 1997. (Ministry of Agriculture, Livestock and Food Supply – MAPA, 'Ordinance No. 352, September 4, 1997', Technical Rules for Identity and Quality of Fresh Minas Cheese. Official Gazette: Brazilia, September 08, 1997).

Brazil (2000) Ministério da Agricultura Pecuária e Abastecimento – MAPA, 'Instrução Normativa N° 37, de

31 de outubro de 2000', Aprova o Regulamento Técnico de Identidade e Qualidade de Leite de Cabra. Diário Oficial da União: Brasília, 08 de novembro de 2000. (Ministry of Agriculture, Livestock and Food Supply – MAPA, 'Ordinance No. 37, October 31, 2000', Technical Rules for Identity and Quality of Goat milk. Official Gazette: Brazilia, November 8, 2000).

Brazil (2001) Ministério da Agricultura Pecuária e Abastecimento – MAPA, 'Instrução Normativa Nº 30, de 26 de junho de 2001', Aprova os Regulamentos Técnicos de Identidade e Qualidade de Manteiga da Terra ou Manteiga de Garrafa; Queijo de Coalho e Queijo de Manteiga. Diário Oficial da União: Brasília, 16 de julho de 2001. (Ministry of Agriculture, Livestock and Food Supply – MAPA, 'Ordinance No. 30, June 26, 2001', Technical Rules for Identity and Quality of "earth Butter" or "bottle Butter"; Coalho Cheese and Butter Cheese. Official Gazette: Brazilia, July 16, 2001).

Brazil (2011a) Ministério da Agricultura Pecuária e Abastecimento – MAPA, 'Instrução Normativa N°62, de 29 de dezembro de 2011', Aprovar o Regulamento Técnico de Produção, Identidade e Qualidade do Leite tipo A, o Regulamento Técnico de Identidade e Qualidade de Leite Cru Refrigerado, o Regulamento Técnico de Identidade e Qualidade de Leite Pasteurizado e o Regulamento Técnico da Coleta de Leite Cru Refrigerado e seu Transporte a Granel. Diário Oficial da União: Brasília, em 30 de dezembro de 2011. (Ministry of Agriculture, Livestock and Food Supply – MAPA, 'Ordinance No. 62, December 29, 2011', Technical Rules for Production, Identity and Quality of Milk Type A, the Technical Rules for Identity and Quality of Refrigerated Raw Milk, the Technical Rules for Identity and Quality of Pasteurized Milk and the Technical Rules for Collection of Refrigerated Raw Milk and Transport in Bulk. Official Gazette: Brazilia, December 30, 2011).

Brazil (2011b) Ministério da Agricultura Pecuária e Abastecimento – MAPA, 'Instrução Normativa N° 57, de 15 de dezembro de 2011'. Diário Oficial da União: Brasília, 16 de dezembro de 2011. (Ministry of Agriculture, Livestock and Food Supply – MAPA, 'Ordinance No. 57, December 15, 2011'. Official Gazette: Brazilia, December 16, 2011).

Brazil (2013) Ministério da Agricultura Pecuária e Abastecimento – MAPA, 'Instrução Normativa N° 30, de 7 de agosto de 2013'. Diário Oficial da União: Brasília, 08 de agosto de 2013. (Ministry of Agriculture, Livestock and Food Supply – MAPA, 'Ordinance No. 30, August 7, 2013'. Official Gazette: Brazilia, Federal District, August 08, 2013).

Brazil (2014) Secretaria da Agricultura, Pecuária e Irrigação – SEAGRI, 'Instrução Normativa Nº 7, de 09 de dezembro de 2014', Aprova o Regulamento Técnico de Identidade e Qualidade do Queijo Serrano. Diário Oficial da União: Rio Grande do Sul, 10 de dezembro de 2014. (Secretariat of Agriculture, Livestock and Irrigation – SEAGRI. Ordinance nº 7, December 09, 2014. Technical Rules for Identity and Quality of Serrano Cheese. Official Gazette: Rio Grande do Sul, December 10, 2014).

Brazil (2017b) Ministério da Agricultura Pecuária e Abastecimento – MAPA, 'Decreto N° 9.013, de 29 de março de 2017', Dispõe sobre o regulamento da inspeção industrial e sanitária de produtos de origem animal, que disciplina a fiscalização e a inspeção industrial e sanitária de produtos de origem animal. Diário Oficial da União: Brasília, 29 de março de 2017. (Ministry of Agriculture, Livestock and Food Supply – MAPA, 'Ordinance No. 9.013, March 29, 2017', Regulation of the industrial and sanitary inspection of products of animal origin, which regulates the industrial and sanitary inspections of products of animal origin. Official Gazette: Brazilia, March 29, 2017).

Brazil (2018) 'Lei Nº 13680, de 14 de junho de 2018', Altera a Lei Nº 1.283, de 18 de dezembro de 1950, para dispor sobre o processo de fiscalização de produtos alimentícios de origem animal produzidos de forma artesanal, Diário Oficial da União: Brazilia, 15 de junho de 2018. (Law No.13,680, June 14, 2018, Modify Law No.1,283, December 18, 1950, to establish about inspection process of animal origin food products from artisanal manufacturing, Official Gazette: Brazilia, June 15, 2018).

Buriti, FCA, da Rocha, JS & Saad, SMI (2005) Incorporation of *Lactobacillus acidophilus* in Minas fresh cheese and its implications for textural and sensorial properties during storage. *Int Dairy J,* 15, 1279-88. [http://dx.doi.org/10.1016/j.idairyj.2004.12.011]

Campagnollo, FB, Margalho, LP, Kamimura, BA, Feliciano, MD, Freire, L, Lopes, LS, Alvarenga, VO, Cadavez, VAP, Gonzales-Barron, U, Schaffner, DW & Sant'Ana, AS (2018a) Selection of indigenous lactic acid bacteria presenting anti-listerial activity, and their role in reducing the maturation period and assuring the safety of traditional Brazilian cheeses. *Food Microbiol,* 73, 288-97. [http://dx.doi.org/10.1016/j.fm.2018.02.006] [PMID: 29526214]

Campagnollo, FB, Gonzales-Barro, U, Cadavez, VAP, Sant'Ana, AS & Schaffner, DW (2018b) Quantitative risk assessment of Listeria monocytogenes in traditional Minas cheeses: The cases of artisanal semi-hard and fresh soft cheeses. *Food Control,* 92, 370-9. [http://dx.doi.org/10.1016/j.foodcont.2018.05.019]

Caminho (2018), 'Caminho do queijo artesanal paulista', viewed 4 October, 2018, https://www.caminhodoqueijopaulista.com/.

Capril Virtual (2018) *Censo Agropecuário IBGE 2017 – Caprinos e Ovinos.*https://www.caprilvirtual.com.br/censo-caprinos-leite.php.

Canastra, IG (2018) *Indicações Geográficas Brasileiras', viewed 10 October, 2018.* http://datasebrae. com.br/ig-canastra/.

Cardoso, VM, Borelli, BM, Lara, CA, Soares, MA, Pataro, C, Bodevan, EC & Rosa, CA (2015) The influence of seasons and ripening time on yeast communities of a traditional Brazilian cheese. *Food Res Int,* 69, 331-40. [http://dx.doi.org/10.1016/j.foodres.2014.12.040]

Cavalcante, ABD & Costa, JMC (2005) Padronização da tecnologia de fabricação do queijo Manteiga. *Cienc Agron,* 36, 215-20.

Cavicchioli, VQ, Camargo, AC, Todorov, SD & Nero, LA (2017) Novel bacteriocinogenic *Enterococcus hirae* and *Pediococcus pentosaceus* strains with antilisterial activity isolated from Brazilian artisanal cheese. *J Dairy Sci,* 100, 2526-35. [http://dx.doi.org/10.3168/jds.2016-12049] [PMID: 28189328]

Chalita, MAN, Silva, ROP, Petti, RHV & Silva, CRL (2009) Algumas considerações sobre a fragilidade das concepções de qualidade no mercado de queijos no Brazil. *Inf Econ,* 39, 77-88.

Chaves, KS & Gigante, ML (2016) Prato cheese as suitable carrier for *Lactobacillus acidophilus* La5 and *Bifidobacterium* Bb12. *Int Dairy J,* 52, 10-8. [http://dx.doi.org/10.1016/j.idairyj.2015.08.009]

Costa, RGB, Alves, RC, Cruz, AG, Sobral, D, Teodoro, VAM, Junior, LCGC, Paula, JCJ, Landin, TB & Miguel, EM (2018) Manufacture of reduced-sodium Coalho cheese by partial replacement of NaCl with KCl. *Int Dairy J,* 87, 37-43. [http://dx.doi.org/10.1016/j.idairyj.2018.07.012]

Cruz, FT & Menasche, R (2014) Tradition and diversity jeopardised by food safety regulations? The Serrano Cheese case, Campos de Cima da Serra region, Brazil. *Food Policy,* 45, 116-24. [http://dx.doi.org/10.1016/j.foodpol.2013.04.014]

Cruz, BEV, Dias, ALC, Souza, ACS, Santos, J & Pinheiro, GS (2015) A Identificação Geográfica para o Queijo do Marajó com Estratégia de Desenvolvimento Territorial para a Microrregião do Arari-Marajó, PA. *Cadernos Prospecção,* 8, 158-68. [http://dx.doi.org/10.9771/S.CPROSP.2015.001.018]

Datamark (2018) Cheese Production Data. Datamark Brazil Focus, Market Intelligence Brazil', viewed 31 July, 2018 http://www.datamark.com.br.

Domingos, LD (2015) *'A influência da redução de gordura e da adição de concentrado proteico de soro nas características do queijo Prato durante a maturação', Tese de doutorado.*Universidade Estadual de Campinas, Campinas.

Dores, MT & Ferreira, CLLF (2012) Queijo Minas Artesanal, tradição centenária: Ameaças e desafios. *Rev Bras Agropecu Sustent,* 2, 26-34.

Dores, MT, Nobrega, JE & Ferreira, CLLF (2013) Room temperature aging to guarantee microbiological safety of brazilian artisan Canastra cheese. *Food Sci Technol (Campinas),* 33, 180-5.

[http://dx.doi.org/10.1590/S0101-20612013005000003]

EMATER (2002) *Empresa de Assistência Técnica e Extensão Rural – EMATER, 'Documento de caracterização da região do Serro/MG como produtora de queijo Minas artesanal', viewed 12 September, 2018.* http://www.emater.mg.gov.br/doc/intranet/upload/queijo_historico/dossi%C3%AA%20do%20serro %20def2.pdf

Figueiredo, EL, Lourenço Júnior, JB, Toro, MU & Lima, SCG (2011) 'Queijo do "Marajó" Tipo Creme: Parâmetros Físico-Químicos e Sensoriais', *Revista do Instituto de Laticínios. Cândido Tostes,* 66, 26-33.

Filho, CG & Silva, PCG (2014) Indicação Geográfica, Uma Certificação Estratégica para os Produtos de Origem Animal da Agricultura Familiar do Semiárido. *Rev Econ Nordeste,* 45, 114-23. [Suplemento Especial].

FoodNews (2018) *Produção de queijos no País deve crescer em 2018.* http://www.foodnewsoficial.com. br/mercado/producao-de-queijos/

Fontenele, MA, Bastos, MD, Dos Santos, KM, Bemquerer, MP & do Egito, AS (2017) Peptide profile of Coalho cheese: A contribution for Protected Designation of Origin (PDO). *Food Chem,* 219, 382-90. [http://dx.doi.org/10.1016/j.foodchem.2016.09.171] [PMID: 27765241]

Fox, PF, Guinee, TP, Cogan, TM & McSweeney, PLH (2000) *Fundamentals of Cheese Science.*Aspen Publication, Maryland.

Furtado, MM (1980) Queijo do Serro: tradição na história do povo mineiro. *Rev Inst Laticínios Cândido Tostes,* 35, 33-6.

Furtado, MM & Lourenço Neto, JPM (1994) *Tecnologia de queijos Manual técnico para produção industrial de queijos* . Dipemar, São Paulo.

Furtado, MM (2005) *Quesos Típicos de Latinoamérica.* Fonte Comunicação e Editora, São Paulo.

Furtado, MRA (2008) *Caracterização histórica, tecnologia de fabricação, características físico-químicas, sensoriais, perfil de textura e comercialização do queijo Reino.*Tese de Doutorado, Universidade Federal de Lavras, Lavras.

Guerra, TMM & Guerra, NB (2003) Influência do sorbato de potássio e do tipo de embalagem sobre a vida útil do Queijo de Manteiga (Requeijão do Norte)'. *Braz J Food Technol,* 6, 259-65.

Guillier, L, Stahl, V, Hezard, B, Notz, E & Briandet, R (2008) Modelling the competitive growth between *Listeria monocytogenes* and biofilm microflora of smear cheese wooden shelves. *Int J Food Microbiol,* 128, 51-7.
[http://dx.doi.org/10.1016/j.ijfoodmicro.2008.06.028] [PMID: 18778863]

Guinee, TP (2017) Pasteurized Processed and Imitation Cheese Products.*Cheese: Chemistry, Physics and Microbiology.*Elsevier Academic Press, London.
[http://dx.doi.org/10.1016/B978-0-12-417012-4.00046-6]

IBGE (2016) https://www.ibge.gov.br/estatisticas-novoportal/economicas/agricultura-e-pecuaria/9-07-producao-da-pecuaria-municipal.html?=&t=resultados Instituto Brazileiro de Geografia e Estatística – IBGE, 'Tabelas de dados de Produção de origem animal, por tipo de produto em 2016', viewed 14 September, 2018,

IMA (2001) Instituto Mineiro de Agropecuária – IMA, 'Portaria N° 546, de 29 de outubro de 2002', Identifica a microrregião do Serro. Diário Oficial de Minas Gerais: 30 de outubro de 2002. (Agriculture Institute of Minas Gerais state, 'Ordinance No. 546, October 29, 2002', Identifies the Serro micro-region. Official Gazette: Minas Gerais, October 30, 2002).

IMA (2004) Instituto Mineiro de Agropecuária – IMA, 'Portaria Nº 694, de 17 de novembro de 2004', Identifica a microrregião da Canastra. Diário Oficial do Estado de Minas Gerais: 18 de novembro de 2004. (Agriculture Institute of Minas Gerais state, 'Ordinance No. 694, November 17, 2004', Identifies the Canastra micro-region. Official Gazette: Minas Gerais, December 20, 2006).

IMA (2006) Instituto Mineiro de Agropecuária – IMA, 'Portaria Nº 818, de 12 de dezembro de 2006,'

Regulamento técnico de produção do queijo Minas artesanal e dá outras providências, Anexo I. Diário Oficial do Estado de Minas Gerais: 20 dezembro de 2006. (Agriculture Institute of Minas Gerais state, 'Ordinance No. 818, December 12, 2006', Standards for the production Minas artisanal cheese and provides other measures – Annex 1. Official Gazette: Minas Gerais, December 20, 2006).

IMA (2013) Instituto Mineiro de Agropecuária – IMA, 'Portaria N° 1305, de 30 de abril de 2013', Estabelece Diretrizes para a produção do Queijo Minas artesanal. Diário Oficial do Estado de Minas Gerais: 01 maio de 2013. (Agriculture Institute of Minas Gerais state, 'Ordinance No. 1,305, April 30, 2013', Standards for the production Minas artisanal cheese. Official Gazette: Minas Gerais, April 30, 2013).

IMA (2016) Instituto Mineiro de Agropecuária -IMA, 'Legislação Certificação Queijo Minas Artesanal', viewed 10 September, 2018, http://www.ima.mg.gov.br/material-curso-cfo-cfoc/doc_details/ 3029-legislacao-certificacao-queijo-minas-artesanal.

Ligabue, LH (2008) Ele inventou o requeijão cremoso', viewed 06 October, 2018, https://www.estadao.com.br/noticias/geral,ele-inventou-o-cremoso,128153.

INPI (2012) Instituto Nacional da Propriedade Intelectual, 'Lista das Indicações de Procedência Concedidas. RPI n° 2149, de 13 de março de 2012', viewed 04 October, 2018, http://www.inpi.gov.br/menu-servicos/indicacao-geografica/pedidos-de-indicacao-geografica-no-brazil.

IPHAN (2014) Instituto do Patrimônio Histórico e Artístico Nacional, 'Dossiê Iphan - Modo Artesanal de Fazer Queijo de Minas: Serro, da Serra da Canastra e Serra do Salitre (Alto Parnaíba)', viewed 14 October, 2018 http://portal.iphan.gov.br/uploads/publicacao/Dossie_Queijo_de_Minas_web.pdf.

Lima, TLS, Alves, RN, Cavalcanti, MT, Rocha, TC & Gonçalves, MC (2017) Padronização do processamento de queijo Coalho caprino condimentado com cumaru produzido por agroindústria na Paraíba. *Rev Verde Agroecol Desenvolv Sustent,* 12, 562-7.
[http://dx.doi.org/10.18378/rvads.v12i3.5051]

Machado, EC, Ferreira, CLLF, Fonseca, LM, Soares, FM & Júnior, FNP (2004) Características físico-químicas e sensoriais do Queijo Minas Artesanal produzido na região do Serro, Minas Gerais. *Food Sci Technol (Campinas),* 24, 516-21.
[http://dx.doi.org/10.1590/S0101-20612004000400006]

Martins, JM, Galinari, É, Pimentel-Filho, NJ, Ribeiro, JI, Jr, Furtado, MM & Ferreira, CLLF (2015) Determining the minimum ripening time of artisanal Minas cheese, a traditional Brazilian cheese. *Braz J Microbiol,* 46, 219-30.
[http://dx.doi.org/10.1590/S1517-838246120131003] [PMID: 26221111]

Mata, GMSC, Martins, E, Machado, SG, Pinto, MS, de Carvalho, AF & Vanetti, MCD (2016) Performance of two alternative methods for *Listeria* detection throughout Serro Minas cheese ripening. *Braz J Microbiol,* 47, 749-56.
[http://dx.doi.org/10.1016/j.bjm.2016.04.006] [PMID: 27268116]

Matera, J, Luna, AS, Batista, DB, Pimentel, TC, Moraes, J, Kamimura, BA, Ferreira, MVS, Silva, HLA, Mathias, SP, Esmerino, EA, Freitas, MQ, Raices, RSL, Quitério, SL, Sant'Ana, AS, Silva, MC & Cruz, AG (2018) Brazilian cheeses: A survey covering physicochemical characteristics, mineral content, fatty acid profile and volatile compounds. *Food Res Int,* 108, 18-26.
[http://dx.doi.org/10.1016/j.foodres.2018.03.014] [PMID: 29735047]

Mato Grosso do Sul (2004) Assembleia Legislativa do Estado de Mato Grosso do Sul, 'Lei N° 2.820, de 04 de maio de 2004', Dispõe sobre o processo de produção do Queijo Artesanal Caipira, e dá outras providências. Diário Oficial do Estado do Mato Grosso do Sul: 05 de maio de 2004. (Legislative Assembly of the Mato Grosso do Sul State, 'Law No. 2,820, May 04, 2004'. Standards for the production of Caipira artisanal cheese and provides other measures. Official Gazette: Mato Grosso do Sul, May 05, 2004).

Medeiros, RS, Araújo, LM, Queiroga Neto, V, Andrade, PP, Melo, MA & Gonçalves, MMBP (2016) Identification of lactic acid bacteria isolated from artesanal Coalho cheese produced in the Brazilian Northeast. *CYTA J Food,* 14, 613-20.
[http://dx.doi.org/10.1080/19476337.2016.1185468]

Medeiros, JMS, Câmpelo, MCS & Silva, JBA (2017) Good manufacturing practices of artisanal products in Northeastern Brazil. *Food Res,* 1, 103-8. [http://dx.doi.org/10.26656/fr.2017.4.050]

Melo, ACM, Alves, LMC & Costa, FN (2009) Avaliação da qualidade microbiológica do Queijo tipo Minas Padrão comercializado na cidade de São Luis, MA. *Arq Inst Biol (Sao Paulo),* 76, 547-51.

Merheb-Dini, C, Chaves, KS, Gomes, E, da Silva, R & Gigante, ML (2016) Coalho Cheese Made with Protease from *Thermomucor indicae-seudaticae* N31: Technological Potential of the New Coagulant for the Production of High-Cooked Cheese. *J Food Sci,* 81, C563-8. [http://dx.doi.org/10.1111/1750-3841.13217] [PMID: 26799929]

Milkpoint (2017) *'MG: produção de queijo minas artesanal ganha reforço com o atendimento de produtores familiares', viewed 15 September, 2018.* https://www.milkpoint.com.br/noticias-e-mercado/giro-noticias-mg-producao-de-queijo-minas-artesanal-ganha-reforco-com-o-atendimento-de-produtor-s-familiares-104224n.aspx

Minas Gerais (2012) Assembleia Legislativa do Estado de Minas Gerais, 'Lei Nº 20549, de 18 de dezembro de 2012', Dispõe sobre a produção e a comercialização dos queijos artesanais de Minas Gerais. Diário Oficial do Estado do Minas Gerais:19 de dezembro de 2012. (Legislative Assembly of the Minas Gerais State, 'Law No. 20,549, December 18, 2012'. Standards for the production and marketing of artisanal cheese in Minas Gerais, Official Gazette: Minas Gerais December 19, 2012).

Minim, VPR, Silva, RCSN, Milagres, MP, Martins, EMF, Sampaio, SCS & Vasconcelos, EMF (2010) Contribuição dos atributos sensoriais para aceitabilidade de requeijão cremoso. *Rev Inst Laticínios Cândido Tostes,* 65, 34-42.

Monteiro, AA, Pires, ACS & Araújo, EA (2011) *Tecnologia de Produção de Derivados do Leite* Editora UFV, Viçosa.

Moreno, I, Dender, AGFV, Costa, GAN, Vialta, A, Lerayer, ALS, Silva, AT & Destro, MT (2002) Propriedades físicas e composição química e bioquímica durante a maturação de queijo Prato de diferentes origens. *Rev Inst Laticínios Cândido Tostes,* 57, 270-3.

Nassu, RT, Araújo, RS, Guedes, CGM & Rocha, RGA (2003) *Diagnóstico das Condições de Processamento e Caracterização Físico-Química de Queijos Regionais e Manteiga no Rio Grande do Norte.* viewed 02 October, 2018 https://www.embrapa.br/busca-de-publicacoes/-/publicacao/425891/diagnostico-das-condicoes-de-processamento-e-caracterizacao-fisico-quimica-de-queijos-regionais-e-m-nteiga-no-rio-grande-do-norte.

Nóbrega, JE (2012) *'Biodiversidade microbiana, descritores físico-químicos e sensoriais dos queijos artesanais fabricados nas regiões da Serra da Canastra e do Serro, Minas Gerais', Tese de doutorado.*Universidade Federal de Viçosa, Viçosa.

Netto, MM (2011) *'A Geografia do Queijo Minas Artesanal', Tese de doutorado.*Universidade Estadual Paulista Júlio de Mesquita Filho, Rio Claro.

Ordónez, JA (2005) *Tecnologia de alimentos: alimentos de origem animal* Artmed, Porto Alegre.

Ortolani, MBT, Moraes, PM, Perin, LM, Viçosa, GN, Carvalho, KG, Silva Júnior, A & Nero, LA (2010) Molecular identification of naturally occurring bacteriocinogenic and bacteriocinogenic-like lactic acid bacteria in raw milk and soft cheese. *J Dairy Sci,* 93, 2880-6. [http://dx.doi.org/10.3168/jds.2009-3000] [PMID: 20630205]

Pará (2011) Agência de Defesa Agropecuária do Estado do Pará – ADEPARA, 'Lei Nº 7.565, de 25 de outubro de 2011', Dispõe sobre normas para licenciamento de estabelecimentos processadores, registro e comercialização de produtos artesanais comestíveis de origem animal e vegetal no Estado do Pará, e dá outras providências. Diário Oficial do Estado do Pará: 26 de outubro de 2013. (Agricultural Defense Agency of Pará State). 'Law No. 7,565, October 25, 2011', Standards for the licensing of processing establishments, registration and marketing of edible artisanal products of animal and plant origin in the State of Pará and provides other measures. Official Gazette: Pará, October 26, 2011).

Pará (2013) Agência de Defesa Agropecuária do Estado do Pará – ADEPARA, 'Portaria ADEPARA Nº 418, de 26 de fevereiro de 2013', Aprova o Regulamento Técnico de Produção do Queijo do Marajó e dá outras providências. Diário Oficial do Estado do Pará: 07 de março de 2013. (Agricultural Defense Agency of Pará State), 'Ordinance No. 418, February 26, 2013', Technical Rules for Production of the Marajó Cheese and provides other measures. Official Gazette: Pará, March 07, 2013).

Perin, LM, Savo Sardaro, ML, Nero, LA, Neviani, E & Gatti, M (2017) Bacterial ecology of artisanal Minas cheeses assessed by culture-dependent and -independent methods. *Food Microbiol,* 65, 160-9. [http://dx.doi.org/10.1016/j.fm.2017.02.005] [PMID: 28399998]

Perry, KSP (2004) Queijos: aspectos químicos, bioquímicos e microbiológicos. *Quim Nova,* 27, 293-300. [http://dx.doi.org/10.1590/S0100-40422004000200020]

Picoli, SU, Bessa, MC, Castagna, SMF, Gottardi, CPT, Schmidt, V & Cardoso, M (2006) Quantificação de Coliformes, *Staphylococcus aureus* e mesófilos presentes em diferentes etapas da produção de Queijo Frescal de leite de cabra em laticínios. *Food Sci Technol (Campinas),* 26, 64-9. [http://dx.doi.org/10.1590/S0101-20612006000100011]

Pinto, MS, Carvalho, AF, Pires, ACS, Souza, AAC, Silva, PHF, Sobral, D, Paula, JCJ & Santos, AL (2011) The effects of nisin on Staphylococcus aureus count and the physicochemical properties of Traditional Minas Serro cheese. *Int Dairy J,* 21, 90-6. [http://dx.doi.org/10.1016/j.idairyj.2010.08.001]

Queiroga, RCRE, Santos, BM, Gomes, AMP, Monteiro, MJ, Teixeira, SM, Souza, EL, Pereira, CJD & Pintado, MME (2013) 'Nutritional, textural and sensory properties of Coalho cheese made of goats', cows' milk and their mixture'. *Lebensm Wiss Technol,* 50, 538-44. [http://dx.doi.org/10.1016/j.lwt.2012.08.011]

Rapacci, M, Van Dender, AGF & Mori, EEM (1999) Avaliação das principais características sensoriais de requeijão cremoso fabricado com massa obtida por fermentação lática e acidificação direta a quente. *Rev Inst Laticínios Cândido Tostes,* 57, 270-3.

Rio Grande do Sul (2014) Secretaria de Estado, da Agricultura, Pecuária e do Agronegócio, 'Instrução normativa Nº 7 de 09 de dezembro de 2014', Aprova o Regulamento Técnico de Identidade e Qualidade do Queijo Serrano. Diário Oficial do Estado do Rio Grande do Sul: 10 de dezembro de 2014. (State Secretary of Agriculture, Livestock and Agribusiness, 'Ordinance No. 7, December 09, 2014', Technical rules for production of the Serrano Cheese. Official Gazette: Rio Grande do Sul, December 10, 2014).

Sangaletti, N, Porto, E, Brazaca, SGC, Yagasaki, CA, Dalla Dea, RC & Silva, MV (2009) Study of Minas cheese shelf life. *Food Sci Technol (Campinas),* 29, 262-9. [http://dx.doi.org/10.1590/S0101-20612009000200004]

Santos, AS (2010) *'Queijo Minas Artesanal da Microrregião do Serro - MG: Efeito da sazonalidade sobre a microbiota do leite cru e comportamento microbiológico durante a maturação', Dissertação de mestrado, Universidade Federal dos Vales do Jequitinhonha e Mucuri.*Diamantina.

Santos, BM, Oliveira, MEG, Sousa, YRF, Madureira, ARMFM, Pintado, MME, Gomes, AMP, Souza, EL & Queiroga, RCRE (2011) Caracterização físico-química e sensorial de queijo de coalho produzido com mistura de leite de cabra e de leite de vaca. *Rev Inst Adolfo Lutz,* 70, 302-10.

Seixas, VNC, Félix, MR, Silva, GM, Perrone, IT & Carvalho, AF (2015) Caracterização do Queijo do Marajó tipo manteiga produzido em duas estações do ano. *Cienc Rural,* 45, 730-6. [http://dx.doi.org/10.1590/0103-8478cr20140463]

Sertão Bras (2018) *Fazendo Carnaúba – Taperoá – PB', viewed 15 October 2018.* https://www.sertaobras.org.br/2018/06/06/fazenda-carnauba-taperoa-pb/.

Sheehan, JJ, Drake, MA & Mcsweeney, PLH (2009) Effect of partial or total substitution of bovine for caprine milk on the compositional, volatile, non-volatile and sensory characteristics of semi-hard cheeses. *Int Dairy J,* 19, 498-509. [http://dx.doi.org/10.1016/j.idairyj.2009.03.009]

Shinohara, NKS, Santos, MCG, Medeiros, STC, Padilha, MRF & Matsumoto, M (2015) Cultural importance of cheese type for kingdom Pernambuco *Holos year,* 31, 62-71.

Silva, FT (2005) *'Queijo Prato', Embrapa Informação Tecnológica.*Agroindústria Familiar, Brasília.

Silva, MCD, Ramos, ACS, Moreno, I & Moraes, JO (2010) Influência dos procedimentos de fabricação nas características físico-químicas, sensoriais e microbiológicas de queijo de coalho. *Rev Inst Adolfo Lutz,* 69, 214-21.

Silva, RA, Lima, MSF, Viana, JBM, Bezerra, VS, Pimentel, MCB, Porto, ALF, Cavalcanti, MTH & Lima Filho, JL (2012) Can artisanal "Coalho" cheese from Northeastern Brazil be used as a functional food? *Food Chem,* 135, 1533-8. [http://dx.doi.org/10.1016/j.foodchem.2012.06.058] [PMID: 22953890]

Silva, RA, Bezerra, VS, Pimentel, MdoC, Porto, AL, Cavalcanti, MTH & Filho, JLL (2016) Proteomic and peptidomic profiling of Brazilian artisanal 'Coalho' cheese. *J Sci Food Agric,* 96, 4337-44. [http://dx.doi.org/10.1002/jsfa.7640] [PMID: 26801736]

Sousa, AZB, Abrantes, MR, Sakamoto, SM, Silva, JBA, Lima, PO, Lima, RN, Rocha, MOC & Passos, YDB (2014) Aspectos físico-químicos e microbiológicos do queijo tipo coalho comercializado em estados do nordeste do Brazil. *Arq Inst Biol (Sao Paulo),* 81, 30-5. [http://dx.doi.org/10.1590/S1808-16572014000100006]

Souza, CHB & Saad, SMI (2009) Viability of *Lactobacillus acidophilus* La-5 added solely or in co-culture with a yoghurt starter culture and implications on physico-chemical and related properties of Minas fresh cheese during storage. *Lebensm Wiss Technol,* 42, 633-40. [http://dx.doi.org/10.1016/j.lwt.2008.07.015]

Souza, TB, Cruz, AG, Moura, MRL, Vieira, APM & Sant'Ana, AS (2008) Microscopic quality indicators of Minas Frescal cheese. *Food Control,* 19, 71-5. [http://dx.doi.org/10.1016/j.foodcont.2007.02.004]

Spadoti, LM, Roig, SM, Zacarchenco, PB, Moreno, I & Alves, ATS (2011) Queijo Prato. *Leite & Derivados,* 128, 92-102.

Tulini, FL, Winkelströter, LK & De Martinis, ECP (2013) Identification and evaluation of the probiotic potential of *Lactobacillus paraplantarum* FT259, a bacteriocinogenic strain isolated from Brazilian semi-hard artisanal cheese. *Anaerobe,* 22, 57-63. [http://dx.doi.org/10.1016/j.anaerobe.2013.06.006] [PMID: 23792229]

Van Dender, AGF, Moreno, I, Alves, ATS & Spadoti, LM (2014) Requeijão e Outros queijos fundidos: Tecnologia de Fabricação e Parâmetros de Controle de Processo In: Van Dender, AGF (Ed) 'Requeijão Cremoso e Outros Queijos Fundidos – Tecnologia de Fabricação, Controle do Processo e Aspectos de Mercado', 2nd ed., Setembro Editora, São Paulo.

Van Dender, AGF, Spadoti, LM, Zacarchenco, PB, Gallina, DA & Alves, ATS (2012) Requeijão cremoso. *Leite & Derivados,* 133, 86-97.

Van Dender, AGF, Zacarchenco, PB & Alves, ATS (2016a) *Requeijão: tecnologias de fabricação da massa básica – Parte 1 . viewed 09 October 2018.* https://www.milkpoint.com.br/artigos/industria/requeijao-tecnologias-de-fabricacao-da-massa-basica-parte-1-91254n.aspx.

Van Dender, AGF, Zacarchenco, PB & Alves, ATS (2016b) *Requeijão: tecnologias de fabricação da massa básica – Parte 2 viewed 09 October 2018.* https://www.milkpoint.com.br/artigos/industria/requeijao-tecnologias-de-fabricacao-da-massa-basica-parte-2-92006n.aspx.

CHAPTER 5

Manufacturing Method and Physicochemical Properties of Lighvan and Koozeh (Koopeh) Cheeses

Mostafa Soltani[1,2,*]

[1] *Department of Food Sciences and Technology, Faculty of Pharmacy, Tehran Medical Sciences, Islamic Azad University, Tehran, Iran*

[2] *Nutrition and Food Sciences Research Center, Tehran Medical Sciences, Islamic Azad University, Tehran, Iran*

Abstract: Ewes' milk and goats' milk and dairy products manufactured from them have an important role in the life of the people living in local areas of Iran. Lighvan and Koozeh (Koopeh) are the most known Iranian traditional cheeses manufactured respectively in East Azarbaijan, West Azarbaijan, Kordestan and Qazvin provinces, in the north-west of Iran. These cheeses are manufactured traditionally from raw ewe or goats' milk or an appropriate mixture of them without using a starter-culture. Using raw milk for the manufacturing of Lighvan and Koozeh (Koopeh) cheeses provides a specific taste and flavor and increases their marketability. During manufacturing of these types of cheese, lamb or commercial calf rennet is used for coagulating the milk to form the curd. While Lighvan cheese is ripened in the cube form in metal tins containing 10-12% of brine for 3-4 months, Koozeh (Koopeh) cheese is ground after dry salting and put into jugs, and then is ripened underground for 2-3 months. In this chapter, besides discussing the properties of ewes' and goats' milk, the manufacturing methods and physicochemical properties of these Iranian traditional cheeses are also explained.

Keywords: Dairy products, Ewes' milk, Goats' milk, Iranian traditional cheeses, Koozeh (Koopeh) cheeses, Lighvan cheeses.

INTRODUCTION

Nowadays, pasteurized milk is preferred for manufacturing of the most types of cheeses in order to minimize the risk of food-borne pathogenic bacteria. However, using raw milk for cheese manufacturing generally provides more intense flavor

* **Corresponding author Mostafa Soltani**: Department of Food Sciences and Technology, Faculty of Pharmacy, Tehran Medical Sciences, Islamic Azad University, and Nutrition and Food Sciences Research Center, Tehran Medical Sciences, Islamic Azad University, Tehran, Iran; Tel: +98-21-22640051; Fax: +98-21-22602059; E-mail: m.soltani@iaups.ac.ir

Celile Aylin Oluk & Oya Berkay Karaca (Eds.)

and faster ripening in the final product (Navidghasemizad *et al.*, 2009; Hayaloglu *et al.*, 2010). Cheese ripening including proteolysis, lipolysis, and glycolysis has an important role in the quality of texture and formation of volatile compounds in cheese (McSweeney, 2007; Soltani *et al.*, 2016). Formation of characteristic volatile compounds in cheese depends on the presence of an adequate number of native and starter microorganisms and indigenous milk enzymes play a major role in the formation of small peptides and free amino acids (Hayaloglu *et al.*, 2010). Pasteurization of raw milk prior to cheese manufacturing influences proteolysis and has a negative effect on the formation of small peptides and free amino acids (Hayaloglu & Brechany, 2007). Due to the inactivation of a large amount of indigenous lipoprotein lipase in the pasteurized milk, it has also been reported that the level of lipolysis in the cheeses manufactured using pasteurized milk is lower than in those manufactured using raw milk (McSweeney, 2007; Soltani *et al.*, 2016). On the other hand, the texture is an important quality parameter in cheese influenced by some factors such as milk type, processing technique, and ripening time (Tsigkros *et al.*, 2003; Chevanan *et al.*, 2006). In this context, Pappa *et al.* (2007) and Aminifar *et al.* (2013) in their studies have reported about the changes in texture properties of cheese manufactured using different types of milk.

Milk produced from goat and ewe is an important part of the national economy in the Mediterranean and Middle-East countries. These types of milk are usually used for manufacturing of the traditional dairy products particularly cheese varieties in local scales. Different parameters such as breed, season, feeding, environmental conditions, health status, and lactation period may influence the composition of ewes' milk and goats' milk. Moreover, the chemical composition of ewes' milk and goats' milk is different from cows' milk and influences the physicochemical properties, proteolysis, texture, and volatile composition of cheese (Park, 2006; Park *et al.*, 2007).

According to the statistics published by FAO, almost 40 million heads of ewe and 15.7 million heads of goat are producing 479000 and 311000 tons of fresh raw milk, respectively, in Iran (FAO, 2017). These types of raw milk are mainly processed by small scale factories or households for the manufacturing of the local and traditional dairy products, especially popular types of cheese such as Lighvan and Koozeh (Koopeh) cheeses manufactured in the East-Azarbaijan, West-Azarbaijan, Kordestan, and Qazvin provinces located in the north-west of Iran. The locations of the manufacturing of these cheeses in Iran are given in Fig. (**1**). The characteristics of ewes' milk and goats' milk used for the manufacturing of cheese as well as the manufacturing method and quality properties of Lighvan and Koozeh cheeses are discussed as follows.

QUALITY CHARACTERISTICS OF EWES' MILK AND GOATS' MILK

Cows' milk dominates the worlds' milk production with about 652 million tons of fresh raw milk as reported by statistics in 2017. However, producing ewes' milk and goats' milk and using them for the manufacturing of various dairy products are important parts of local economies in the Mediterranean and Middle -East countries (FAO, 2017; Balthazar *et al.*, 2017).

Fig. (1). Location of the manufacturing of Lighvan and Koozeh cheeses in Iran.

Ewes' milk has high concentrations of protein, fat, minerals, and vitamins. Higher content of fat fraction in ewes' milk compared to cows' milk has provided favorable flavor and greater consistency to the dairy products manufactured from this type of milk. Moreover, the absence of agglutinin as a significant factor in clustering and coalescence of fat globules provides better digestibility for ewes' milk-manufactured products compared to dairy products manufactured from cows' milk. On the other hand, high protein content involving caseins (4.2-5.2 g/100g) and whey proteins (1.02-1.3 g/100g) can make ewes' milk a proper raw material for the manufacturing of various types of cheese. Additionally, the casein fraction in ewes' milk is rich in terms of calcium content. Therefore, technologically, no calcium chloride is needed to assist cheese manufacturing from heat-treated ewes' milk. In general, renneting properties of ewes' milk for cheese manufacturing including the rate and time of coagulation, the amount of required rennet, and the curd firmness are influenced by some physicochemical properties of milk i.e. pH, the size of casein micelles, concentrations of calcium, and other minerals. However, several parameters including species and age of animal, season, nutrition, environmental conditions, lactation stage, and efficiency may cause changes in chemical composition of ewes' milk over time (Park *et al.*, 2007; Claeys *et al.*, 2014; Gantner *et al.*, 2015; Balthazar *et al.*, 2017).

Goats' milk has different casein content and proportions of four main caseins including α-$_{S1}$ casein, α_{S2}-casein, β-casein, and κ-casein from cows' milk. Despite naturally lower content of α-$_{S1}$ casein in goats' milk, it has higher proportion of α_{S2}-casein, β-casein, and κ-casein compared to cows' milk. The poor cheese-manufacturing ability of goats' milk is mainly related to low casein content and composition, size, and hydration of casein micelles have a significant negative effect on the quality of final product (Park *et al.*, 2007; Clark & Mora Garcia, 2017). From the nutritional aspect, lower allergenic properties are important characteristics of goats' milk making it an alternative to cows' milk in the human diet. The genetic polymorphism of α_{S1}- casein has been confirmed to be the main factor influencing the allergenicity of goats' milk, and the types of milk containing null α_{S1}-casein genotypes have lower activity of allergenicity (Ballabio *et al.*, 2011; Clark and Mora Garcia, 2017). However, Lisson et al (2014) reported that due to cross-reactivity of cows' milk caseins with possible IgE antibodies of goats' milk, the dairy products manufactured from goats' milk should be consumed cautiously by cows' milk-allergic patients.

In addition to health-related features, the strong and attractive flavor is one of the famous properties of goats' milk. Several studies have demonstrated the unique flavor property of goats' milk. In this context, Chilliard *et al.* (2003), reported that fatty acids composition, triglyceride structure, and characteristics of lipoprotein lipase influence goats' milk flavor. Although the beneficial effects of goats' milk on health maintenance and physiological functions have been proven by the nutritionists and health authorities, using this type of milk for the manufacturing of dairy products is limited probably due to the seasonal production and negative public perception toward "goat-like" flavor (Park, 2005; Ribeiro & Ribeiro, 2010). However, the quality and property of goats' milk make it a potential source of manufacturing of dairy products, especially various types of cheese. The protein content of goats' milk may vary between 2.6 - 4.1 gL^{-1}, which is less than protein content present in ewes' milk or cows' milk. Furthermore, goats' milk has less α-$_{S1}$ casein content compared to other ruminants' milk. Therefore, goats' milk is generally used in combination with other types of milk for the manufacturing of cheese (Raynal-Ljutovac *et al.*, 2008). Fat content and properties may be considered as the most important quality variable of goats' milk influencing the texture and flavor of final product. The content of short and medium-chain fatty acids has been reported to play an important role in the production and release of flavor compounds in goats' milk fat, which is at least two times more than the cows' milk fat (Chilliard *et al.*, 2006).

In terms of microbial contamination, D'Amico *et al.* (2008) demonstrated that the incidence of *Staphylococcus aureus* was lower in goats' milk than cows' or ewes' milk collected from Vermont State in the USA. The authors also reported that *Escherichia coli* O157:H7, *Salmonella* spp., and *L. monocytogenes* were not found at food-borne infection levels in goats' milk. In a similar study, Rahimi & Alian (2013) reported that the prevalence of *Staphylococcus aureus* was lower in raw goats' milk than raw cows' milk or ewes' milk analyzed in Iran. However, they emphasized that, regardless of milk type, raw milk should not be used for the manufacturing of dairy products because of high potential risk of staphylococcal food poisoning. In this regard, the results of a research on microbial quality of cheese manufactured from different types of milk showed that cheese manufactured from goats' milk had higher risk of coliform detection compared to the cheeses manufactured from cows' milk. Accordingly, some reasons such as natural properties of goats' milk, manufacturing method and the facilities used for the manufacturing of goats' milk cheese can be considered (Trmcic *et al.*, 2016).

Considering the mentioned issues, ewes' milk and goats' milk are used for manufacturing of the Lighvan and Koozeh cheeses in Iran. These types of cheese have specific characteristics and are manufactured traditionally in local dairy plants.

THE MANUFACTURING METHOD AND SOME QUALITY PROPERTIES OF LIGHVAN CHEESE

The manufacturing procedure of Lighvan cheese and the final cheese consumed are shown in Figs. (**2** and **3**), respectively. As seen in Fig. (**2**), for the manufacturing of cheese, first, raw milk is heated to 28-30 °C and is coagulated using the lamb rennet (chymosin). After 2 hours, the curd is cut into pieces and is transferred to rectangular cotton bags for whey drainage. Then, a weight is placed on the bags for better whey drainage. After about 2 hours, the curd is cut into blocks and is salted with brine (22%) for 6 hours. At the next stage, the cheese blocks are removed from brine and are salted with coarse-grained salt on the surface for 24 hours. For better whey removal, at this stage cheese blocks are turned upside down for about 10-15 times. Finally, the cheese blocks are put in tins filled with brine (12%) and are ripened for 90-120 days at a cold room with an average temperature of 9±1 °C (Aminifar & Emam-Djome, 2016)

Milk (usually is a mixture of raw ewes' milk and goats' milk)
⇩
Heat treatment until 28-30 °C
⇩
Addition of lamb rennet for coagulation
⇩
Cutting the curd into cube pieces
⇩
Transferring to rectangular cotton bags
⇩
Placing a weight on the cotton bags for compression
⇩
Cutting the curd into blocks
⇩
Placing the blocks in the brine (22 % for 6 hours)
⇩
Removing the blocks from brine and dry-salting them for 24 hours
⇩
Keeping the cheese blocks in the brine (12%) for 3-4 months

Fig. (2). Flow chart of manufacturing of Lighvan cheese.

Lighvan cheese is a semi-hard and white-brined cheese manufactured traditionally from raw ewes' milk and goats' milk or an appropriate mixture of them without using starter-culture in the Lighvan village located in East-Azarbaijan province, Iran. Due to the desired texture and excellent natural flavor, Lighvan cheese has high popularity among Iranian consumers. However, seasonal production of ewes' milk and goats' milk causes some limitations regarding the manufacturing of this type of cheese. Therefore, cows' milk is sometimes used in combination with ewes' milk or goats' milk for the manufacturing of Lighvan cheese in order to meet the consumer demand (Lavasani *et al.*, 2012; Aminifar & Emem-Djome, 2016).

Fig. (3). Lighvan cheese manufactured from raw ewe and goats' milk.

Aminifar *et al.* (2012) reported that the chemical properties of Lighvan cheese manufactured from ewes' milk are as follows: 4.7 pH, 62.1% moisture content, 8.0% salt content, and 21.5% fat content after 90 days of ripening. In a similar research, Aminifar *et al.* (2013) reported that the chemical properties of Lighvan cheese manufactured from ewes' milk or a mixture of ewes' milk and cows' milk are as follows: 4.65 and 4.49 pH, 61.63 and 62.15% moisture content, 13.61% and 14.05% salt content, and 19.11% and 17.80% WSN/TN content, respectively after 90 days of ripening. In other research, Attar *et al.* (2018) investigated the hypothesis of the presence of inherent lactic acid bacteria with antibiotic properties in Lighvan cheese based on lack of food born disease report related to consumers of this type of cheese. The results of their research showed that due to manufacturing of Lighvan cheese from unpasteurized raw milk, high number of indigenous lactic acid bacteria are existed in it which have an important role in maintaining the sanitary quality of final product. In similar research, Nayyeri *et al.* (2017), reported that lactic acid bacteria isolated from Lighvan cheese can be classified as natural bio-preservatives. They proved that *Lactobacillus Plantarum* isolated from different stages of manufacturing of Lighvan cheese has antifungal effect on *Penicillium expansum* and *Rhodotorula mucilaginosa* as main indicators for fungal spoilage in fruit juices.

THE MANUFACTURING METHOD AND SOME QUALITY PROPERTIES OF KOOZEH CHEESE

The manufacturing procedure of Koozeh cheese and the final cheese consumed are shown in Figs. (**4** and **5**), respectively. As seen in Fig. (**4**), for the manufacturing of cheese, first, the raw milk is heated to 33-34 °C and is coagulated using the commercial calf rennet. After 45-60 min, the curd is mashed and is put on a cloth bag and is hanged for about 14-15 hours for whey drainage.

Then, the curd is transferred into a large cloth bag and is dry-salted. Finally, the cheese curd is ground and is put into pottery jugs, and then is ripened for 60-90 days in the shade or underground (Edalatian *et al.*, 2012).

Milk (prepared from raw ewes' milk or a mixture of ewes' milk and goats' milk)

⇩

Heat treatment until 33-34 °C

⇩

Addition of calf rennet for coagulation (45-60 min)

⇩

Mashing the curd and transferring it to a cloth bag

⇩

Hanging the curd for whey drainage (14-15 hours)

⇩

Transferring the curd to a large cloth bag and dry-salting

⇩

Grinding the curd and putting it into the pottery jugs

⇩

Ripening in the shade or underground for 2-3 months

Fig. (4). Flow chart of manufacturing of Koozeh cheese.

Fig. (5). Koozeh cheese manufactured from raw ewe and goats' milk.

Koozeh or Koopeh cheese is a semi-soft cheese manufactured from ewes' milk or a mixture of ewes' milk and goats' milk without adding starter-culture in the West-Azarbaijan, Kordestan, and Qazvin provinces, Iran. "Koozeh" or "Koopeh" means the pottery jug in "Persian" and "Azarbaijani Turkish", respectively. Because of keeping the cheese in the pottery jugs, this type of cheese is called Koozeh cheese. Possibly, using ewes' milk and goats' milk for manufacturing and spending 2-3 months in the jug as ripening time leads to the formation of unique taste and flavor in Koozeh cheese (Edalatian *et al.*, 2012; Hassanzadazar *et al.*, 2014). Shahbazi *et al.* (2017) reported that Koozeh cheese is one of the most commonly consumed Iranian traditional cheeses, and is classified as white-brined, semi-soft cheese with crumbly texture, sour taste, and pleasant flavor. However, they emphasized that due to potential risks of Aflatoxin M_1 concentration, this type of cheese should be consumed cautiously. In a similar research, Hasanzadazar *et al.* (2014) reported that the chemical properties of Koozeh cheese manufactured from a mixture of ewes' milk and goats' milk are as follows: 6.20 pH, 36.72% of total solid content, 14.76% fat content, 12.45% protein content and 3.24% salt content, respectively after 60 days of ripening in pottery jugs.

CONCLUSION

Lighvan cheese and Koozeh cheese are classified as starter-free cheeses manufactured from raw ewes' milk or goats' milk or a mixture of them in the north-west provinces of Iran. Both cheeses have high consumption rate because of unique texture and flavor. The manufacturing methods and ripening conditions of these cheeses are different from each other. Calf rennet and cloth bags are used for coagulation of milk and whey drainage of both kinds of cheese, respectively. While Lighvan cheese is salted in brine, and is ripened in tins for 90-120 days, Koozeh cheese is salted using the dry-salting method and is ripened in pottery jugs for 60-90 days' underground. Due to the manufacturing of these types of cheese from raw milk, potential precautionary measures of microbial contamination should be taken into account. Considering the high consumption of these cheeses, further studies should be conducted in order to identify the physicochemical, microbiological, texture, and sensory properties of these cheeses and modify their manufacturing methods.

CONSENT FOR PUBLICATION

Not applicable.

CONFLICT OF INTEREST

The author confirms that this chapter contents have no conflict of interest.

ACKNOWLEDGEMENTS

Declared none.

REFERENCES

Aminifar, M, Hamedi, M, Emam-Djomeh, Z & Mehdinia, A (2014) Investigation on proteolysis and formation of volatile compounds of Lighvan cheese during ripening. *J Food Sci Technol,* 51, 2454-62. [http://dx.doi.org/10.1007/s13197-012-0755-3] [PMID: 25328184]

Aminifar, M, Hamedi, M, Emam-Djome, Z & Mehdinia, A (2013) The effect of ovine and bovine milk on the textural properties of Lighvan cheese during ripening. *Int J Dairy Technol,* 66, 45-53. [http://dx.doi.org/10.1111/1471-0307.12009]

Aminifar, M & Emam-Djome, Z (2016) Investigation on the microstructural and textural properties of Lighvan cheese produced from bovine milk fortified with protein and gum tragacanth during ripening. *Int J Dairy Technol,* 69, 225-35. [http://dx.doi.org/10.1111/1471-0307.12243]

Attar, MA, Yavarmanesh, M, Mortazavi, A, Edalatian Dovem, MR & Habibi Najafi, MB (2018) Antibacterial effects of Lactococcus lactis isolated from Lighvan cheese regarding the recognition of Nisin, Lacticin and Lactococcin structural genes. *Lebensm Wiss Technol,* 89, 186-91. [http://dx.doi.org/10.1016/j.lwt.2017.10.044]

Ballabio, C, Chessa, S, Rignanese, D, Gigliotti, C, Pagnacco, G, Terracciano, L, Fiocchi, A, Restani, P & Caroli, AM (2011) Goat milk allergenicity as a function of αs_1-casein genetic polymorphism. *J Dairy Sci,* 94, 998-1004. [http://dx.doi.org/10.3168/jds.2010-3545] [PMID: 21257068]

Balthazar, CF, Pimentel, TC, Ferrao, LL, Almada, CN, Santillo, A, Albenzio, M, Mollakhalili, N, Mortazavian, AM, Nascimento, JS, Silva, MC, Freitas, MQ, Sant'Ana, AS, Granato, D & Cruz, AG (2017) Sheep Milk: Physicochemical Characteristics and Relevance for Functional Food Development. *Compr Rev Food Sci Food Saf,* 16, 247-62. [http://dx.doi.org/10.1111/1541-4337.12250]

Chevanan, N, Muthukumarappan, K, Upreti, P & Metzger, LE (2006) Effect of calcium and phosphorus, residual lactose and salt-to-moisture ratio on textural properties of cheddar cheese during ripening. *J Texture Stud,* 37, 711-30. [http://dx.doi.org/10.1111/j.1745-4603.2006.00080.x]

Chilliard, Y, Ferlay, A, Rouel, J & Lamberet, G (2003) A review of nutritional and physiological factors affecting goat milk lipid synthesis and lipolysis. *J Dairy Sci,* 86, 1751-70. [http://dx.doi.org/10.3168/jds.S0022-0302(03)73761-8] [PMID: 12778586]

Chilliard, Y, Rouel, J, Ferlay, A, Bernard, L, Gaborit, P, Raynal-Ljutovac, K, Lauret, A & Leroux, C (2006) Optimising goat's milk and cheese fatty acid composition. *Improving the Fat Content of Foods* .Woodhead Publishing Ltd, Cambridge, UK 281-312. [http://dx.doi.org/10.1533/9781845691073.2.281]

Claeys, WL, Verraes, C, Cardoen, S, De Block, J, Huyghebaert, A, Raes, K, Dewettinck, K & Herman, L (2014) Consumption of raw or heated milk from different species: an evaluation of the nutritional and potential health benefits. *Food Control,* 42, 188-201. [http://dx.doi.org/10.1016/j.foodcont.2014.01.045]

Clark, S & Mora García, MB (2017) A 100-Year Review: Advances in goat milk research. *J Dairy Sci,* 100, 10026-44. [http://dx.doi.org/10.3168/jds.2017-13287] [PMID: 29153153]

D'Amico, DJ, Groves, E & Donnelly, CW (2008) Low incidence of foodborne pathogens of concern in raw milk utilized for farmstead cheese production. *J Food Prot,* 71, 1580-9. [http://dx.doi.org/10.4315/0362-028X-71.8.1580] [PMID: 18724751]

Edalatian, MR, Habibi Najafi, MB, Mortazavi, A & Mayo, B (2012) The biodiversity and evolution of lactic flora during ripening of the Iranian semi-soft Lighvan cheese. *Int J Dairy Technol,* 65, 81-9. [http://dx.doi.org/10.1111/j.1471-0307.2011.00738.x]

FAO (2017) Food and Agriculture Organization of the United Nations. Official Data.

Gantner, V, Mijic, P, Baban, M, Zoran, S & Alka, T (2015) The overall and fat composition of milk of various species. *Mljekarstvo,* 65, 223-31. [http://dx.doi.org/10.15567/mljekarstvo.2015.0401]

Hayaloglu, AA & Brechany, EY (2007) Influence of milk pasteurization and scalding temperature on the volatile compounds of Malatya, a farmhouse Halloumi-type cheese. *Lait,* 87, 39-57. [http://dx.doi.org/10.1051/lait:2006025]

Hayaloglu, AA, Deegan, KC & McSweeney, PLH (2010) Effect of milk pasteurization and curd scalding temperature on proteolysis in Malatya, a Halloumi-type cheese. *Dairy Sci Technol,* 90, 99-109. [http://dx.doi.org/10.1051/dst/2009052]

Hassanzadazar, H, Ehsani, A & Mardani, K (2014) Antibacterial activity of *Enterococcus faecium* derived from Koopeh cheese against *Listeria monocytogenes* in probiotic ultra-filtrated cheese. *Vet Res Forum,* 5, 169-75. [PMID: 25568714]

Lavasani, AS, Ehsani, MR, Mirdamadi, S & Ebrahim Zadeh Mousavi, MA (2012) Changes in physicochemical and organoleptic properties of traditional Iranian cheese Lighvan during ripening. *Int J Dairy Technol,* 65, 64-70. [http://dx.doi.org/10.1111/j.1471-0307.2011.00724.x]

Lisson, M, Novak, N & Erhardt, G (2014) Immunoglobulin E epitope mapping by microarray immunoassay reveals differences in immune response to genetic variants of caseins from different ruminant species. *J Dairy Sci,* 97, 1939-54. [http://dx.doi.org/10.3168/jds.2013-7355] [PMID: 24485684]

McSweeney, PLH (2007) *Effect of milk pasteurization and curd scalding temperature on proteolysis in Malatya, a Halloumi-type cheese* Dairy Science & Technology 99-109.

Navidghasemizad, S, Hesari, J, Saris, P & Nahaei, MR (2009) Isolation of lactic acid bacteria from Lighvan cheese, a semihard cheese made from raw ewe milk in Iran. *Int J Dairy Technol,* 62, 260-4. [http://dx.doi.org/10.1111/j.1471-0307.2009.00462.x]

Nayyeri, N, Edalatian Dovom, MR, Habibi Najafi, MB & Bahreini, M (2017) A Preliminary study on antifungal activity of lactic acid bacteria isolated from different production stages of Lighvan cheese on *Penicillium expansum* and *Rhodotorula mucilaginosa. J Food Meas Charact,* 11, 1734-44. [http://dx.doi.org/10.1007/s11694-017-9554-x]

Pappa, EC, Kandarakis, I & Mallatou, H (2007) Effect of different types of milks and cultures on the rheological characteristics of Teleme cheese. *J Food Eng,* 79, 143-9. [http://dx.doi.org/10.1016/j.jfoodeng.2006.01.038]

Park, YW (2005) Goat milk products: quality, composition, processing, marketing.*Encyclopedia of Animal Science* CRP Press 478-81.

Park, YW (2006) Goat milk-chemistry and nutrition. *Handbook of Milk of Non-bovine Mammals,* Blackwell Publishing Professional, Oxford, UK/Ames, Iowa 34-58. [http://dx.doi.org/10.1002/9780470999738.ch3]

Park, YW, Juarez, M, Ramos, M & Haenlein, GFW (2007) Physico-chemical characteristics of goat and ewe milk. *Small Rumin Res,* 68, 88-113. [http://dx.doi.org/10.1016/j.smallrumres.2006.09.013]

Rahimi, E & Alian, F (2013) Presence of enterotoxigenic *Staphylococcus aureus* in cow, camel, sheep, goat, buffalo bulk tank milk. *Vet Arh,* 83, 23-30.

Raynal-Ljutovac, K, Lagriffoul, G, Paccard, P, Guillet, I & Chilliard, Y Composition of goat and sheep milk products: An update. *Small Rumin Res,* 79, 57-72. [http://dx.doi.org/10.1016/j.smallrumres.2008.07.009]

Ribeiro, AC & Ribeiro, SDA (2010) Specialty products made from goat milk. *Small Rumin Res,* 89, 225-33. [http://dx.doi.org/10.1016/j.smallrumres.2009.12.048]

Shahbazi, Y, Nikousefat, Z & Karami, N (2017) Occurrence, seasonal variation and risk assessment of exposure to aflatoxin M1 in Iranian traditional cheeses. *Food Control,* 79, 356-62.

[http://dx.doi.org/10.1016/j.foodcont.2017.04.021]

Soltani, M, Boran, OS & Hayaloglu, AA (2016) Effect of various blends of camel chymosin and microbial rennet (*Rhizomucor miehei*) on microstructure and rheological properties of Iranian UF White cheese. *Lebensm Wiss Technol,* 68, 724-8.
[http://dx.doi.org/10.1016/j.lwt.2016.01.028]

Trmčić, A, Chauhan, K, Kent, DJ, Ralyea, RD, Martin, NH, Boor, KJ & Wiedmann, M (2016) Coliform detection in cheese is associated with specific cheese characteristics, but no association was found with pathogen detection. *J Dairy Sci,* 99, 6105-20.
[http://dx.doi.org/10.3168/jds.2016-11112] [PMID: 27289158]

Tsigkros, D, Folland, E, Moate, R & Brennan, CS (2003) Feta cheese texture: the effect of caprine and ovine milk concentration. *Int J Dairy Technol,* 56, 233-6.
[http://dx.doi.org/10.1046/j.1471-0307.2003.00112.x]

CHAPTER 6

Traditional Cheeses of Marmara Region in Turkey

Nayil Dinkci*

Faculty of Agriculture, Department of Dairy Technology, Ege University, 35100 Bornova-Izmir, Turkey

Abstract: The Marmara Region is situated in the north-western part of Turkey. It is a point of juncture between Europe and Asia, thousands of civilizations passed from one continent to the other and left some parts of their cuisines in the region. Some cheese varieties are real artworks; they have to be formed and ripened in the hands of the cheese master. Mihaliç Cheese is one of these cheeses. It is an aged cheese made from high-fat sheep milk in the north western provinces of Balıkesir and Bursa. It's been made in the region since more than 200 years. Also, the best Lor cheese is made from the whey of Mihaliç cheese. Çerkez Füme is a Circassian smoked cheese produced in the eastern Marmara region. This cheese is usually made from low fat milk, and is fumed with pinewood smoke. Abaza is a semi-hard cheese that is infused with various herbs, the most popular version being with nigella seeds. Ezine, is famed for its high-quality Beyaz Peynir variations which are produced from full fat sheep-goat and cow milk mixtures and ripened for at least 6 months. Kirli Hanım cheese, "dirty lady," is a type of Sepet cheese variety which is produced in Ayvalık region. It is a hard cheese made from a blend of goat, sheep and cow's milk, that is shaped into a basket. Beyaz Peynir and Kaşar cheese are specific cheeses for the Thracian region (Edirne, Tekirdağ) which are now produced conventionally in many parts of Turkey.

Keywords: Abaza cheese, Beyaz cheese, Çerkez fume cheese, Ezine cheese, Thrace Kaşar cheese, Lor cheese, Mihalic cheese, Sepet cheese.

INTRODUCTION

Geography

Turkey is situated in Anatolia between the 39°00'N and 35°00'E geographic coordinates. The area is about 783,562 km^2 while the European part known as Thrace is about 3% of the total area. The population is about 80,000,000 (Wikipedia, 2018). Anatolia is geographically on the crossroads of Europe, Asia and partially Africa, and also the ancient Egypt and Mesopotamia. Many civilisations passed through Anatolia to Asia or Europe by using the Silk Road.

* **Corresponding author Nayil Dinkci:** Faculty of Agriculture, Department of Dairy Technology, Ege University, 35100 Bornova-Izmir, Turkey; Tel/Fax: + 90 232 311 13 94; E-mail nayil.dinkci@ege.edu.tr

Celile Aylin Oluk & Oya Berkay Karaca (Eds.)

Anatolia itself has been also the home for thousands of years for many civilisations. Every civilisation left some parts of his cuisine in this ancient part of the world. This is the main reason for the formation of the huge amount of traditional products (Durlu-Özkaya & Gün, 2008). Therefore, it is not a surprise that it offers also such a diverse range of cheese varieties.

Ancient Origins of Cheese - Anatolia

Approximately, more than 2000 varieties of cheese exist worldwide. It is not exactly known where and when cheese was made for the first time. Cheese is a part of our ancient cultural heritage. Historically these foods enabled populations to survive periods of famine, since they provide nutritional elements; they are good for vital health and are also desirable in the daily diet. According to Kosikowski (1977); the evolution and world movement of fermented foods began many centuries before Christ (BC), while the first of these products must be discovered by accident, probably in the warm climate of the Mediterranean Sea Basin. It can be seen from archaeological records that domestication of plants begun in the Fertile Crescent (the comparatively watery land from the Jordan River Valley of the southern Levant northward through Lebanon and Syria to the south-eastern part of Anatolia and Zagros Mountains, parallel to the alluvial soils of Euphrates and Tigris Rivers) (Fig. **1**) around 9000 to 8500 BC (Barker, 2006; Simmons, 2007). It is also known that the first evidence for cattle domestication occurred in central Anatolia, around 7000 BC (Cauvin, 2000). The origin of cheese production goes also back to the beginnings of agriculture. One of the theories for the first cheese is that shortly after 7000 BC, with the development of ceramics and pottery, that the collected milk has been stored in ceramic pods in the warm climate of south-east Asia, the milk fermented and coagulated spontaneously by the bacteria present in milk. It was only a matter of time to discover that adults were able to consume the curds (Kindstedt; 2012). It is known from the analysis of organic residues in pottery shards corresponding to around 6500 – 6000 BC that the pods contained dairy products like cheese and butter (Evershed *et al.* 2008). Another theory is that a nomadic traveller filled his bag, made from the stomach of a mammal (goat – sheep), with fresh milk and realized after a while when he wanted to drink milk, that the milk has coagulated. The discovery of cheese making during the 7. millennium BC came at the right time for the Neolithic man, since the Fertile Crescent was failing catastrophically due to rising populations and environmental degradation. It transformed the nutritional value of milk into a vital food that could be stored for later use. During the mass movements from the Fertile Crescent the people take the knowledge of cheese making with them and continue it wherever they journeyed and settled during many centuries. Migrations to the south-east let to the settlements of Mesopotamia (6000 BC). Migrations to the south-west reached Egypt (5000 BC), the ones moved to the east reached the Indus River Valley (6500 BC). And the migrations to the north and west was trough Turkey into Thrace and the Balkan peninsula they reached England 4500 BC. And another route proceeded westward from the sea to Greece, Italy, France and Spain. Archaeological founding's confirmed that the Neolithic farmers brought a Near East culture, based on mixed farming which included milk production and cheese - butter making (Kindstedt; 2012). The Sumerians were the earliest settlers of Mesopotamia, who recorded the making of 20 different types of cheese on clay tablets in 3000 BC (L.Y. Ergil, 2014). Once a cheese making principle was established it must have been easier to create related products. The variety of cheese comes from the traditional adaptations of a few cheese making techniques. The roots of cheese production in Anatolia must be as deep as the migrations and settlements to north-west Anatolia 6500 BC. Anatolia has been also the home for many civilisations in the next thousands of years and many civilisations passed through Anatolia and left some of their culture in this land. It is possible to find in each region different cheeses with their distinct flavour. Ünsal (1997) reported 230 different names of cheese varieties in Turkey.

Fig. (1). Map of the region (Max Pixel; 2018).

However, Kamber (2015) indicated that this number includes the different names by which the same cheese is known and also the names of regions renowned for their cheeses. It was mentioned that in Turkey more than 130 varieties of cheese are to be found while Swan (2009) reported about 160 different traditional cheeses in Turkey. Turkish cheese is one of the most delectable delights of Turkish cuisine, it is served usually at the breakfast table, it is used as meze staple or also featured in a number of unique desserts (Ergil; 2014). Dishes like börek (flaky layered phyllo), gözleme (savoury crepe with various fillings) and poğaca (savory) are made also with cheese (Donelly; 2016).

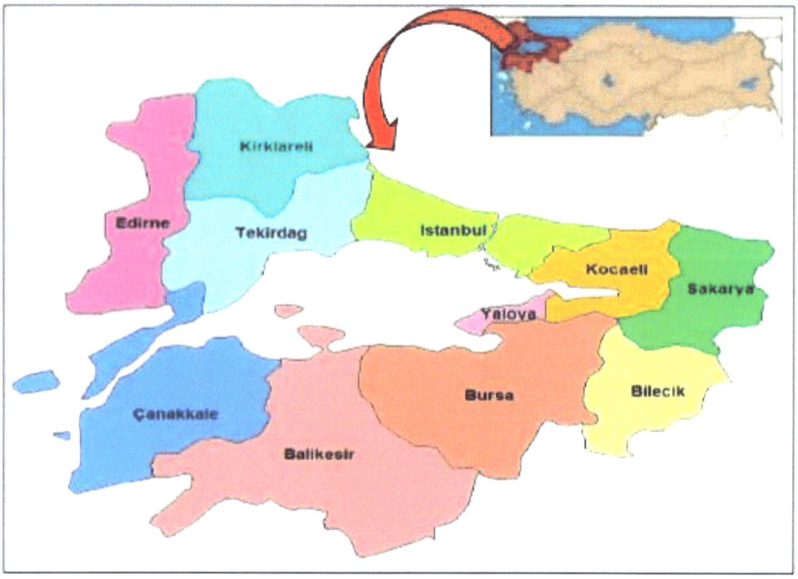

Fig. (2). The Marmara Region.

The Marmara Region

Turkey is separated in 7 geographical regions according to their climate, location, flora and fauna. The Marmara Region is the northwest part of Turkey (Fig. **2**) (Yenen; 2018). This region gets its name from the sea it surrounds; the Sea of Marmara. The Sea of Marmara is the inland sea that connects the Black Sea to the Aegean Sea, thus separating Turkey's Asian and European parts. It separates also Istanbul into its Asian and European parts; it is the geographical border between Europe and Asia: northern coasts of it are in Europe, while southern/eastern coasts are in Asia (Wikitravel; 2017). The Marmara region is the second smallest region of Turkey in size, it is only a bit smaller than Ireland or Netherlands and Belgium combined. But it is also the most populated and industrialized part of Turkey. The region has a moderate climate and is quite rich in terms of its flora. Large parts of the region are used for agricultural farming, most of the modern dairies and also small traditional cheese producers are found in this region. It's no wonder that only in the province Balıkesir are more than 50 different traditional cheeses been produced (Bal-Onur, Aksoy-Biber; 2017). Commercially the most important traditional cheese of the region can be listed as; Beyaz Peynir, Kaşar Peyniri, Mihaliç Peyniri, Sepet Peyniri and Lor (Fig. **3**). Beyaz Peynir, KaşarPeyniri and Lor varieties are seen more regularly in whole Turkey with some regional variations.

Fig. (3). Some of the traditional cheeses of the region (breakfast plate).

TRADITIONAL CHEESES OF THE REGION

Beyaz Peynir (White Cheese)

Turkish White Cheese is a brined cheese variety. It is one of the most produced traditional cheeses with an annual production amount of more than 60% of the total cheese production in Turkey (USK, 2014). Its varieties can be found; with some physicochemical, textural and sensory differences according to the region, in whole Turkey and also many neighbour countries, with the names Feta, Telemes (Greece), Telemea, Branza de Braila (Romania), Bjalosalamurenosirene, Bjalo sirene (Bulgaria), Bienosirenje (FYROM), Mohant (Slovenia), Sjenicki, Homoljski, Zlatarski, Svrljiŝki (Serbia), Pljevaljski, Polimosko-Vasojevaski, Ulcinjski (Montenegro), Travnicki, Vasicki (Bosnia-Herzegovina), Liqvan, Iranian White (Iran), Brinza (Israel), Akawi (Lebanon) and Domiati, Mish (Egypt). Traditionally it was mainly made from sheep's and goat's milk. While the milk of these species does not contain carotenoids the resulting cheese is in white colour, which is giving the name to this cheeses (Alichanidis and Polychroniadou, 2008).

Since the lactation period of sheep / goats is short and the milk amount is not enough for the production, cow's milk or a combination with sheep's / goat's milk is also used. Traditionally white cheese is produced with raw milk and its natural microbial flora. Therefore, starter culture is not used. The curd is handled extensively by the cheese master and the cheeses are matured for at least 6 months in cold rooms (+5°C) (Turantas *et al.*, 1989, Erkmen, 1995). Turkish White Cheese is a rindless cheese with a soft to semi-hard, close texture (Hayaloğlu, Güven and Fox; 2002). Usually it has a salty acid taste, it may have also a piquant flavour (Dinkci and Gönc; 2000) especially when made from sheep's milk and matured for more than 6 months. It's ripened at least for one month if produced with pasteurized milk otherwise it has to be matured for at least 4 months according to the Turkish Cheese Codex (TGK, 2015). But the best taste occurs in sheep – goat cheeses after 1 year of ripening. According to the Turkish regulations fresh and matured white cheese; has to be at least 35%, 40% Dry Matter and a max concentration of 6,5%, 6,5% Salt (NaCl) in dry matter respectively. And it can be produced in four different fat in dry matter (FDM) levels; Full-fat Cheese (45 ≤ FDM %), Half-fat Cheese (25≤FDM%<45), Low-fat Cheese (10 ≤ FDM % < 25) and as Non-fat Cheese (10 > FDM %) (TGK, 2015). The manufacturing steps of White Cheese are described in Fig. (**4**). Chemical properties of White Cheese samples (from cow's milk, ripened for 6 months) collected from different geographical regions of Turkey are summarized in Table **1**.

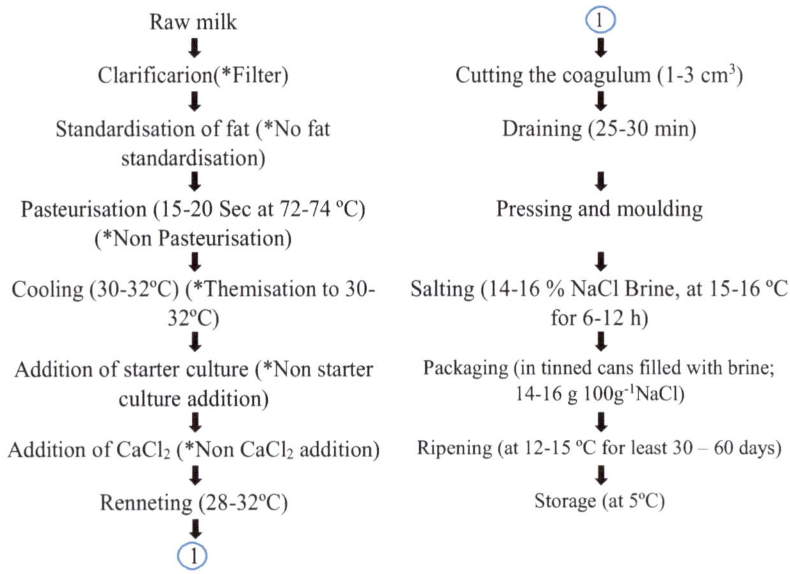

Fig. (4). Flow diagram for the manufacture of White Cheese (*Traditional White Cheese production) (Adapted from Hayaloğlu, 2002).

Table 1. Composition of ripened White Cheeses from different regions (Salum *et al.*, 2018).

Province	Moisture (%)	Protein (%)	Fat in DM (%)	Salt in DM (%)	Titratable Acidity (%)	pH	Ash (%)
	50.0±0.4	18.7±0.5	51.3	5.45	0.86±0.02	5.11±0.10	3.9±0.0
Çanakkale	52.5±0.4	18.3±0.3	47.7	6.49	0.88±0.00	4.81±0.06	5.3±0.0
Kars	51.2±0.2	18.3±0.2	50.2	6.40	0.79±0.01	4.97±0.02	4.3±0.0
Antalya	50.7±0.5	17.9±0.2	52.4	5.50	1.00±0.01	4.77±0.09	3.8±0.1
Denizli	50.0±0.2	17.1±0.5	56.3	5.20	0.79±0.05	5.02±0.18	3.7±0.1
Mersin	54.1±0.3	18.0±0.4	49.5	5.81	0.76±0.02	4.78±0.09	3.9±0.0
Balıkesir	51.3±0.5	17.9±0.3	51.0	7.38	0.81±0.02	4.87±0.07	4.2±0.0
Aksaray	50.1±0.3	18.7±0.1	53.1	6.68	0.90±0.02	4.90±0.07	4.2±0.1
Edirne	54.6±0.1	17.1±0.5	51.8	6.56	0.83±0.04	5.10±0.09	4.1±0.1

DM: Dry matter, Values are means ± SD

Edirne Peyniri (Edirne Type White Cheese)

Edirne Cheese is one of the most popular white cheese varieties in Turkey. White cheese was called as Edirne Cheese in the markets for many years. The main features of Edirne cheese are that it is made from the milk of only one species; sheep, goat or cow's and not from a mixture. It's traditionally produced from full fat fresh milk of the animals which are fed by the diversified vegetation in the delta formed by the Maritsa, Tundzha, Arda and Ergene rivers (Oraman, 2016). Since 2004, Edirne Cheese is protected under the Turkish Patent Agency's Protection of Geographical Indications (PGI) regulatory classification system (Karaca, 2016). Usually just one hour after the milking process begins the production. This cheese gets it ideal taste after a specific ripening period in cold rooms (+4°C - +6°C); for at least 6, 8 or 12 months, respectively for cow's, goats and sheep's milk. It is bright and white in colour and does not contain pores (Fig. 5).

Ezine Peyniri (Ezine Type White Cheese)

Ezine Cheese is probably one of the most popular white cheese varieties in Turkey. Two towns in Turkey are famous for their white cheeses, Edirne and Ezine (near Çanakkale). These white cheeses are known and sold under the name of the cities "Edirne" or "Ezine" (Cervantes, 2011). Ezine cheese is a full-fat, white pickled cheese ripened in tinplate containers like Edirne Cheese. But in contrast to Edirne Cheese, Ezine Cheese is produced from the mixture of goat, sheep and cow milk 40, 50 and 10% respectively, without any additional starter culture (Karagül-Yüceer; 2007). The cheese making season is from March to July.

The manufacturing procedure is similar as Beyaz Peynir but the geographical origin of the milk contributes some specialities to Ezine cheese. Since 2006, Ezine Cheese is protected under the Turkish Patent Agency's Protected Designation of Origin regulatory classification system (Karaca, 2016). Only the cheeses produced from the milk of animals which are grazing on natural plant sources on the north-west part of the mountain Kaz Dağları in the villages Ezine, Ayvacık, Bayramıç and a part of Çan can use the name Ezine Cheese.

Fig. (5). Edirne Type White Cheese.

The mountain is also known as mount Ida in the poem Iliad from Homer (ancient Greek mythology) and is famous with its thousands of natural springs. Many aromatic plants, including sweet marjoram (*Origanummarjorana L.*) oregano (*Origanumvulgare*), garden sage (*Salvia officinalis L.*), mint (*Menthalonifolia L*), melisa (*Melisa officinalis L.*), thyme (*Thymus vulgaris L.*) and hundreds of other plant species grow on the Kaz Dağları. This specific diversity of the plants plays a crucial role on the formation of the specific microflora of the raw milk and of course on the aroma of Ezine Cheese (Hayaloğlu *et al.*, 2008, Uymaz; 2015). Ezine Cheese has its own characteristic taste and aroma that differ from other white cheeses. Karagül-Yüceer (2007) reported that Ezine cheese is characterized by a rancid, cooked, creamy, whey-like, goaty, salty and sour flavour and taste, and a soft or semi-hard texture (Fig. **6**). The chemical and sensory properties of 22 Ezine Cheese samples were reported as; 48.2 - 56.9%, 0.73 - 2.13%, 49.3 - 58.8% and 5.6 - 11.9%, moisture, titratable acidity, fat-in-dry matter and salt-in-dry matter, respectively.

Fig. (6). Ezine Type White Cheese.

Trakya Kaşarı (Kashar Cheese of Thrace)

Kashkaval type cheeses are the well-known pasta-filata type cheeses, which are made for centuries in the Mediterranean region. In many countries of the area they are called Kashkaval or with some differences in spelling like Kaşar in Turkey, Kashar in Albania or Kasseri in Greece (Alichanidis & Polychroniadou, 2008). Kashar cheese is the second most consumed cheese after white cheese in Turkey. It is especially common in Thrace, Central Anatolia and East Anatolia regions. However, it is being produced in all regions of Turkey with some physico-chemical and textural differences today. The main production differences include the kind of milk used, it's thermal treatment, the use of cheese culture, the curd scalding step before acidification, the way of salting and the final cheese ripening.

The main properties of Thrace Kashar can be listed as below (Durlu-Özkaya, 2014; Kamber, 2008);

- Traditional Thrace Kashar Cheese is made with (raw) sheep's milk or it's mixtures with goat and cow's milk, the use of cow's milk is more common in other regions.
- Kashar Cheese belongs to the hard or midle-hard cheeses and is produced in cylindrical form. The cheese mould is about 10 cm in height with a diameter of 30 cm and weighs about 11 - 12 kg (Figs. **8 - 10**).
- Kashar cheeses produced in the eastern cities like Kars, Erzurum are up to 16 cm high while Thrace Kashar is flatter.

- The surfaces of Kashar cheeses produced in Thrace (Kırklareli, Edirne, Tekirdağ) are not mouldy and have faintly yellow colour, the other ones are mouldy and has an amber color.
- The crust of Thrace Kashar is thinner than the other ones.

The manufacture of Kashar Cheese consists of two distinct, independent stages:

a. Production of the curd and its acidification.
b. Texturizing of the acidified curd, this involves heating, kneading and stretching by soaking in hot brine.

The manufacturing steps of Kashar Cheese are described in Fig. (7).

Raw milk
↓
Filtration
↓
Pasteurisation (15-20 Sec at 72-74 °C) (*Non Pasteurisation)
↓
Cooling (30-32°C) (*Themisation to 30-32°C)
↓
Addition of $CaCl_2$ (*Non $CaCl_2$ addition)
↓
Renneting (Summer 30°C, Winter 35-40°C, ~ 45 min)
↓
Cutting the curd (6-8 mm cubes)
↓
Pressing for one hour, cutting into large blocks
↓
(1)

(1)
↓
Leafe to acidify (for 2 h summer and 4 h
↓
Checking the desired consistency and cutting the curd into long thin slices
↓
Immers the slices in baskets in hot water (75-80°C, 8 % NaCl, ~ 3 min)
↓
Kneading, salting inside the dough and kneading until thoroughly mixed
↓
Transfer into circular moulds for 12 h
↓
Transfer to the salting room (<18°C), keep on coarse grained salted shelves for one day
↓
After 2-3 days moulds are removed and the cheeses salted again for 2-3 weeks
↓
Ripened for 2-3 months (12°C), and stored for at least 4-6 months in cool rooms (2-4°C)

Fig. (7). Flow diagram for the manufacture of Thrace Kashar Cheese (*Traditional) (Kamber, 2008; Alichanidis and Polychroniadou, 2008; Durlu-Özkaya & Gün, 2007).

Fig. (8). Thrace Kashar Cheese ripened for more than 1 year.

Fig. (9). Slice of Thrace Kashar Cheese.

Besides traditional Kashar Cheese, fresh (unripened) kashar cheeses are also being produced in Turkey. The fresh cheese is known as taze kaşar or fresh kashar. Fresh kashar is a smooth, firm, light yellow cheese usually made from cow's milk. It is a soft tasting cheese that has often been compared to mozzarella (Donelly, 2016). It is a very versatile cheese good for slicing, melting, grating or eating straight up. It most often accompanies white cheese at breakfast. It is also used on pizzas, in sandwiches and salads and as the main ingredient in toast. The gross composition of Thrace Kashar Cheese is shown in Table **2**.

Fig. (10). Surface moulded Kashar Cheese from Cunda Island – Ayvalık.

Table 2. Chemical composition of Thrace Kashar Cheese (%).

Moisture Content	Fat-free Dry Matter	Fat	Protein	Ash	Salt	Acidity*	References
44.8	30.2	25.0	22.6	4.5	3.21	52.2	Hamzacebi & Anter (1978)
42.72	33.18	24.10	26.42	3.05	2.82	-	Demirci & Diraman (1990
43.0	29.0	28.0	27.0	-	2.84	2.00**	Kurultay, Öksüz & Simsek (2000)
44.0	31.36	24.64	-	-	3.0	-	Durlu-Özkaya & Gün, (2014)

*Soxhelet Henkel, **Titratable acidity

According to the Turkish regulations matured Kashar Cheese; has to be at least 60% Dry Matter and a max concentration of 4,0% Salt (NaCl) in dry matter. And it can be produced in four different fats in dry matter (FDM) levels; Full-fat Cheese (45 ≤ FDM %), Half-fat Cheese (25 ≤ FDM % <45), Low-fat Cheese (10 ≤ FDM % < 25) and as Non-fat Cheese (10 > FDM %) (TGK, 2015).

Mihaliç Peyniri (Mihalic Cheese)

According to a story; there was a small migrant village along the coast of the Marmara Sea, during the Roman period. They had a way of making their milk last throughout the year. They emptied their vessels of milk and boiled it. They strained the curds with cloths that they hung anywhere they could. The hanging cheeses were so abundant that the site inspired the Romans who later passed on the cheese making methods throughout Europe. Cheese, along with so many preserves, was made so that travellers could easily transport their food and eat it, too (Donnelly, 2016). Was this cheese may be the ancestor of Mihalic Cheese?

Some cheeses varieties are real artworks; they cannot be produced in fabrication style with continuous machines. They have to be formed one by one and step by step in the hands of the cheese master and have to be ripened in special conditions. Mihalic Cheese is one of these cheeses (Fig. **11**). It is a traditional Turkish cheese, mainly produced in the villages Karacabey, Gönen, Bandırma, Manyas, Mustafakemalpaşa of the provinces Balıkesir and Bursa. It is also known as Mağlıç, Mahlıç and Kelle cheese (Dinkci *et al.*, 2013). "Mihaliç" is also the old name of the village Karacabey. It has been told that this cheese was also made by migrant Albanians in the region during the Ottoman Empire, more than 200 years ago. It is usually made from whole fat, raw, sheep's milk.

Fig. (11). Mihalic Cheese.

The milk is filtered through a cotton cloth, coagulated at 32 - 35°C using home-made calf rennet or commercial rennet in wooden barrels of 100 - 250 L capacity. The coagulation time requires 60 to 90 min., afterwards the coagulum is being cut into very small pieces (the size of a grain of rice) using a cross-shaped wooden stick. The curds are stirred continuously and rested for 10–15 min to settle down. The curds are cooked gradually by adding boiling water slowly to the broken curd until the curd–water mixture reaches 45°C. It's mixed and shaken for 5-10 min again. At this stage the curd becomes elastic. It's rest again for another 10–15 min. During this time the curd precipitates. Then, the curds are taken to a cotton cloth for whey drainage. A cloth is tied up and hung for complete draining which takes 3–8 h. By this way the curd takes the form of the tied cloth. It is usual now that, the curd is pressed using 50 kg weights for each 250 kg of cheese milk, to minimize the draining period. The cloth containing the curd is pierced periodically with a needle to accelerate the whey time. The formation of regular openings in the cheese

occurs at this stage and this is one of the unique characteristics of Mihalic. The drained curd is cut into 3–5 kg blocks and brined at first in 15% NaCl for 2 days, then the blocks are taken to a medium brine 18% NaCl for 2-3 days. At this stage the blocks absorb brine and expand a little. And finally they are taken in 20-22% NaCl brine for 15–30 days. At this stage a crust develops on the cheese moulds. The cheese blocks are placed carefully on the bottom of a wooden barrel containing a layer of coarse salt; the cheese blocks are covered with 20–22% NaCl brine and the barrels are closed with a wooden lid (Hayaloğlu *et al.*, 2008). The cheese is ripened in the barrels at 15-25° Cabout 3 months (Kamber, 2008; Hayaloğlu *et al.*, 2008; Aday & Karagül Yüceer, 2014). The ripening period is very important, the cheese moulds have to be formed one by one and step by step in the hands of the cheese master. The best taste occurs after one year of ripening. The yield in Mihalic is 20–22% from sheep's milk or 10–12% from cows' milk (Hayaloğlu *et al.*, 2008).

Physicochemical properties of 15 Mihalic Cheese samples, collected from towns of Balıkesir (Gönen, Havran, Ivrindi), Bursa (Karacabey) and Çanakkale (Yenice) are given in Table **3**. All of the cheese samples are produced from raw sheep's milk and ripened for at least six months.

Table 3. Physicochemical properties of Mihalic Cheese (Aday and KaragülYüceer, 2014).

Sample No	pH	Acidity (%)	(%) DryMatter	Salt (%)	Fat (%)	Ash (%)	Protein (%)	WSN (%)	TCA (%)	PTA (%)
1	5.79	0.37	64.1	6.66	30.5	8.42	24.6	0.35	0.10	0.03
2	5.92	0.55	60.4	3.38	30.2	4.93	22.2	0.84	0.26	0.11
3	5.60	0.59	57.8	6.77	25.0	8.14	22.8	0.30	0.06	0.03
4	5.85	0.48	58.7	6.31	25.2	5.87	24.6	0.34	0.08	0.06
5	5.83	0.50	63.9	7.25	26.2	9.76	23.7	0.26	0.05	0.03
6	5.63	0.64	62.2	6.07	28.2	7.75	22.0	0.60	0.11	0.06
7	5.54	0.64	56.7	4.90	26.2	6.98	21.8	0.48	0.11	0.06
8	5.67	0.52	56.7	4.44	25.0	5.96	21.2	0.44	0.15	0.03
9	5.72	0.44	62.4	5.84	33.0	7.38	18.6	0.42	0.13	0.06
10	5.47	0.68	58.3	4.67	27.5	6.11	21.2	0.52	0.11	0.03
11	5.22	0.41	61.3	3.27	28.0	5.66	22.7	0.65	0.14	0.08
12	5.13	0.55	59.5	6.30	25.2	7.78	22.1	0.44	0.10	0.08
13	5.09	0.50	60.2	6.77	26.5	8.62	21.6	0.51	0.13	0.08

(Table 3) cont.....

Sample No	pH	Acidity (%)	(%) DryMatter	Salt (%)	Fat (%)	Ash (%)	Protein (%)	WSN (%)	TCA (%)	PTA (%)
14	5.10	0.55	63.6	7.95	27.0	10.09	21.6	0.50	0.03	0.03
15	5.16	1.04	59.8	8.18	27.0	9.58	19.5	0.42	0.09	0.03
Min.	5.09	0.37	56.7	3.27	25.2	4.93	18.6	0.26	0.03	0.02
Max.	5.92	1.04	64.1	8.18	33.0	10.09	24.6	0.84	0.26	0.11
Aver.	5.51	0.56	60.4	5.92	27.4	7.54	22.0	0.47	0.11	0.05

Acidity: Titratable acidity (g lactic acid / 100 g), WSN: Water Soluble Nitrogen, TCA: Trichloracetic acid soluble nitrogen, PTA: Phosphotungustic acid soluble nitrogen

Its colour ranges from cream to light yellow, roundish holes and 3 – 4 mm in diameter pores are decreasing from the centre to the sides in its cross section (Fig. **12**). It has a characteristically sharp taste and odour, is rather salty and has a hard and crusty structure with a 3 – 4 mm thick crust. The characteristic flavour comes from the natural microflora of the raw milk and the contaminating microorganisms that grow within the cheese and contribute to the maturation during the ripening process (Özer & Kesenkaş, 2017). Mihalic is sold nowadays in vacuum-packages as pieces weighing about 250 to 300 g (Fig. **13**).

Fig. (12). Slice of Mihalic Cheese.

Fig. (13). Mihalic Cheese.

Sırvatka Loru (Sırvatka Lor Cheese – Whey Cheese)

Lor Cheese is an interesting cheese because it is not made from strained milk curds (casein) like the other ones. It is made from the leftover that's strained out of milk curds (whey proteins) (Donelly, 2016). It is the whey cheese of Anatolia. It is produced almost in every region in Turkey and also in most East-Mediterranean countries (Durlu-Özkaya & Gün, 2014). Most of them are heat coagulated cheese. Their production method varies according to the region in respect of the used raw material (kind of milk, cheese whey), the addition of milk and/or cream and the traditional recipe. Traditionally these cheeses are manufactured mostly from the whey of sheep's and goat's milk cheeses, because this whey is much richer in protein and fat as compared with cow's milk cheese whey. But since the production of sheep's and goat's milk is seasonal, whey cheeses are produced also from cow's milk cheeses throughout the year. The yield of whey cheese, independent of the kind of whey, is higher when it is made from the whey of pasta filata cheeses than that of soft cheeses (Alichanidis & Polychroniadou, 2008).

Sırvatka Lor Cheese is manufactured in the same region as Mihalic Cheese. The whey obtained during the production of Mihalic Cheese is known as "Sırvatka" therefore the whey cheese has also the same name. Sırvatka Lor Cheese, which is produced from the whey of Mihalic Cheese, has a higher quality than the other whey cheeses. The manufacturing steps of Sırvatka Lor Cheese are described in Fig. (**14**). Chemical properties of Lor Cheese samples are summarized in Table **4**.

Filtration ofMihalic Cheese whey through cotton cloth
↓
Heating until it curdles (80-82°C)
↓
Leave to coagulate (88-92°C-45 min)
↓
Collecting the curd that gathered on the surface
↓
Place into a cotton cloth, hang and leave it for 1 day to filter in a cold room
↓
Salting with thin grained salt (10-15g L^{-1})
↓
(1)

(1)
↓
Press inside a small wooden cask
↓
Cover the cask with oily paper
↓
Closing their lids tight
↓
Storage (2 days)
↓
Ready for consumption

Fig. (14). Flow diagram for the manufacture of Sırvatka Lor Cheese (Adapted from Kamber, 2008; Alichanidis & Polychroniadou, 2008; Durlu-Özkaya & Gün, 2014).

Table 4. Chemical properties of Lor Cheese (%).

Moisture	Fat Free DM	Fat	Protein	Ash	Salt	Acidity	Reference
74.01	20.65	5.34	13.50	1.24	-	58.23*	Kamber (2015)***
65.0-70.0		7.0-15.0	9.0-14.0	-	0.6-0.8	5.5-6.0**	Alichanidis & Polychroniadou (2008)****

*Soxhelet Henkel, **pH, *** Average, ****Min-Max

The resulting product, Sırvatka Lor, has an odor and taste like milk cream, it's color ranges between pure white and off-white. The shape is that of the container used. The texture is soft, smooth and creamy (Fig. **15**). They have no rind and no holes caused by gas (Kamber, 2008; Alichanidis & Polychroniadou, 2008; Durlu-Özkaya & Gün, 2014). It is preferred fresh at breakfast on bread and is also used in pastry, pies and many desserts. According to it's is low-salt, low fat and high whey protein content it is known as healthy and is also preferred by body-builders.

There are also other whey cheeses in different parts of the Marmara region. Their production amount is less than Sırvatka Lor Cheese. They have special production techniques according to the region. The main of these whey cheeses can be described as follows.

a. Trakya Loru (Thrace whey cheese) is mainly produced in the Thrace region. It is made by the whey of Kashar cheese.

b. Ayvalık Kirli Hanım Peyniri (Dirty Lady Cheese) is a gourmet taste, white cheese variety, usually made from a blend of goat, sheep and cow's milk, that is shaped, salted and matured in reed baskets (Gökçe, 2018). This kind of basket cheese varieties are mainly produced in the Aegean region (Ergil, 2014).
c. Ayvalık Sepet Loru (Ayvalık Basket Whey Cheese) is a traditional cheese of Ayvalık (Fig. **16**). It is mainly produced in the South-Western parts of the Marmara region and the North-Western parts of the Aegean region. It is made by draining the whey cheese in baskets. While it is being consumed fresh, it is also used for the production of Ayvalık Kirli Hanım cheese. Besides these major varieties of Ayvalık Sepet cheese there are also Sepet cheeses with walnuts (Fig. **17**), thyme (Fig. **18**), nigella seeds and red pepper.

Fig. (15). Sırvatka Lor Cheese.

Çerkez Peyniri (Circassian Cheese)

Traditional circassian cheese, named Adyghe Koaye, is a full fat cheese originated in the Northern Caucasus. It is mainly produced from goats', sheep' and cows' milk or a mixture of them, skimmed buffalo milk is also being used (Güneser & Karagül-Yüceer, 2011; Gül *et al.*, 2015). In Turkey it is traditionally made by Circassian families mainly in in the Eastern Marmara Region (Balıkesir, Bursa, Adapazarı, Çanakkale, Gönen, Biga, Hendek) and Western Black Sea Region (Bolu, Düzce and Sinop) (Kamber, 2008). It is also made by circassian families in Jordan, Syria, Israel and in the Adygey, Kabardino-Balkaria and Karachi Cherkess republics in the Russian Federation. While it is being produced for centuries in the Northern Caucasia there is not much information about its history

(Uysal *et al.*, 2010). Circassian cheese is an acid coagulated cheese and has different characteristics than enzyme coagulated cheeses but it's production has also similarities with Kashar Cheese and partially with Mihalic Cheese (Güneser & KaragülYüceer, 2011; Kamber, 2008). Its texture shows similarities to Hellim cheese. Some varieties of Circassian cheese are sprinkled with different herbs or fumed types are also produced. Circassian cheese is consumed fresh or matured. The fumed type has a very long shelf life in relative to the fresh one. It is usually fumed with pinewood or beech and oak trees (Ergil, 2014; Güneser & Karagül Yüceer, 2011). It is usually formed in the basket during draining of the whey, therefore the out surface has the form and figure of the basket. The moulds are about 1 kg, the color is varying from light yellow in the fresh ones to white - cream in the dried version (Fig. **19 & 20**). It is a slightly salty cheese with a specific aroma and a soft consistency. The smoked type (Figs. **21 & 22**) has a light brown exterior with a thin crust and a light yellow – cream interior.

Fig. (16). Traditional basket for filtering (left) and AyvalıkSepetLor Cheese (right).

Fig. (17). Sepet Cheese with walnut (Cunda-Ayvalık region).

Fig. (18). Sepet Cheese with thyme (Cunda-Ayvalık region).

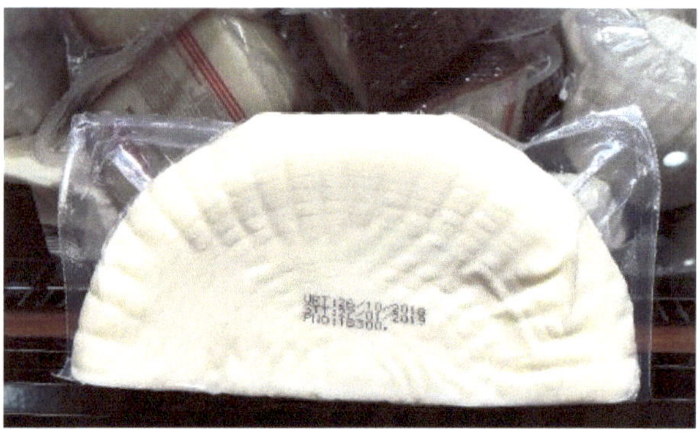

Fig. (19). Vacuum packaged Circassian Cheese (½).

Fig. (20). Vacuum packaged Circassian Cheese (¼).

Fig. (21). Fumed Circassian cheese (1/1).

Fig. (22). Fumed-Vacuum Packaged Circassian Cheese (¼).

The manufacturing steps of Circassian Cheese are described in Fig. (**23**).

Filtration of raw milk (Sheep's, Goat's, Cow's milk or a mixture of them)
⬇
Heating (~95 - 100°C)
⬇
Addition of sour whey / yoghurt whey (~400mL whey / 10 kg milk)
⬇
Stirring and heating until coagulation
⬇
Collection of the curd by perforated ladles
⬇
Draining of the whey in cheese baskets
⬇
①

①
⬇
Salting of the curds upper part / turning over for salting the bottom part
⬇
Ready for fresh consume / for maturation
⬇
Hanging in the baskets in to the sun. Turning upset down frequently
⬇
Drying the cheese under sun (3 - 7 days) / fuming with pinewood, beech or oak trees
⬇
Ready for consumption

Fig. (23). Flow diagram for the manufacture of Circassian Cheese (Adapted from Güneser, KaragülYüceer, 2011, Uysal *et al.*, 2010).

Circassian cheese can be offered for consumption as fresh (until 2 weeks) or in sun dried version. A fumed variety, which can be kept for long periods, is being produced (Kamber, 2008). The chemical properties of different varieties of Circassian Cheese samples from different regions of Turkey are summarized in Table **5**. As it can be seen from Table **5**, there are significant differences among the cheese samples according to the cheese variety and region.

Table 5. Chemical properties of Circassian Cheese (Uysal *et al.*, 2010; Güneser & KaragülYüceer, 2011).

(%) DM	Fat (%) in DM	Protein (%)	Salt (%) in DM	Ash (%)	Acidity (%LA)	pH	Sample
44.06	48.05	16.74	4.81	-	0.66	5.64	Fresh Circassian Cheese
66.96	42.26	30.24	6.27	-	1.09	5.60	Sun dried Circassian cheese
69.82	41.89	32.42	5.78	-	0.98	5.50	Stove dried Circassian cheese
65.53	45.78	24.95	6.44	4.90	0.58	5.52	Kayseri Region
43.07	53.13	16.21	3.61	2.32	0.55	5.37	Biga-Çanakkale Region
47.38	52.75	19.40	7.28	3.82	0.54	5.34	Susurluk - Balıkesir Region
53.33	48.75	25.20	4.99	3.96	1.06	5.27	Çanakkale Region
55.67	44.90	25.58	3.92	4.75	1.26	5.66	Fumed Circassian Cheese (Çanakkale)
54.02	49.98	22.46	4.50	6.77	1.28	5.45	Fumed Circassian Cheese (Çanakkale)

DM: Dry Matter, LA: Lactic Acid

Abaza Peyniri (Abkhasian Cheese)

Abaza Cheese has a similar story like Circassian's Cheese. It is made by Abkhazian's, who migrated from Caucasia after the war with the Russians in the mid-19th century to Turkey and settled mainly in the Marmara (Adapazarı, Kocaeli, Bilecik, Bursa, Çanakkale) and Western Black Sea (Düzce, Bolu, Hendek, Sinop) regions. It is also known as Apsuhas cheese. It's often confused with Circassian Cheese but it differs totally in the production method, preservation conditions and its gross composition. The manufacturing steps of Abaza Cheese are described in Fig. (**24**).

Filtration of raw milk (Sheep's, Goat's, Cow's, Buffalo milk or a mixture of them)

↓

Heating (~30 - 33°C)

↓

Rennet addition (~10 g rennet / 100 kg milk)

↓

Coagulation (2 – 2,5 h)

↓

Cutting of the curd into small pieces and pressing for draining of the whey

↓

Curd ripening in cotton cloths at 35°-40°C until 5.1 – 5.4 pH (~ 24 h)

↓

(1)

(1)

↓

Cutting of the curd into small moulds about 10 cm thick

↓

Throwing the moulds into boiling water and kneading by hand

↓

Shaping by hand into the desired form (braid, wheel or pita bread)

↓

Salting of it's upper surface and bottom (dry) and pouring of a spoon hot water

↓

Ready for consumption fresh (2-4 month refrigerated

↓

Can also be sun/oven dried or fumed consumed

Fig. (24). Flow diagram for the manufacture of Abaza Cheese (Adapted from; Atamer, 1987; Uysal, 1998; Kamber, 2008; Güzeler and Parlak, 2010).

Abaza Cheese is a semi-hard cheese that is mostly infused with various herbs, the most popular version being with nigella seeds (Ergil, 2014). It has usually a cream to light yellow colour and a slightly salty taste. It has the form of pita bread, wheel or is braid (Fig. **25**). Fumed versions are also available. The chemical composition of Abaza Cheeses are summarized in Table **6**.

Table 6. Chemical properties of Abaza Cheese (Uysal *et al.*, 1998; Kamber, 2008; Güzeler & Parlak, 2010).

(%) DM	Fat (%)	Protein (%)	Salt (%)	Acidity (%LA)	pH	Sample
52.90	21.54	23.66	6.36	0.53	5.40	Fresh Abaza Cheese (n=20)*
68.05	28.15	32.79	4.85	1.19	5.10	Sun dried Abaza Cheese (n=10)*
65.05	25.76	30.27	5.42	1.28	5.15	Oven dried Abaza Cheese (n=10)*
56.60	21.76	28.33	4.56	0.94	5.15	Different varieties (n=20)*
40.4-55.7	3.5-29.5	-	0.70-7.25	0.24-1.18	-	Varieties of Abaza Cheese (n=10)**
32.0-44.4	21.5-29.8	23.7-30.3	3.4-6.4	0.64-1.28	-	Varieties of Abaza Cheese**

*Values are the average of n samples, **Min – Max values, DM: Dry Matter

Abaza Cheese is usually consumed fresh or it can be eaten dried in the sun or oven. Fumed varieties with a characteristic flavour are also being produced.

Fig. (25). Fresh braid Abaza Cheese.

Other Traditional Cheeses

Other than these major varieties, there are many other types of Turkish traditional cheeses produced in the region. Most of these cheese varieties are made in villages, at home for the consumption of the family. Some are also sold by these families in the marketplace at the next city. Economically they are not important according to the production amount.

Balkabağı Küp Peyniri (Pumpkin Cube Cheese)

Balkabağı Küp Cheese is also made by Abkhazian families like Abaza Cheese in the region of Adapazarı, Hendek, Arifiye and Doğançay. For the production of this cheese; the inside of a pumpkin is emptied, cleaned and dried at a sunny place. While the pumpkin is dried, raw milk and yoghurt are mixed in a container, salted and left overnight for coagulation. The curdled mixture is transferred into the dried pumpkin and preserved at a cool place. The cheese inside can be gradually consumed. It can be kept unspoiled for 2 years (Kamber, 2008).

Çoban Peyniri (Shepherd's Cheese)

Çoban peyniri is mostly made in the villages of Yenişehir-Bursa. Usually it has been made by shepherds by the same method as White Cheese. The main difference is that during fermentation several herbs are being added to the cheese milk, which were collected from the mountains where the sheep were grazing.

Köy Peyniri (Village Cheese)

Köy Peyniri is made usually in the villages by the farmers for their own consumption. The cheese is filtered from a cotton cloth, heated to the renneting temperature (~ 30°C). After coagulation the curd is collected in a cotton cloth, the whey is drained in a cool place over night. And the obtained cheese is preserved in brine, in a cool room. This fresh unsalted Köy Cheese is also used by the production of Höşmerim dessert (Bal-Onur and Aksoy-Biber, 2015). Höşmerim is mainly known as a dessert of the Balıkesir region.

Taraklı Keş Peyniri (Keş Cheese of Taraklı)

Keş Cheese is made mainly in the region Bolu and Sakarya. Taraklı Keş Cheese is Traditionally made from non-fat yoghurt. The drained yoghurt is pressed for two days, afterwards it is salted and formed by hand by throwing it in a tray. It gets the form of a triangle and the upper side is salted and wrapped with a thin cotton cloth. The wrapped cheeses are taken for a week, on a board in an open-air place which is not much windy for drying. The dried cheeses can be kept for 1 year in refrigerated conditions. Taraklı Keş Cheese can be consumed fried in butter at breakfast, on noodles or by frying with eggs (Bal-Onur and Aksoy-Biber, 2015).

CONCLUSION

The Marmara region, the north-western part of Turkey, for many civilisations a bridge to Europe or Asia. The home and Capital for many civilisations and imperia since thousands of years, hundreds of years the new lands of immigrants from the Balkan, Caucasia. The taste of a multicultural cuisine with the roots at the kitchen of the imperium.

No wonder that nowadays only in the province Balıkesir are more than 50 different traditional cheeses being produced. While some of them are real artworks and have to be formed and ripened in the hands of the cheese master. Some has to be made from the milk of the mammalians grazing on the ancient Homers Ida Mountain. Others from the full fat fresh milk of the animals fed at the diversified vegetation in the delta formed by the Maritsa, Tundzha, Arda and Ergene rivers. Other ones have to be drained through traditional braided baskets or cotton cloths.

Investigation and documentation of the traditional production methods and some

Investigation and documentation of the traditional production methods and some basic parameters like the biochemical, microbiological, textural, rheological and sensory properties of traditional cheeses are important for the protection of the cultural heritage. Extensive studies have been made on the commercially most important varieties including White Cheese or Kashar Cheese but there is still insufficient literature for the other varieties.

CONSENT FOR PUBLICATION

Not applicable.

CONFLICT OF INTEREST

The authors confirm that this chapter contents have no conflict of interest.

ACKNOWLEDGEMENTS

I would like to thank Fahri ŞAHAN (Fahri Şahan Peynir Evi, Bandırma) and İsmail KESEBİR (Kesebir Mandıra, Cunda, Ayvalık) for sharing their experience and providing some of the cheeses for the photographs.

REFERENCES

Aday, S & Karagül Yüceer, Y (2014) Physicochemical and Sensory Properties of Mihalic Cheese. *Int J Food Prop,* 17, 2207-27.
[http://dx.doi.org/10.1080/10942912.2013.790904]

Alichanidis, E & Polychroniadou, A (2008) Characteristics of major traditional regional cheese varieties of East-Mediterranean countries: a review. *Dairy Sci Technol,* 495-510.
[http://dx.doi.org/10.1051/dst:2008023]

Bal-Onur, B & Aksoy-Biber, N (2017) *50 Peynirli Şehir Balıkesir.* Balıkesir Büyükşehir Belediyesi Yayınevi, Balıkesir. (In Turkish)

Bal-Onur, B & Aksoy-Biber, N (2015) *Peynir Aşkına.* Türkiye İş Bankası Kültür Yayınları Sertifika No: 29619. İstiklal Cad. MeşelikSok. No: 2/4 Beyoğlu, 34433 İstanbul. (In Turkish)

Cauvin, J (2000) *The Birth of the Gods and the Origins of Agriculture.*Cambridge University Press, New York.

Cervantes, M (2011) Foods of Turkey. http://www.foodsofturkey.com/Shop/Entries/2011/6/8.

Demirci, M & Diraman, H (1990) Study on Production Technique, Physical, Chemical, Microbiologic Features and Energy Levels of Packed with Vacuumed Fresh Kashar Cheese, Made in Trachea region. *GIDA-The Journal of Food,* 15, 83-8.

Dinkçi, N, Özer, E & Kesenkaş, H (2013) A Real Artwork: Mihalic Cheese. *The 2nd International Symposium on "Traditional Foods from Adriatic to Caucasus" 26-26 October Struga, Macedonia*

Dinkci, N & Gönç, S (2000) Mucormichei'den Elde Edilen Lipaz (Piccantase A) Enziminin Beyaz Peynirin Olgunlaşmasında Kullanılması Üzerinde Araştırmalar. *Ege Univ Ziraat Fak Derg,* 37, 141-8. (In Turkish).

Donelly, C (2016) *Turkish Cheeses a Very Quick Guide.* https://istanbites.com/features /turkish -cheeses-a-very-quick-guide/.

Durlu-Özkaya, F & Gün, I (2007) Anadolu'da Peynir Kültürü. ICANAS, Uluslararası Asya ve Kuzey Afrika Çalışmaları Kongresi,10-15, Ankara In Turkish Durlu-Özkaya, F & Gün, I (2014), Aroma Compounds of

Some Traditional Turkish Cheeses and Their Importance for Turkish Cuisine. *Food Nutr Sci,* 5, 425-34.

Ergil, LY (2014) *A Tour of Turkey's Artisanal Cheeses.* https://www.dailysabah.com/life/2014/08/14/a-tou--of-turkeys-artisanal-cheeses

Erkmen, O (1995) Behaviour of *Staphylococcus aureus* in Turkish Feta cheese during manufacture and ripening. *J Food Prot,* 58, 1201-5.
[http://dx.doi.org/10.4315/0362-028X-58.11.1201] [PMID: 31137313]

Evershed, R (2008) Organic residue analysis in archaeology: The archaeological biomarker revolution. *Archaeometry,* 50, 895-924.
[http://dx.doi.org/10.1111/j.1475-4754.2008.00446.x]

Gökçe, Y (2018) *Turkish Cheese, The Indispensible Ingredient of Turkish Tables: Cheese.* http://www.turkishculture.org/culinary-arts/cheeses-312.htm

Gül, O, Dervişoğlu, M, Mortas, M, Aydemir, O, Ilhan, E & Akşehir, K (2015) Evaluation of Polycyclic Aromatic Hydrocarbons in Circassian Cheese by High-Performance Liquid Chromatography with Fluorescence Detection. *J Food Compos Anal,* 37, 82-6.
[http://dx.doi.org/10.1016/j.jfca.2014.07.004]

Güneser, O & Karagül-Yüceer, Y (2011) Chracterisation of Aroma-Active Compounds, Chemical and Sensory Properties of Acid-Coagulated Cheese: Circassian Cheese. *Int J Dairy Technol,* 64, 517-25.
[http://dx.doi.org/10.1111/j.1471-0307.2011.00703.x]

Güzeler, N & Parlak, Y (2010) Abaza Peyniri Üretimi ve Özellikleri. 1. *Uluslararası "Adriyatik'ten Kafkaslar'a Geleneksel Gıdalar" Sempozyumu, 15-17 Nisan 2010 Tekirdağ,* 63-5.

Hamzaçebi, Y & Anter, C (1978) Investigations on Chemical Composition of Kashar Cheeses Produced in East, central and Trachea Part of Turkey. *J of Etlik Veterinary Micro Pub,* 4, 142-55.

Hayaloğlu, AA, Güven, M & Fox, PF (2002) Microbiological, Biochemical and Technological Properties of Turkish White Cheese "Beyaz Peynir". *Int Dairy J,* 12, 635-48.
[http://dx.doi.org/10.1016/S0958-6946(02)00055-9]

Hayaloğlu, AA, Özer, BH & Fox, PF (2008) Cheeses of Turkey: 2. varieties ripened under brine. *Dairy Sci Technol,* 88, 225-44.
[http://dx.doi.org/10.1051/dst:2007014]

Kamber, U (2008) The traditional cheeses of turkey: "Marmara Region. *Food Rev Int,* 24, 175-92.
[http://dx.doi.org/10.1080/87559120701764613]

Kamber, U (2015) Traditional turkey cheeses and their classification. *Van Veterinary Journal,* 26, 161-71.

Karaca, OB (2016) The Importance of Our Traditional Cheeses in Gastronomy Tourism. *Journal of Tourism and Gastronomy Studies,* 4, 17-39.
[http://dx.doi.org/10.21325/jotags.2016.12]

Karagul-Yuceer, Y, Isleten, M & Uysal-Pala, C (2007) Sensory characteristics of Ezine cheese. *J Sens Stud,* 22, 49-65.
[http://dx.doi.org/10.1111/j.1745-459X.2007.00094.x]

Kindstedt, P (2012) *Cheese and Culture (A History of Cheese and its Place in Western Civilizations)* Chelsea Green Publishing, 85 North Main Street, Suite 120, White River Junction 295-6300.

Kosikowski, F (1977) *Cheese and Fermented Milk Foods* . Printed in U.S.A. and Distributed by Edwards Brothers, Inc., Ann Arbor, Michigan.

Kurultay, S, Oksüz, O & Simsek, O (2000) The effects of hydrocolloids on some physico-chemical and sensory properties and on the yield of Kashar cheese. *Nahrung,* 44, 377-8.
[http://dx.doi.org/10.1002/1521-3803(20001001)44:5<377::AID-FOOD377>3.0.CO;2-L] [PMID: 11075385]

Özer, E & Kesenkaş, H (2017) The effect of using different starter culture combinations on organic and fatty acid composition of mihalic cheese. *Acta Aliment,* 46, 492-500.

[http://dx.doi.org/10.1556/066.2017.46.4.12]

Pixel, M (2018) Russia Turkey Europe globe syria black sea world. https://www.maxpixel.net/Russia-Turkey-Europe-Globe-Syria-Black-Sea-World-2234559.

Swan, S (2009) *Türkiye'nin Peynir Hazineleri.* Boyut Yayıncılık ve Ticaret AŞ, İstanbul In Turkish.

Taviloğlu, E (2018) *Everything You Ever Wanted to Know About Turkish Cheese.* www.thespruceeats.com/turkish-cheese-overview-3274228

TGK (2015) *Türk Gıda Kodeksi Peynir Tebliği (Tebliğ No: 2015/6) In Turkish.*

Turantaş, F, Unlütürk, A & Göktan, D (1989) Microbiological and compositional status of Turkish white cheese. *Int J Food Microbiol,* 8, 19-24.
[http://dx.doi.org/10.1016/0168-1605(89)90076-7] [PMID: 2701567]

USK (2014)

Uymaz, B & Yüksel, Z (2015) Identification of indigenous lactic acid bacteria isolated from ezine cheese (PDO) during the ripening. *Conference Paper, International Dairy Federation World Dairy Summit.* Vilnius, Lithuania

Uysal, H, Akbulut, N, Kavas, G & Kesenkaş, H (1998) An Investigation on properties of Abaza (Abkhaz) cheeses. geleneksel süt ürünleri, 5. *Süt ve Süt Ürünleri Sempozyumu 21-22 Mayıs 1998 Tekirdağ* 126-32.

Uysal, H, Kavas, G, Kesenkaş, H & Akbulut, N (2010) Some Properties of traditional circasian cheese produced in Turkey. *Int J of Dairy Sci,* 5, 150-2.
[http://dx.doi.org/10.3923/ijds.2010.150.152]

Wikipedia (2018) https://en.wikipedia.org/wiki/Geography_of_Turkey

Wikitrave l (2017) https://wikitravel.org/en/Marmara_(region)

Yenen, Ş (2018) http://www.turkishodyssey.com/tag/the-marmara-region

CHAPTER 7

Traditional Turkish Cheeses of Black Sea Region

Selin KALKAN[1,*] and **Evrim GUNES ALTUNTAS**[2]

[1] *Department of Food Engineering, Faculty of Engineering, Giresun University, 28200, Giresun, Turkey*

[2] *Biotechnology Institute 06135, Ankara University, Ankara. Turkey*

Abstract: Cheese is one of the dairy products consumed throughout the world. For this reason, the production, consumption, and variety of cheeses are increasing day by day. Cheese is a valuable dairy product from the aspects of durability and diversification that can meet the expectations of consumers from every segment of the society. Francc, Netherlands, and Italy are the first countries that are well-known for their cheese varieties. Moreover, with its wide range of cheese varieties, Turkey is also among these countries. Although the production methods of more than 130 Turkish cheese varieties produced in Turkey are similar, the differences between the cheeses arise from the materials used in production and the regions, in which they are produced / consumed. Each of these cheese varieties has its own unique chemical and sensory qualities including the taste, texture, and appearance. The Black Sea Region extends eastwards alongside the coastline from the east of Sakarya province to the Georgian border. Among the other regions in the country, this region comes to the forefront with its wide range of cheese varieties. There are more than 30 cheese varieties identified in the Black Sea region. The geographical and climatic conditions of this region have significant effects on these cheese varieties. The civil cheese, Imansiz cheese, Kargı tulum cheese, curry cheese, and Kolete cheese are the most popular varieties in the region. This chapter aims to give information about the structure and physical, chemical, and microbiological properties of all the traditional cheese varieties of the Black Sea region.

Keywords: Black sea region, Cheese process, Traditional cheese, Turkey.

INTRODUCTION

The milk and dairy products are important nutritional sources and they offer significant health benefits for humans. One may encounter with these products in the form of butter, cheese, cream, or yogurt. The composition of dairy products may differ among the product varieties but the main components, such as water, protein, fat, lactose, vitamins, and minerals, remains same. The cheese, which is a

* **Corresponding Author Selin KALKAN:** Department of Food Engineering, Faculty of Engineering, Giresun University, 28200, Giresun, Turkey; Tel: +90 454 310 17 40 fax: +90 454 310 17 49; E-mail:selin.kalkan@giresun.edu.tr, selin9806@hotmail.com

Celile Aylin Oluk & Oya Berkay Karaca (Eds.)

popular fermented dairy product, is produced through the coagulation of caseinmicelles *via* the enzyme chymosin, starter bacteria, or heat treatment (Zheng *et al.*, 2015).

Raw milk contains many microorganisms such as lactic acid bacteria, yeasts, and molds. Generally, some lactic acid bacteria species play dominant role in this biodiversity, and the resting microorganisms play a role in bringing out different cheese species. This may explain why the cheese made of raw milk has more intense and rich flavor than the pasteurized milk cheese (Montel *et al.*, 2014).

The microflora of cheeses is the main determinant of the product characteristics. The starter cultures are generally responsible for the initiation of fermentation, but the nonstarter lactic acid bacteria also play a role in the development of the organoleptic characteristics during the ripening period. In this group, *Lactobacillus* (*Lb.*) *paracasei*, *Lactobacillus plantarum*, and *Lactobacillus curvatus* draw attention and *Pediococcus* and *Leuconostoc* species, as well as enterococci and streptococci, may also be present (Licitra & Carpino, 2014).

Apart from these bacteria, the cheese and raw milk may contain hazardous pathogens and they may also cause various diseases. The listeriosis is an important disease associated with the cheese and the bacterium named *Listeria monocytogenes* is responsible for it. Beside this pathogen, several key pathogens related with the dairy industry such as *Cronobacter* spp., (*Enterobacter sakazakii*), enterohemorrhagic *E. coli*, and EHEC including *E. coli* O157:H7 have been identified (Boor *et al.*, 2017).

Similar to the other animal products, the fat content of dairy products attracts attention because of the effect of LDL cholesterol. Yet, it is estimated that the impact of high- and low-fat dairy foods on LDL cholesterol concentrations is rarely seen. The consumption of a high-fat dairy type diet does not significantly increase the LDL cholesterol when compared with a low-fat dairy diet (Brassard *et al.*, 2016). Kratz *et al.* (2013) observed that the dairy fat or high-fat dairy food does not contribute to obesity or cardio-metabolic risk, and it suggests that high-fat dairy consumption within typical dietary patterns is inversely associated with the obesity risk.

For a while, the foods were classified as the therapeutic or aggravating agents of health. Food consumption is one of the current issues of selection criteria causing Cancer, which is the most widely known diseases of this century. From this aspect, the relationship between cancer and dairy products is explained mainly by the positive interaction, but the opposite results have also been reported. For instance, the casein was reported to be a potential anti-mutagenic and anti-carcinogenic protein (Lu *et al.*, 2016). Among the cancer types, the colorectal cancer is known as the third most common type in the world. The epidemiological studies, which were carried out on the relationship between dairy products and colorectal cancer risk, reported different results. Although the dairy products were reported to provide protection against colorectal cancer risk thanks to the high amount of calcium they contain, some dairy products such as specific cheeses and creams also have a high-fat content that potentially could increase the colorectal cancer risk by increasing the bile acid levels in the colon (Aune *et al.*, 2012).

A meta-analysis study was carried out by Lu *et al.* (2016) in order to reveal the relationship between the consumption of dairy products and the risk of death of cancer. In that study, 11 population-based cohort studies involving 778.929 individuals searched in PubMed and EMBASE were included in the analyses. The authors concluded that the total intake of dairy products has no significant impact on the increased cancer mortality risk, while the low total dairy intake even reduced relative risk based on the nonlinear model. However, whole milk intake significantly contributed to the elevated prostate cancer mortality risk among men.

The regions of Turkey differ from each other by significant nutrition types. A significant portion of the people living in the Black Sea earns their lives from agriculture. Moreover, Black Sea Region is one of the coastal regions of Turkey and the fish takes a great part of daily diet. In our country, 45% of corn production, more than 80% of hazelnut production, and 65% of tea production are carried out in Black Sea Region. It may not be listed in the first ranks but the cheese is an important breakfast food and also the main ingredient of kuymak (mıhlama, muhlama), the most widely known traditional Black Sea Region food (Ertaş & Gezmen-Karadağ, 2013; Sufer & Sezer, 2014). The Minci (Minzi) cheese used in this food and also the other mostly consumed and produced cheeses will be explained in this chapter.

Turkish Cheese in the World

It is seen that the first cheese has been produced in Mesopotamia in B.C. 7000-10000 and the cheese production has spread from this region to the other parts of the world. In “Divanu Lugati’t Turk” written by Kasgarli Mahmut in 1072-1074, the cheese was named “udma” and “udhıtma” and explained by cream of milk (Durlu-Ozkaya & Gun, 2008). In Turkey, various recipes are used for the production of the cheeses. Other than Turkey, the cheese production also dates back to old ages in Greece, Italy and Spain.

The cheese production processes are improved day by day; however, some basic tools remain the same for centuries. The farmers and cheese producers have used wooden tools to collect and transform the milk in this region. Today, the wooden shelves are still used for ripening the cheese. Many cheese producers choose this method in order to improve the organoleptic and typical characteristics of the final product (Lortal *et al*., 2014). The cheese production processes using the raw milk in the past and nowadays are illustrated in Fig. (**1**) (Zheng *et al*., 2015).

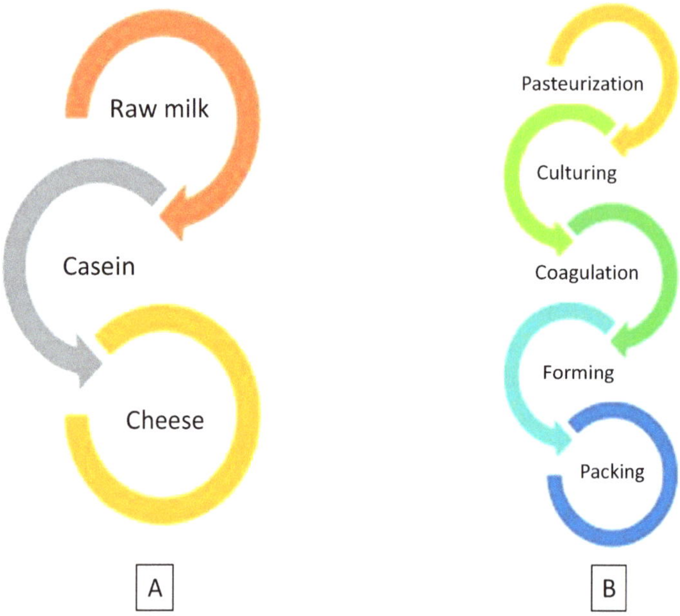

Fig. (1). The cheese production process using raw milk in the past **(A)** and today **(B)** (Zheng *et al*., 2015).

The production and consumption of cheese in Turkey significantly increased in recent years. One of the most important factors playing role in this increase is the consumption in large cities, especially the cities that are distant from the region where the traditional cheeses are produced. This demand laid the foundation of the industrial-scale production of the traditional cheeses, which have been historically produced in some of our regions or in our province. It is necessary to promote the production of the cheese varieties that are becoming increasingly popular to gain economic benefits with them. For this purpose, it is important to examine the status of cheese production in our country, to discuss the results obtained from the studies carried out on this subject, and to prepare new strategies.

More than 50 varieties of cheese are produced in Turkey; the most popular ones are feta, Kashar, and Tulum cheese. According to the data obtained from the Turkish Statistical Institute, the feta cheese constitutes approximately 60% of the total, whereas Tulum cheese constitutes 15%, and Mihaliç constitutes 15%. The remaining 10% consists of the other regional varieties (Unsal, 2018).

The cheeses produced in Turkey are classified by the milk used as cow, sheep and goat cheese or by the fat content as full-fat, medium-fat, and fat-free or tasteless. On the other hand, we can also classify our cheeses as soft (*e.g.* non-edible sediment), semi-hard (*e.g.* pickled feta cheese, Tulum cheese), and hard (*e.g.* Mihalic cheese, Aegean Tulum cheese). Apart from this classification, we can also distinguish the cheeses by their maturation status; fresh consumed ones eaten without maturation (*e.g.* curd cheese), semi-ripened (white-brined cheese), and fully ripened (Tulum cheese, Kashar cheese, Mihalic cheese).

Majority of the salted (brined) cheeses produced in Turkey are known as "feta cheese" or most often as "Edirne cheese" or "Tin cheese". Besides that, there also are 8-10 cheese varieties that are ripened. Some of these cheeses are called with the name of the region or geography such as Urfa, Antep, Malatya, Diyarbakır Örgü, Mihaliç, and Ezine cheeses. The identification of the dominant flora of cheeses ripened in brine and produced in different locations of Turkey plays an important role in the selection of starter cultures during the production of these cheeses on an industrial scale. It is stated that, for the white cheese, *Lactococcus* spp. and *Enterococcus* spp. are the dominant flora during the early maturation, whereas *Lactobacillus* spp. is dominant in the late phases of maturation.

There are new approaches for identifying the products specific to the special places. Indeed, this new perspective is explained with "Protected Designation of Origin (PDO)" or Protected Geographically Indication (PGI) terms. PDO identifies the designation of a product; the production, processing and preparation have to take place in a specific geographical area and the protection logos of these products are provided by the European Community (Regulation EC. No 509/2006 and Regulation EC. No 510/2006). The cheese is one of the main subjects of Protected Geographically Indication (PGI) in the world and Turkey has many cheese varieties that can be listed in this class. Turkey has more than 50 cheese varieties and the processes used in production of many cheeses are specific to our country. Similar to Turkey, the other Mediterranean countries such as Italy, Greece, and Spain have also many cheeses labeled with PGI (Endrizzi *et al.*, 2012; Pirisi *et al.*, 2011).

Cheeses Belonging to Black Sea Region

There are many varieties of cheese produced in Black Sea Region such as Civil Cheese, Imansiz Cheese, Kargi Tulum Cheese, Curry Cheese, and Kolot Cheese. The Circassian cheese is also one of the cheese types merchandised in this region (Gul *et al.*, 2015).

The Black Sea region has always been rich in diversity in terms of local dairy products for many years. A very small portion of these products is produced industrially, and most of them are made locally and sold in the neighborhood markets. The Black Sea cheese varieties hold an important place among these products, which have various flavors and aroma. Giresun dry cottage cheese (Çökelek), Peskütan and Keşi, Rize Kurçi cheese, Çamlıhemşin Minzi cheese, and Trabzon Yayla cheese are the first ones coming to the mind. The most known cheese varieties belonging to the provinces and districts of the Black Sea region are shown in Table **1**.

Table 1. The most popular cheese varieties of Black Sea Region (Kamber, 2015).

Traditional Cheese Name	City/ District Name
Acı Cheese	Giresun
Aho Cheese	Trabzon, Gümüşhane, Bayburt
Ayran Cheese	Rize
Cami boğazı	Trabzon
Ekşimik	Ordu, Giresun, Samsun
Eridik Cheese	Artvin
İmansız Cheese	Giresun, Trabzon
Kadina Cheese	Rize (Çamlıhemşin, Ayder)
Kargı Tulum Cheese	Çorum (Kargı)
Kesik	Ordu, Giresun, Samsun
Kolete (Golot) Cheese	Rize, Artvin, Trabzon, Bayburt
Kurçi	Rize
Küflü köylü Cheese	Artvin (Yusufeli)
Külek (Varella) Cheese	Trabzon, Artvin, Rize, Ardahan
Lorlu kaşar kırığı	Bayburt
Mezele Cheese	Trabzon (Sürmene)
Minzi	Trabzon, Rize, Artvin
Oğma Cheese	Trabzon, Artvin,
Su (Sulu) Cheese	Trabzon
Süt kırması	Trabzon
Şor Cheese	Artvin (Şavşat)
Tekne Cheese	Artvin
Telli Cheese	Trabzon, Artvin
Telli krema Cheese	Trabzon, Artvin (Yusufeli)
Torba Cheese	Ordu
Tulum Kaşarı	Tokat, Trabzon

(Table 1) cont.....

Traditional Cheese Name	City/ District Name
Yayla Cheese	Trabzon, Artvin
Yer Cheese	Artvin, Trabzon
Yumne Cheese	Artvin

As indicated in Table **1**, most of the cheese varieties are concentrated in the Eastern Black Sea Region. Some of the most popular cheese varieties of Black Sea Region are presented in Fig. (**2**).

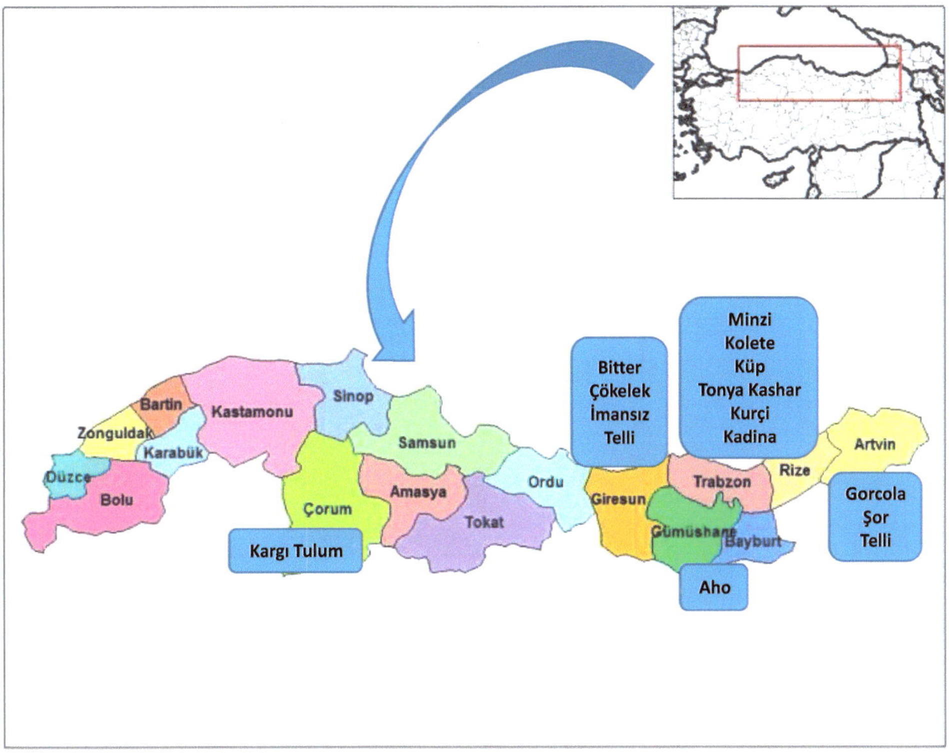

Fig. (2). Some cheeses of Black Sea Region in Turkey.

Chemical and Microbial Parameters of Cheeses of Black Sea Region

The cheese varieties having different quality characteristics are produced locally in the Black Sea region and there is no standard identified for the production of these cheeses. There are few scientific studies carried out on this subject. The microbiological and chemical results of several studies carried out on the cheese produced in the Black Sea Region are summarized in Tables **2** and **3**.

Table 2. Chemical composition of traditional Black Sea Region cheeses.

Variety of Cheeses	Moisture Content	Fat	Fat in Dry Matter	Protein	Ash	Salt	Acidity	Reference
Civil	56.64	2.78	40.58	16.46	9.40	9.15	0.64*	Bakirci, 1999
Civil	63.94	-	5.81	-	2.11	-	1.25*	Sengul *et al.*, 2007
Civil	90.66	0.40	-	3.37	-	-	17.89**	Cambaztepe *et al.*, 2009
Civil	-	2.30	6.50	32.40	-	0.23	2.39*	Yildiz *et al.*, 2010
Gorcola	37.90	26.10	36.00	25.80	5.40	4.80	0.50*	Kamber&Çelik, 2007
Kadel	56.50	3.45	7.93	30.33	-	3.36	1.86*	Yildiz *et al.*, 2008
Kargı Tulum	34.66	20.53	44.81	21.37	-	3.69	0.62*	Dinkci *et al.*, 2012
Keş	27.46	36.50	50.31	-	-	4.68	1.85*	Kirdar, 2004
Keş	39.02	7.50	-	38.85	4.19	3.39	1.75*	Tarakci, 2005
Keş	43.83	8.79	-	-	4.31	3.22	-	Kavas&Kavas, 2017
Kolete	56.49	5.31	38.20	33.64	-	3.12	0.73*	Caglar *et al.*, 1998a
Kolete	51.37	7.41	41.42	35.68	-	2.95	0.93*	Tunçtürk&Özdemir, 2005
Külek	53.64	6.75	39.61	28.22	8.16	7.09	2.16*	Yazici *et al.*, 1998
Külek	46.33	-	-	-	-	5.69	-	Licitra *et al.*, 2018
Külek	-	0.40	6.67	-	-	-	0.12*	Aydemir&Dervisoglu, 2010)
Küp	50.49	9.44	40.07	-	9.10	6.70	0.87*	Pekel&Korukluoglu, 2009
Minzi (Minci)	68.69	7.22	-	9.56	1.68	-	1.24*	Aloglu *et al.*, 2012
Telli	45.60	25.30	46.40	25.10	-	3.97	-	Kesenkas *et al.*, 2012
Yayla	52.85	9.06	19.21	30.18	-	2.88	1.75*	Yildiz *et al.*, 2008

*LA value; **SH value

Table 3. Microbiological properties of traditional Black Sea region cheeses.

Variety of Cheeses	TMAB	LAB	Total Coliform	*Lactococcus* spp.	Proteolytic Bacteria	Lipolytic Bacteria	*Staphylococcus* spp.	Yeast and Mould	Reference
Aho	6.72*	6.42*	<1.00*	-	-	-	<1.00*	4.12*	Kesenkas *et al.*, 2012
Civil	1.70x10^{8}	-	-	-	-	-	-	2.50x10^{5}	Bakirci, 1999
Civil	3.49x10^{8}	3.68x10^{6}	2.36x10^{4}	-	-	-	1.86x10^{4}	4.75x10^{5}	Caglar *et al.*, 1998a
Civil	7.43*	7.35*	5.66*	-	-	-	-	3.61*	Sengul *et al.*, 2007
Civil	7.44*	6.41*	2.00*	-	-	-	-	-	Yildiz *et al.*, 2010

(Table 3) cont.....

Variety of Cheeses	TMAB	LAB	Total Coliform	*Lactococcus* spp.	Proteolytic Bacteria	Lipolytic Bacteria	*Staphylococcus* spp.	Yeast and Mould	Reference
Gorcola	$1.9x10^{7}$	$2.90x10^{4}$	-	$1.00x10^{7}$	-	-	-	$1.40x10^{7}$	Kamber&Çelik, 2007
Kadel	$4.41x10^{8}$	-	$1.20x10^{4}$	-	-	-	-	$3.42x10^{7}$	Yildiz *et al.*, 2008
Kargı Tulum	6.98*	7.39*	-	7.28*	-	-	-	6.10*	Dinkçi *et al.*, 2012
Kolete	2.99*	3.07*	-	-	-	-	-	2.35*	Tunçtürk&Özdemir, 2005
Külek	$5.90x10^{5}$	-	-	-	-	-	-	$1.30x10^{6}$	Yazici *et al.*, 1998
Külek	4.96*	-	-	-	2.45*	2.67*	-	4.10*	Licitra *et al.*, 2018
Küp	7.31*	-	-	-	-	-	-	7.57*	Pekel&Korukluoglu, 2009
Minci (Minzi)	8.39*	6.74*	3.64*	6.34*	-	-	2.65*	2.89*	Aloğlu *et al.*, 2012
Pesküten	$3.70x10^{6}$	-	-	-	$1.20x10^{5}$	$1.00x10^{5}$	-	$1.40x10^{6}$	Kurt&Çağlar, 1998
Yayla	$2.35x10^{8}$	-	$4.60x10^{3}$	-	-	-	-	$1.63x10^{7}$	Yildiz *et al.*, 2008

*log cfu/g; TMAB: Total mesophilic aerobic bacteria; LAB: Lactic acid bacteria

Properties and Production Methods of Traditional Black Sea Region Cheeses

Aho Cheese

Aho Cheese is produced by small family farms in Trabzon, Bayburt, and neighboring towns located in the Black Sea region. Aho cheese was named after the Ayvadere (formerly known as Aho) Village of Araklı, Trabzon. The villagers living in Aho, who have been living in this region for hundreds of years and engaged in animal husbandry, store the excess of Aho Cheese, which they produce for their own needs in winter when they migrate to plateaus and the milk is abundant, and then they consume it after maturation. The surplus product is also sold in the local markets for commercial purposes. Aho Cheese is mostly produced in the highlands in May and August. This is because the hot air in this season offers the best fusion between Golot Cheese and Minzi (Minci) Cheese. Moreover, the milk production increases in these months. Aho Cheese can be stored for longer than a year. It has a dirty cream-color and a granular structure. This cheese variety is produced in different sizes, has a throat-burning aroma and is very peppery.

Aho Cheese, formerly made with Minzi, is now made of the mixture of two cheeses by adding Golot Cheese. Aho Cheese has a unique flavor with the combination of these two cheeses and is very popular in this region. Aho Cheese contains four components namely Minzi Cheese, Golot Cheese, salt, and red pepper for the spicy aroma. At the manufacturing sites, the floor is wet and slightly sloped and, thus, a nylon or cloth cover is laid on it. A fair amount of unsalted and fresh Minzi (Minci) cheese is spread on the ground with 23-25cm thickness. The cheese warms up and its color turns into yellow from top to bottom under the effect of summer temperature. This process is named bleaching. After the bleaching process, yellowing is seen in approx. 15cm from the top of Minzi cheese and the lower part remains white. The white portion of cheese is turned upside down and the milled rock salt is sprinkled. Subsequently, a sufficient amount of fresh Golot Cheese is sliced on top the Minzi Cheese. The aim of adding Golot Cheese is to increase the value of Minzi cheese, which is worthless, as well as accelerating the absorption of salt. Then, the red pepper is added to the cheese in a sufficient amount. The cheeses are blended with shovels and collected in the middle of the cloth on the floor. The brine of the cheese is leaked out for 24h. Then, they are put into special non-nylon sacks for a week. The cheese releasing its remaining water in this way is then pounded with the help of the mallet, and pressed into the sacks in order to remove all the bubbles. The cheese sacks are stacked on top of each other and they are kept for a month. After a month, the bottom sack is repositioned to the top and they are kept for one more month. The cheese is matured for consumption after two months of storage period. The longer the ripening time, the better the flavor of cheese (Çetinkaya & Özçakmak, 2017). A sample of Aho cheese was shown in (Fig. **3**).

Fig. (3). A sample of Traditional Aho Cheese.

Artvin-Yusufeli Wormy Curd Cheese

"Rotten cheese", which is made of milk with increased acidity in the highlands of Tevek village of the Yusufeli district of Artvin, is dried. The dried cheese is crumbled and broken into pieces. After a while, the cheese becomes moldy and the helminthiasis begins. These worms seem similar to the walnut worms, which have a black head and white body (Unsal, 2018).

Civil Cheese

The Civil cheese is also very popular in Turkey. It has a filamentous structure, as well as satisfactory melting and stretching properties (Kindstedt, 1993). It is produced by the small family business for their needs and by small dairy companies for commercial purposes. This variety of cheese is usually produced between May and September. The producers claim that it is very difficult to obtain a high-quality product in other months. There is no standard method for the production of Civil cheeses. Because of the production conditions and procedures, significant differences can be seen between Civil cheese samples produced by traditional methods. Traditionally, the fat-free, low-fat, or full-fat cow's milk, sheep's milk, or mixtures of these raw milks are used in the production. In producing the Civil cheese, the skim milk is divided into two equal portions. The first portion is allowed to rest for a night at 15 °C in order to increase acidity with natural microflora. Then, after this process, it is mixed with the second portion. Then, the mixture is heated to approx. 30 °C and the rennet is added. The starter cultures are not used in producing the traditional Civil Cheese. The heat is gradually increased with slow agitation and a clot begins at approx. 50 °C. Heating is continued up to approx. 70 °C by stirring a shovel. The cheese block is transferred cleanly. It is spread on the surface by using the shovel and kneaded by hand. The cheese block is hung from a platform for stretching by means of its own weight. Curd cheese has a smooth, plastic and fibrous character during the stretching process. Later, the civil cheese is cut into pieces and a saline solution is added to it (Cambaztepe *et al.*, 2009; Yildiz *et al.*, 2010; Caglar *et al.* 1998a; Sengül, 2006).

Civil cheese can be freshly consumed after the production but, when the cheese production is abundant, it is generally stored in the saline solution. This cheese is consumed without maturation by many people living in this region. The possibility of pathogenic microorganisms arising from the use of raw milk may cause certain health problems (Kara *et al.*, 1999). It is very difficult to produce a high-quality product using this traditional method. The milk acidification is the key step in the production of Civil cheese. Improper pH adjustment may cause a cheese texture that does not have a fibrous character.

The chemical and biochemical properties of commercial Turkish cheeses are well known only for some of the varieties including the White, Kashar, and Tulum cheeses. Other traditional cheeses, such as civil cheese, have drawn little attention, when compared to the other types of cheese. As known, the quality of cheese is significantly affected by the levels of peptides derived from proteolysis and lipolysis, as well as amino acids, and free fatty acids. The proteolysis is likely the most important biochemical conversion that has a great influence on the taste and texture of most cheese varieties. The cheese quality is also influenced by the enzymatic catabolism of amino acids. There are few studies carried out on the proteolysis in Civil cheese. Only a few researchers have investigated the chemical and biochemical properties of Civil cheese (Caglar *et al.*, 1998b; Sengül, 2006; Yazici & Dervisoglu, 2003).

Giresun Bitter Cheese

Giresun Bitter Cheese is made of fresh sheep cheese. The cheese is salted, and then placed in a plastic bag and filtered. It becomes slightly bitter when kept in cool conditions. After 3-4 days, the cheese is crumbled and pressed into plastic drums (Özdemir, 2018).

Giresun Çökelek Cheese

During the production, the Giresun Çökelek Cheese is left for the growth of lactic acid bacteria in "Ayran", which is made by diluting yogurt. Thus, the clot formation is ensured. Then, the clot is filled into a cloth bag and all the bags are pressed with stones in order to remove the whey. This cheese is similar to the "peasant cheese" in appearance and has plenty of bubbles. This variety of cheese is produced in Ordu region and at the same time used in cooking the traditional black sea pita (Özdemir, 2018). A sample of Giresun Çökelek Cheese is shown in Fig. (**4**).

Giresun İmansız Cheese

Giresun İmansız Cheese is a fresh cheese made of fat-free cow's milk. The peasant cheese (imansız) made of sheep milk on the Kazıkbeli plateau of Tirebolu in Giresun is also delicious. This cheese can be made of full-fat cow's milk and no salting is done. It is a type of cheese that is used in cooking the traditional meal named "Yağlıaş". A sample of Giresun İmansız cheese is shown in Fig. (**5**) (Özdemir, 2018).

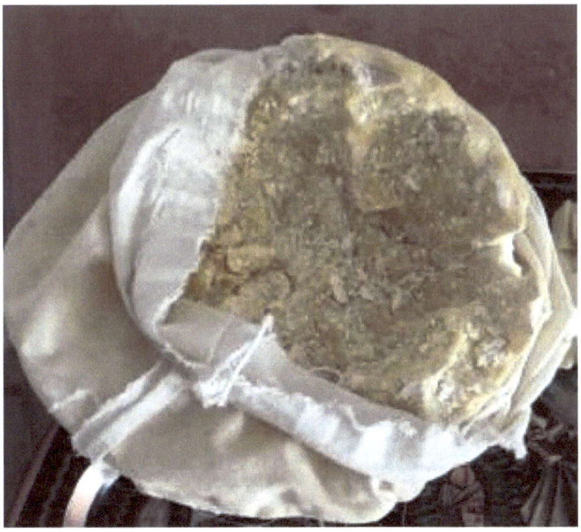

Fig. (4). An example of Giresun Çökelek Cheese.

Fig. (5). An example of Giresun İmansız Cheese.

Gorcola Cheese

The Gorcola cheese produced in Posof region of Ardahan and Şavşat district of Artvin is a local variety of cheese (Kamber, 2005). Gorcola cheese is tasty and fatty. It is usually made of skimmed cow's milk. This cheese is mostly made in summer (June-July). During the maturation process, the cheese develops mold and becomes green; it is locally known as "göğermiş gorcola" (matulase gorcola).

For the production of Gorcola cheese, raw milk is filtered with a cotton bag and then left for the fermentation about 15 SH. Then, it is poured into a large container and boiled until it coagulates. The cooled clot is filled into a cloth bag and it is filtered under pressure for 7-8 hours. Following the filtration process, the curd cheese is crumbled, spread on a fabric at 2-3 cm depth, and left for drying for 2-3 days in the open air. During this period, the whey evaporates and a very sharp odor is achieved. Then, the product obtained is called "gorcola" cheese. The Gorcola is crumbled by hand and put on a cloth that is folded in two, and then placed in a deep bowl. The cheese is kept there for 1.5 - 2 days and crumbled by hand once every 2-3 hours.

Then, "approx. 1 handful of finely ground salt is added per 1 kg of cheese" and the cheese is pressed into the animal skin by using a hammer. The pressed cheese is covered with a finger-thick salt and then tied up. Then, the filled skin is turned upside down, buried in moist soil at a depth of approx. 0.5-1 meters, and left for maturation for approx. 2-3 months. At the end of this period, the cheese is taken out of the soil and kept at room temperature until consumed. The Gorcola cheese becomes ready for consumption at the beginning of winter and can be kept untouched until the beginning of the next cheese season (Kamber & Çelik, 2007).

Kargı Tulum Cheese

This cheese produced in the summer season is introduced to the market in the autumn season. Its annual production is approx. 25 tons, and it is consumed especially in Çorum, Kastamonu, Samsun, and Ankara. This cheese, which is a very sharp and salty cheese and wrapped in sheepskin, is sold in 500 g, 1 kg, and 1.5 kg packages (Anonymous, 2003). Kargı Tulum cheese, which is produced by traditional methods on a small scale in houses by in family-type enterprises and dairy firms, is a traditional cheese variety.

Kargı Tulum cheese is made using the traditional production method in the flatland of Kargı district of Çorum. In the production process, the raw milk (8-10 kg) is first filtered and then heated to 24-25 °C. 100 ml of cheese water and 5-6 drops of microbial rennet (Trakya Cheese Yeast, Büyükçekmece, Istanbul) (1: 8000 strength) are added in order to coagulate the milk. The clot is transferred to a special filtration bag made of cloth after resting at 15-18 °C for 24 hours. These bags are hung for 24 hours in a room at 5-7 ° C temperature in order to separate the whey. Then, the bags are placed in wooden gutters in order to accelerate the discharge of whey in the same room and a pressure is applied for 24 hours. The clot is broken when the pH reaches 5.0; thus, the dry salted (2-2.5%) product is made and hand-kneaded at 15-18°C. The cheese, which is produced on daily basis, is salted and kneaded and it is squeezed firmly by using large cloth bags of 15-20 kg and stored in the same room again. When these cloth bags are completely full, the cheeses are pressed so that there is no more air in the bag with a capacity of 50 kg. The cheeses are matured during the summer months by transferring them to another bag every 15 to 20 days. The cheeses are then filled tightly as 0.5, 1 or 1.5 kg to sheepskins, which are produced using the traditional methods, and stored in these sheepskins until they are consumed (Dinkçi *et al.*, 2012; Kirdar *et al.*, 2015). A sample of Kargı Tulum cheese is shown in Fig. (**6**).

Fig. (6). An example of Kargı Tulum Cheese.

Keş Cheese

Keş cheese, which is a traditional Turkish dairy product, is manufactured commonly in different regions of Turkey. These cheeses are generally produced by small family businesses for their own needs. Commercial milk processing plants also produce them. Generally, the cow's milk is used for producing the Keş cheese. When the sheep and buffalo milks are available, these milks are mixed with cow's milk. The Keş cheese is generally produced in summer (Tarakci, 2005).

For the production of Keş cheese, the raw milk is filtered through the cheesecloth before heat treatment in order to remove unwanted ingredients. After the separation, the raw milk is heated to 90-100 °C for approx. 15-20 minutes. The milk is cooled to 40-45 °C after the heat treatment and then inoculated with yogurt culture (2-3%). The resultant clot is heated to 90-100°C for 10 minutes and then transferred to the cheesecloth. The pressure is applied until the whey is removed. The obtained clot is salted by salt rate of 2-3% and filled into the cotton bags. The bags are hung to a platform for 2-3 days in a warm room. The cheese bags are sometimes turned for homogeneous dryness. In recent years, 20-30% added white cheese to the curd cheese for making the product more attractive, resulting in a mosaic-like structure of the final product. Keş cheeses reach a total solids content of up to 60 to 70%. They are stored in a cool place for 4-5 months (Tarakçı, 2005; Tarakçı *et al.*, 2001; Dervisoglu *et al.*, 2009).

Kolete Cheese

Kolete cheese is generally made of fat-free cow or sheep milk in Trabzon, Rize, and Bayburt provinces. This cheese is locally named "Kolete", "Koleti", "Golot", and "Kolot". In production of Kolete cheese, the raw milk is filtered after milking and passed through the cream separating machine for separating as cream and fat-free milk. Skim milk is kept in enamel or tinned boilers for one day. After the same treatments, the raw milk obtained the next day is mixed with previous milk and heated to 40-50°C. During this heating, the milk coagulates due to improved acidity. The raw cheese obtained in this way forms the raw material of Kolete cheese. In another production method, the fat-free milk is heated to 30 °C and fermented. The rennet is usually added by diluting at a rate 2 tablespoons for 15-20 liters of milk and clotting is completed in 1 - 1.5 hours. After coagulation, the clot is collected on one side of the vessel by a scoop and is allowed to adhere to each other. Then the clot is taken into a cloth bag and a large part of whey is removed. This period lasts 10-12 hours. 1 kg of raw cheese is obtained from 11-14 liters of fat-free milk. Raw cheeses are sold in the weekly markets of the region. Producers of Kolete cheese processing collect these raw cheeses coming from the market from peasant producers and process them into Kolete cheese. The raw cheeses purchased from the marketplace are finely chopped into small pieces. The cheeses that are divided into small pieces are taken to the vessels and left for fermentation for 1 - 3 days according to the air temperature. Then, it is boiled at 80-90 °C. Boiling process, which is executed usually in the boiling brine boilers in a basket, is performed for 2-4 minutes. The boiled cheese is kneaded for a while like Kashar cheese, and the desired consistency is achieved. Then, it is divided into 1-1.5 kg pieces. Each part is attached to the hub and is placed on a flat surface in order to spread spontaneously and to form a round, slightly domed shape. Kolete cheeses are marketed after 1-2 days of ripening. If it will be stored, the cheese is taken to the tree barrels called "kurun" and salted using dry salting method.

Considering the production technique of Kolete cheese, it can be understood that this variety of cheese is similar to the Kashar cheese. However, the fat-free milk processing and coagulation, mostly through natural acidification, constitute the aspects distinguishing from the Kashar cheese. These differences distinguish the composition of Kolete cheese from the Kashar cheese, especially in terms of the fat content (Tunçtürk & Özdemir, 2005).

Kulek Cheese

Kulek cheese is a kind of cheese ripened by adding cheese yeast. It is known as one of the most consumed cheese varieties in the Black Sea region of Turkey. Kulek cheese is similar to cottage cheese (acid curd-uncreamed). However, since it is consumed after a maturation period of 3 months, it has a more compact curd structure (Dervisoglu & Yazici, 2001). Yazıcı *et al.* (1998) reported traditional production process and some characteristics of this cheese. Besides that, there is no standard processing technology for this variety. Dervisoglu & Yazıcı (2001) suggested the use of an appropriate initial culture in order to produce a Kulek cheese with a good quality.

Kulek cheese is ripened for 1-3 months traditionally in goatskin, wood, or clay containers (Yazici *et al.*, 1998). For this reason, the organoleptic and rheological properties of Kulek cheese vary depending on both the region and the production seasons. The cheese starter cultures are used to produce lactic acid and to change the organoleptic properties of cheese. However, the textural, microbiological and sensory properties of cheeses produced using commercial cheese starter cultures may be different from the traditional cheeses. It may be possible to obtain a market product with standard properties by using an appropriate commercial starter culture.

In the production of Kulek cheese, firstly the raw milk is filtered through a cotton cloth. The milk, fat of which is separated, is left to rest for an hour for casein precipitation after the warming process. The clot achieved is pressed for 7-8 hours using a specific weight in the cheesecloth (called the American cloth). The curd is laid on cotton cloth with 2 - 3 cm thickness and left to dry in the open air for 2 - 3 days. In this phase, the curd is traditionally called "Gorcolo". The Gorcolo is crumbled by hand and placed in a deep bowl. Then, the curd is stirred in this bowl once every 2-3 hours for 1.5 to 2 days, and then salted using rock salt. The cheese is crushed with a wooden stick in goatskin, wood, or soil containers. Once the surface of the containers is covered with salt, they are closed and placed 50 cm below the floor to ripe for 1 - 3 months. After the ripening period, the salty surface layer is separated and the Kulek cheese is stored at room temperature for consumption. There is no scientific explanation for burying cheese in the soil, but it is believed that the soil is easier to use and cheaper and it contributes to the formation of a better odor-structure when the cheese is ripened. A high-quality product can be produced by using this conventional method, because most of the very difficult production steps, especially the maturation conditions, are not standardized (Temiz & Kılıç, 2015).

Küp (Jar) Cheese

The Küp (Jar) Cheese is produced by different methods from the milk of sheep, goat and cow in the Central and North Anatolia. According to the regions, fatty or skimmed milk is left to itself and clotted by souring or sometimes rennet is used for clotting. Fatty, fresh village cheeses taken from the market in North Anatolia (Sürmeli, Trabzon) are put into a cloth and stones are used for pressing, thereby, removing the whey. The resulting curd is crushed by hand after being kept for a while. For the clot obtained by using rennet, the curd is slightly heated after cutting, filtered in cloth bags, pressed, and then crushed by hand. The curd, which is produced with different methods, is mixed by adding salt, and then is filled by pressing firmly in wide-mouthed unglazed jars. In some regions, the cream is added to the curd, which was obtained from skimmed milk, after filling into the jars. On top of this, sufficient amount of soil is poured onto the bag. Thus, the contact between air and cheese is blocked and the growth of mold is prevented. However, in some regions, the upper surface of the cheese is allowed to contact with the air due to which the cheese becomes moldy and gains a special flavor. The jars are buried into a cool ground such as cellar. The cheese gradually loses its moisture during the formation period within the jars, and the biochemical events occurring during this period create a unique odor and aroma. The ideal storage temperature for Küp (Jar) Cheese, which should be stored below room temperature, is 6-10 °C (Unsal, 2018).

Minci (Minzi) Cheese

Minci (Minzi) cheese is characterized as a kind of cheese, which is made of “ayran”, “yoğurt”, or “raw milk”. Minzi is the name given to the Çökelek cheese in the Black Sea region and used in cooking the “kuymak”, which is a traditional food. The protein and calcium-rich sediment cheese was a highly consumed foodstuff. Çökelek cheese is also important as an inexpensive animal protein/nutrient source for the people with insufficient economic sources, as well as the use of dairy residues. If it is produced under hygienic conditions, it can be possibly used as an important dietary food. Minzi is generally produced from “ayran”, although it is possible to use different production methods. The production process of Minzi made of “ayran” is as follows; a tin of yogurt is mixed with half tin of water and “ayran” is churned on the wooden churn until the oil and water are separated. The oil layer is removed from the ayran and the remaining part is transferred to an oven, then it is heated to the boiling temperature on the hob and filtered through the cloth bags. During this period, the precipitate (clot) is formed. This clot is called Minci (Minzi) in the Black Sea region. The production of Çökelek cheese by using "ayran" is very important for decreasing the animal protein deficiency in Turkey. Moreover, some of the dairy farm owners use yogurt, which they cannot sell, in making Çökelek. This cheese is widely consumed in Turkey and seen as a very important food product in terms of health (Unsal, 2018; Aloğlu *et al.*, 2012).

Telli Cheese

Telli cheese is one of the popular traditional cheese varieties in the Eastern Black Sea region of Turkey. It is usually made of cow's milk obtained from Trabzon and Artvin, as well as the provinces of Sürmene and Akçaabat in Trabzon. Yellowish colored Telli cheese is similar to the Macaroni Filata cheese (especially kashar cheese), and its production method is also similar to Örgü cheese produced around Diyarbakır and its appearance is similar to Erzurum Civil Cheese (Kamber& Terzi, 2007). However, Telli cheese is more acidic (pH 4.8 to 5.1) and has a more elastic texture than Kashar cheese (pH 5.2). The kneading process of the fresh clot in salted hot water does not only give the product its fibrous structure but also its melting and stretching properties. Although there is no standard method of producing Telli cheese, there are few differences between the above-mentioned cities, where the producers try to use the modern processing methods in the production. Telli cheese can be consumed fresh (usually within 30 days after the production) and after maturation by mixing with Minzi cheese in wooden tubs (Kamber & Terzi, 2007). In the production of Telli cheese, the fat content of the cow's milk is standardized to approximately 3.5% and the raw milk is heated to 35°C. A sufficient amount of cheese yeast is then added in order to achieve the coagulation in the milk within approx. 35 minutes. The resulting clot is cut into 1 x 1 x 1 cm^3 cubes and then left to rest in whey for 2-3 minutes. Then, the clot is transferred to the molds, covered with cheesecloth, and pressed for draining the whey for 30 ± 5 minutes. The pressure is set at a weight of 150 kg per 100 kg clot. Then the clot is fermented at the ambient temperature until 5.0-5.10 pH. The clot is then cut into slices (5 cm wide, 5 mm thick) and boiled in hot brine (72-75 °C, 12-13% salt) for 2-3 minutes. The obtained cheese is cooled by arranging a cord that is 30-45 cm long and 6 cm diameter. At the end of these processes, Telli cheese is packed with vacuum and stored at 4 °C (Kesenkas *et al.*, 2012). A sample of Giresun Telli Cheese is shown in Fig. (**7**).

Fig. (7). An example of Giresun Telli Cheese.

Rize "Kadina" Cheese

Kadina cheese is a fat-free cheese pressed in pine barrels called "Kadina" in Ayder Yukarı Kavrun Plateau, Rize. The families prepare their cheese daily and keep them in these wooden barrels. This cheese variety is consumed for breakfast, and is also used in cooking the local food called "mıhlama" or "muhlama" as well. Kadina Cheese is white and non-porous and has a slightly elastic structure. For producing Kadina cheese, the raw milk cream is mechanically removed in a centrifuge and then boiled. Then, the milk is brought to a moderately warm temperature and fermented using a bottle of rennet. The clot formation occurs within half an hour. The clot is collected and transferred to another container. Then, the cheese is squeezed in order to remove the whey. After applying pressure, the dry salting step is done. Then, it is placed into the barrels, which are named "Kadina". As cheese is made on a daily basis, a new one is added on the "Kadina" and the cheeses in maturation period are kept in a cool part of the house (Unsal, 2018; Kamber & Terzi, 2007). A sample of Kadina Cheese is shown in Fig. (**8**).

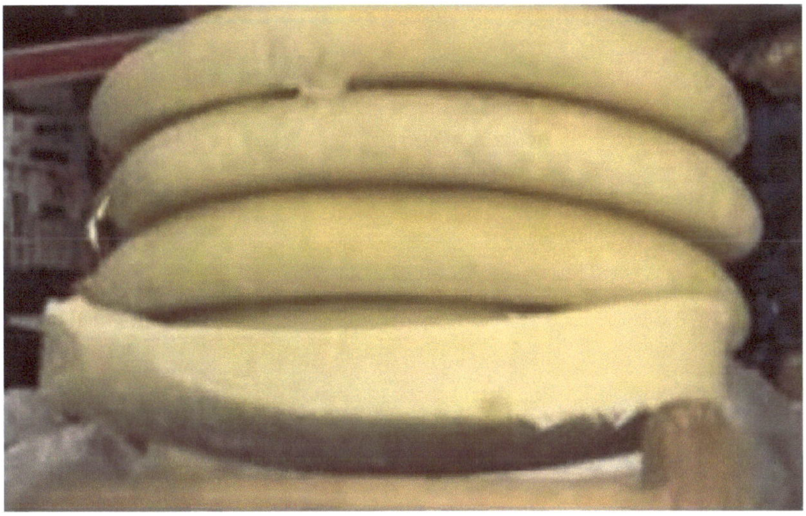

Fig. (8). An example of Kadina Cheese.

Rize Kurçi Cheese

"Kurçi" or "Kurç" cheese is another type of Çökelek made by letting the acidity to increase in the buttermilk. It is crumbled by hand. In the same region, a kind of skimmed cheese made of skim milk is also called "Kurçi" (Unsal, 2018).

Şor Cheese

Şor cheese is produced in Şavşat, Artvin. The traditional production method of this cheese is similar to that of Aho cheese. The cheese was named after the word "Şor" meaning "bitter" in the region. Its color is dark yellow and has a bitter taste since it is very salty. In the production of Şor cheese, the raw milk is poured in a large barrel and then acidified. The coagulated milk is boiled and the curd on the surface is transferred to another boiler using a scoop. Then, the clot is taken into a large cloth bag and pressed using large stones in order to remove the whey. The cheese obtained is cut into small pieces, spread over a fabric, and dried for 2-3 days. When the cheese starts to get a yellowish color, it is mixed with rock salt. Thus, the salt completely penetrates into the cheese. In the past, the salted cheese has been placed into an animal skin. However, nowadays, it is packed in cloth bags or wooden barrels for the consumption (Kamber & Terzi, 2007). An example of Şor Cheese is shown in Fig. (**9**).

Fig. (9). A sample of Şor Cheese.

Trabzon Tonya Kashar Cheese

Tonya Kashar Cheese is made of full-fat cow's milk. The clot is pressed using a hydraulic press and the whey is removed for 3-5 hours. The mass is chopped using the mincing machine and is left for 1 day for acidification. Then, they are put into perforated buckets, boiled in boiling water for 3-5 minutes, and then stirred. The cheese dough taken out from the boiler is salted and kneaded thoroughly. Then the cheese is poured into rectangular or round molds and vacuumed. A sample of Trabzon Tonya Kashar Cheese is shown in Fig. (**10**) (Mutlu, 2018).

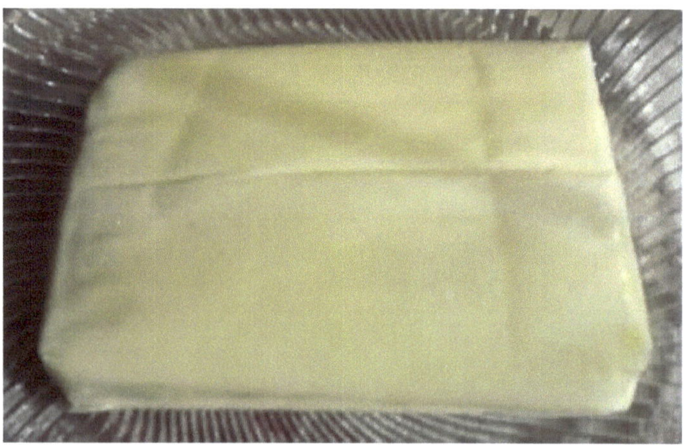

Fig. (10). An example of Trabzon Tonya Kashar Cheese.

Trabzon Yayla (Scouring) Cheese

This cheese is made in Hamsiköy plateau, Trabzon. For the production of this cheese, Minzi and white cheese made of fat-free cow's and goat's milk are mixed. The mixture is crushed and "scrubbed" by hand. The crumbled cheese mix is pressed to the barrels (called as Kurun or Külek) made of pine wood. Cheese-filled barrels are buried in well-covered soil wells and maturated for 4 months (Mutlu, 2018).

FUTURE EXPECTATIONS

In the past, the people cared about "what should not be eaten". However, this opinion has turned into "what should we eat". So, the "functional food" term is the key for the future food demands. The cheese is a strong component of the daily diet. This food has many special characteristics such as being a protein source, containing beneficial bacteria (lactic acid bacteria), and is made of milk. The new aspect of the cheese industry is to manufacture the probiotic products in order to address the people demanding the functional foods.

The probiotic cultures found in the food should be minimum 10^6-10^7 CFU/g or CFU/mL per 100 g or 100 mL. It is not easy to keep the microorganisms alive during the entire process and to ensure the transportation to gastrointestinal canal in an appropriate manner (Castro *et al.*, 2015). At this point, the fermented dairy products come to the forefront and the cheese is a good option.

The production of cheese is a dynamic process. The raw milk, the sterilization technique, processing technology, and the starter culture are the main parameters influencing the quality and safety of product. Thus, the molecular genetic methods are the key to settle the suitable cultures. The genetic engineering increases the functionality of starter bacteria in fermentation process, which enables feasible automation in the production (Johnson, 2017).

Another aspect of the dairy products is the relationship between obesity / being overweight and the consumption of dairy products. In the past, it was assumed that the dairy products have an undesirable effect on overweight and obese individuals due to a high level of saturated fat content. However, recent studies have revealed the effects of high level of dairy fat consumption. Moreover, the high level of calcium content in the dairy products may facilitate weight loss by suppressing lipogenesis, stimulating lipolysis, increasing fecal fat excretion, and fat oxidation. Similarly, it was reported that the dairy products have a positive effect on cancer, type 2 diabetes, and hypertension (Zheng *et al.*, 2015).

Some researchers reported adhesive effects of high calcium intake. On the other hand, an opposite opinion also exists. For instance; Yang *et al.* (2016) declared that the intake of dairy products or calcium was not statistically associated with the risk of lung cancer.

Around the globe, the authorities tend to label the domestic products as "Geographically Marked Products (Geographical Indication: GI)". According to WIPO (World Intellectual Product Organization), this term refers to "*a sign used on the products that has a specific geographical origin and possesses qualities or a reputation arising from that origin. In order to function as a GI, a sign must identify a product as originating from a given place. In addition, the qualities, characteristics or reputation of the product should be essentially arising from the place of origin. Since the qualities depend on the geographical place of production, there is a clear link between the product and its original place of production.*" Many types of foods are included in this classification. However, it cannot be ignored that the cheeses are strong candidates of this labeling system. On the other hand, while modern and standard efforts are being made for the cheeses, the local characteristics, traditional taste, aroma, and structures should not be sacrificed.

In fact, the European Union (EU) supports three types of quality labels: Protected Designation of Origin (PDO), Protected Geographical Indication (PGI) and Traditional Speciality Guaranteed (TSG) (European Commission, 2013). Dias & Mendes (2018) reported four clusters: "Protected Geographical Indication", "Certification of Olive Oil and Cultivars", "Certification of Cheese and Milk", and "Certification and Chemical Composition" according to the bibliometric analysis. The "Certification of Cheese and Milk" cluster is the development of authentication methods for certified traditional products. The percentage of the cheeses among all certificated products is 17%, which is a good value.

Turkey has many products, which can be commercially introduced to the world. As it can be seen in this chapter, even the cheeses produced in a single region of Turkey are strong candidates. The authorities should focus on labeling these products and taking a position in the global market.

CONCLUSIONS

In conclusion, the consumption of dairy products in Turkey has a very long historical background thanks to the favorable geographical characteristics. Although the cheese production gradually increases as an industry in Turkey, approximately 90% of cheeses are still being produced by the small enterprises, dairies, or family businesses. This also applies to the majority of the cheeses produced in the Black Sea Region. The production of milk and cheese in our country and Black Sea Region is not sufficiently evaluated since it is mostly produced by primitive materials and methods. The local and traditional Turkish cheese production methods should be modernized using modern production technologies with pasteurized milk and cultures in order to ensure production in healthy conditions. Thus, while traditional cheeses meet the modern technology, the risk of uniformity of large enterprise cheeses preserved at local characteristics will be avoided. It can be concluded that Turkey has a great potential of becoming the leader in cheese production. But we need to standardize our processes and label our products according to the global labeling system.

CONSENT FOR PUBLICATION

Not applicable.

CONFLICT OF INTEREST

The authors confirm that this chapter contents have no conflict of interest.

ACKNOWLEDGEMENTS

Declared none.

REFERENCES

Aloğlu, HŞ, Turhan, İ & Öner, Z (2012) Minci (Minzi) Peynirinin Özelliklerinin Belirlenmesi. *GIDA-The Journal of Food,* 37, 349-54. (In Turkish).

Anonymous, (2003) *Dünden Bugüne Kargı Tarihi, Kültürü, Coğrafyası, Ekonomisi ve İnsanı ile Kargı Türkiye Cumhuriyetinin Kuruluşunun 80.*Yılı Anısına Hazırlanan Yıllık, Kargı-Çorum. In Turkish.

Aune, D, Lau, R, Chan, DSM, Vieira, R, Greenwood, DC, Kampman, E & Norat, T (2012) Dairy products and colorectal cancer risk: a systematic review and meta-analysis of cohort studies. *Ann Oncol,* 23, 37-45. [http://dx.doi.org/10.1093/annonc/mdr269] [PMID: 21617020]

Aydemir, O & Dervişoğlu, M (2010) The effect of heat treatment and starter culture on colour intensity and sensory properties of Kulek cheese. *Int J Dairy Technol,* 63, 569-74. [http://dx.doi.org/10.1111/j.1471-0307.2010.00610.x]

Bakirci, I & Andic, S (1999) Muş Bulanik Yöresinde Üretilen Cecil Peyniri Üzerinde bir Araştirma. *Yüzüncü Yıl Üniversitesi Veteriner Fakültesi Dergisi,* 10, 67-71.

Boor, KJ, Wiedmann, M, Murphy, S & Alcaine, S (2017) A 100-Year Review: Microbiology and safety of milk handling. *J Dairy Sci,* 100, 9933-51. [http://dx.doi.org/10.3168/jds.2017-12969] [PMID: 29153181]

Brassard, D, Tessier-Grenier, M, Desroches, S & Couture, P (2016) Comprehensive Review of the Impact of Dairy Foods and Dairy Fat on Cardiometabolic Risk 1 – 3. *Adv Nutr,* 1041-51. [http://dx.doi.org/10.3945/an.115.011619.1041]

Caglar, A, Kurt, A, Ceyhan, G & Hursit, S (1998) Different prevention methods of Civil cheese. *Traditional Dairy Products* MPM 65-78. a

Caglar, A, Kurt, A, Ceylan, ZG & Hursit, S (1998) The Research on Conservation in Different Shape of Civil Cheese. V National Productive Center Publication No: 621 *Milk and Milk Products Symposium,* Ankara, Turkey. 65-78. b

Cambaztepe, F, Çakmakçı, S & Dağdemir, E (2009) Effect of some technological parameters on microbiological, chemical and sensory qualities of Civil cheese during ripening. *Int J Dairy Technol,* 541-8. [http://dx.doi.org/10.1111/j.1471-0307.2009.00536.x]

Castro, JM, Tornadijo, ME, Fresno, JM & Sandoval, H (2015) Biocheese: a food probiotic carrier. *BioMed Res Int,* 2015723056 [http://dx.doi.org/10.1155/2015/723056] [PMID: 25802862]

Çetinkaya, A & Özçakmak, S (2017) Specification of Mineral Composition, Heavy Metal Content, Microbiological Characteristics of a Traditional Anatolian Cheese-Aho Cheese. *Sch. Bull.*; Vol-3. *Iss,* 12, 655-61.

Dervisoglu, M, Tarakcı, Z, Aydemir, O, Temiz, H & Yazici, Y (2009) A survey on some chemical, biochemical and sensory properties of Kes cheese, a traditional Turkish cheese. *Int J Food Prop,* 12, 358-67. [http://dx.doi.org/10.1080/10942910701784688]

Dervisoglu, M & Yazici, F (2001) Ripening changes of kulek cheese in wooden and plastic containers. *J Food Eng,* 48, 243-9. [http://dx.doi.org/10.1016/S0260-8774(00)00164-3]

Dias, C & Mendes, L (2018) Protected Designation of Origin (PDO), Protected Geographical Indication (PGI) and Traditional Speciality Guaranteed (TSG): A bibiliometric analysis. *Food Res Int,* 103, 492-508. [http://dx.doi.org/10.1016/j.foodres.2017.09.059] [PMID: 29389640]

Dinkçi, N, Ünal, G, Akalin, AS, Varol, S & Gönç, S (2012) Chemical and microbiological properties of Kargi Tulum Cheese. *Ege Univ Ziraat Fak Derg,* 49, 287-92.

Durlu-Ozkaya, F & Gun, I (2008) Anadoluda peynir kültürü. ICANAS 38. *Uluslararası Asya ve Kuzey Afrika çalışmaları Kongresi, Eylül* Ankara-Turkey. In Turkish

Endrizzi, I, Fabris, A, Biasioli, F, Aprea, E & Franciosi, E Poznanski, Cavazza A & Gasperi, F (2012), The effect of milk collection and storage conditions on the final quality of Trentingrana cheese: Sensory and instrumental evaluation. *Int Dairy J,* 105-14. [http://dx.doi.org/10.1016/j.idairyj.2011.10.004]

Ertaş, Y & Gezmen-Karadağ, M (2013) Sağlıklı Beslenmede Türk Mutfak Kültürünün Yeri. *Gümüşhane Üniversitesi Sağlık Bilimleri Dergisi / Gümüşhane University. Journal of Health Sciences,* 2, 117. [In Turkish].

Franciosi, E, Settanni, L, Cologna, N, Cavazza, A & Poznanski, E (2011) Microbial analysis of raw cows' milk used for cheese-making: influence of storage treatment on microbial composition and other technological traits. *World J Microbiol Biotechnol,* 27, 171-80. a [http://dx.doi.org/10.1007/s11274-010-0443-2]

Gul, O, Dervisoglu, M, Mortas, M, Aydemir, O, Ilhan, E & Aksehir, K (2015) Evaluation of polycyclic aromatic hydrocarbons in Circassian cheese by high-performance liquid chromatography with fluorescence detection. *J Food Compos Anal,* 37, 82-6. [http://dx.doi.org/10.1016/j.jfca.2014.07.004]

Johnson, EA, Nelson, JH & Johnson, M (1990) Microbiological safety of cheese made from heat-treated milk, Part 1. Execu- tive summary, introduction and history. *J Food Prot,* 53, 441-52. [http://dx.doi.org/10.4315/0362-028X-53.5.441] [PMID: 31018298]

Johnson, ME (2017) A 100-Year Review: Cheese production and quality. *J Dairy Sci,* 100, 9952-65. [http://dx.doi.org/10.3168/jds.2017-12979] [PMID: 29153182]

Kamber, U (2015) Traditional Turkey Cheeses and Their Classification, *Van. Vet J,* 26, 161-71.

Kamber, U & Çelik, H (2007) Some Microbiological And Chemical Characteristics Of Gorcola Cheese. *Yüzüncü Yıl Üniversitesi Veteriner Fakültesi Dergisi,* 18, 87-92.

Kamber, U (2005) *Geleneksel Anadolu Peynirleri* Miki Matbaacılık AS, Ankara 174. In Turkish

Kamber, U & Terzi, G (2007) The Traditional Cheeses of Turkey: Middle and Eastern Black Sea Region. *Food Rev Int,* 24, 95-118.
[http://dx.doi.org/10.1080/87559120701764555]

Kara, AA, Algur, OF & Kaya, M (1999) Isolation and identification of Listeria species in Beyaz and Civil cheese gained from Erzurum market. *Turk J Biol,* 23, 331-7.

Kavas, N & Kavas, G (2017) Use of turmeric *(Curcuma longa L.)* essential oil added to an egg white protein powder-based film in the storage of çökelek cheese. *J of Food Chem and Nanotech,* 3, 105-10.
[http://dx.doi.org/10.17756/jfcn.2017-045]

Kesenkas, H, Dinkci, N, Seckin, K, Gursoy, O & Kinik, O (2012) Physicochemical, biochemical, textural and sensory properties of telli cheese - a traditional turkish cheese made from cow milk. *Bulg J Agric Sci,* 18, 763-70.

Kindstedt, PS (1993) Mozzarella and Pizza cheese. *Cheese: chemistry, physics and microbiology* Chapman and Hall, London.
[http://dx.doi.org/10.1007/978-1-4615-2648-3_12]

Kirdar, SS (2004) A reasearch on Kes cheese. *International Dairy Symposium,* May 24-28 2004: 239-41.

Kirdar, SS, Kose, Ş, Gun, I, Ocak, E & Kursun, Ö (2015) Do consumption of Kargi Tulum cheese meet Daily requirements for minerals and trace elements? *Mljekarstvo,* 65, 203-9.
[http://dx.doi.org/10.15567/mljekarstvo.2015.0307]

Kratz, M, Baars, T & Guyenet, S (2013) The relationship between high-fat dairy consumption and obesity, cardiovascular, and metabolic disease. *Eur J Nutr,* 52, 1-24.
[http://dx.doi.org/10.1007/s00394-012-0418-1] [PMID: 22810464]

Kurt, A & Çağlar, A (1988) Investigations on chemical and microbiologic features of Pesküten. *GIDA-The J of Food,* 13, 341-7.

Lan, CH, Son, CK, Ha, HP, Florence, H, Binh, LT, Mai, LT & Yves, W (2013) Tropical traditional fermented food, a field full of promise. Examples from the Tropical Bioresources and Biotechnology programme and other related French-Vietnamese programmes on fermented food. *Int J Food Sci Technol,* 48, 1115-26.
[http://dx.doi.org/10.1111/ijfs.12064]

Leitner, G, Silanikove, N, Jacobi, S, Weisblit, L, Bernstein, S & Merin, M (2008) The influence of storage on the farm and in dairy silos on milk quality for cheese production. *Int Dairy J,* 18, 109-13.
[http://dx.doi.org/10.1016/j.idairyj.2007.09.001]

Licitra, G & Carpino, S (2014) *The Micro floras and Sensory Profiles of Selected Protected Designation of Origin Italian Cheeses,* 1-12.
[http://dx.doi.org/10.1128/microbiolspec.CM-0007-2012.]

Licitra, G, Caccamo, M, Valence, F & Lortal, S (2018) Traditional Wooden Equipment Used for Cheesemaking and Their Effect (Chapter 7), Global Cheesemaking Technology. In: Papademos, Photis, Bintsis, Thomas, (Eds.), *Cheese Quality and Characteristics* John Wiley & Sons, Inc, NJ 07030, USA 157-69.

Lortal, S, Licitra, G & Valence, F (2014) Wooden Tools: Reservoirs of Microbial Biodiversity in Traditional Cheesemaking. *Microbiol Spectr,* 2, CM-0008-2012.
[http://dx.doi.org/10.1128/microbiolspec.CM-0008-2012] [PMID: 26068076]

Lu, W, Chen, H, Niu, Y, Wu, H, Xia, D & Wu, Y (2016) Dairy products intake and cancer mortality risk: a meta-analysis of 11 population-based cohort studies. *Nutr J,* 15, 91.
[http://dx.doi.org/10.1186/s12937-016-0210-9] [PMID: 27765039]

Malacarne, M, Summer, A, Formaggioni, P, Franceschi, P, Sandri, S, Pecorari, M, Vecchia, P & Mariani, P (2008) Dairy maturation of milk used in the manufacture of Parmigiano-Reggiano cheese: effects on physico-chemical characteristics, rennet-coagulation aptitude and rheological properties. *J Dairy Res,* 75, 218-24.
[http://dx.doi.org/10.1017/S0022029908003221] [PMID: 18474140]

Montel, MC, Buchin, S, Mallet, A, Delbes-Paus, C, Vuitton, DA, Desmasures, N & Berthier, F (2014) Traditional cheeses: rich and diverse microbiota with associated benefits. *Int J Food Microbiol,* 177, 136-54. [http://dx.doi.org/10.1016/j.ijfoodmicro.2014.02.019] [PMID: 24642348]

Mutlu, C (2018) Personal Communication, Trabzon Country.

Özdemir, G (2018) Personal Communication, Giresun Country.

Pekel, M & Korukluoğlu, M (2009) Microbiological cube of cheese produced in the Sivas region chemical quality and determination of mold flora. *Anadolu Tarım Bilimleri Dergisi,* 24, 1-7.

Pirisi, A Comunian, Urgeghe, PP & Scintu, MF (2011), Sheep's and goat's dairy products in Italy: Technological, chemical, microbiological, and sensory aspects. *Small Rumin Res,* 101, 102-12. [http://dx.doi.org/10.1016/j.smallrumres.2011.09.030]

Scallan, ERM, Hoekstra, FJ, Angulo, RV & Tauxe, MA Wid- dowson, SL, Roy, JL, Jones, PM & Griffin (2011) Foodborne illness acquired in the United States—Major pathogens. *Emerg Infec Dis,* 17, 7-15. [http://dx.doi.org/10 .3201/ eid1701 .091101p1.]

Sengül, M (2006) Microbiological characterization of Civil cheese, a traditional Turkish cheese: microbiological quality, isolation and identification of its indigenous Lactobacilli. *World J Microbiol Biotechnol,* 22, 613-8.
[http://dx.doi.org/10.1007/s11274-005-9079-z]

Sengul, M, Gurses, M, Dervisoglu, M & Yazici, F (2007) A Survey on the Some Chemical and Biochemical Properties of Civil Cheese, a Traditional Turkish Cheese. *Int J Food Prop,* 9, 791-801. [http://dx.doi.org/10.1080/10942910600596415]

Sufer, O & Sezer, YÇ (2014) *Bir Karadeniz Klasiği: Kuymak 4 Geleneksel Gıdalar Sempozyumu* 17-9. Nisan 2014 Adana-Turkey. In Turkish

Tarakçı, Z, Küçüköner, E & Yurt, B (2001) Ordu ve yöresinde imal edilen kesin yapılışı ve bazı özellikleri üzerinde bir araştırma. *GIDA- J Food,* 26, 295-300. [In Turkish].

Tarakci, Z (2005) The Chemical and sensorial properties of Kes produced in Ordu. *ITAFE'05-International Congress on Information Technologies in Agriculture, Proceedings volume II,* 846-849, 12-4. Adana-Turkey

Temiz, H & Kılıç, S (2015) A survey of the chemical, biochemical, microbiological and sensorial quality of Aho cheese, a traditional cheese from Eastern Black Sea Region, Turkey. *Int J Dairy Technol,* 69, 209-16. https://onlinelibrary.wiley.com/doi/epdf/10.1111/1471-0307.12237.
[http://dx.doi.org/10.1111/1471-0307.12237]

Tunçtürk, Y & Özdemir, M (2005) Production technique of production and consumption in the Eastern Black Sea Region offered Golotte Cheese, some chemical, biochemical and microbiological properties. *GIDA- J Food,* 30, 167-72.

Unsal, A (2018) When Milk Slept "Turkish Cheeses. *Bank of Yapi Kredi Publications* (8 th edition) :Istanbul 216-44.

Weisbecker, A (2007) A legal history of raw milk in the United States. *J Environ Health,* 69, 62-3. [PMID: 17450957]

Yang, Y, Wang, X, Yao, Q, Qin, L & Xu, C (2016) Dairy Product, Calcium Intake and Lung Cancer Risk: A Systematic Review with Meta-Analysis. *Sci Rep,* 6, 20624.
[http://dx.doi.org/10.1038/srep20624] [PMID: 26877260]

Yazici, F, Dervisoglu, M & Temiz, H (1998) Sensory, physical, chemical, and microbiological properties of Kulek cheese. In: Demirci , M., (Ed.), *Traditional dairy products,* May;21-22621, 133-44.

Yazici, F & Dervisoglu, M (2003) Effect of pH adjustment on some chemical, biochemical, and sensory properties of Civil cheese during storage. *J Food Eng,* 56, 361-9.
[http://dx.doi.org/10.1016/S0260-8774(02)00208-X]

Yildiz, F, Sirt, S & Yetisemeyen, A (2008) Some Properties of Yayla and Kadel Cheese Produced in the

Black Sea Region. *10 Gida Kongresi, Mayis* 21-3. Erzurum, Turkey: Gıda Teknolojisi Derneği Yayın No: 37 Ankara.

Yildiz, F, Yetisemeyen, A, Senel, E, Ozkaya, F, Oztekin, S & Sanli, E (2010) Some properties of Civil Cheese. *Int J Dairy Technol,* 63, 575-80.
[http://dx.doi.org/10.1111/j.1471-0307.2010.00612.x]

Zheng, H, Clausen, MR, Dalsgaard, TK & Bertram, HC (2015) Metabolomics to explore impact of dairy intake. *Nutrients,* 7, 4875-96.
[http://dx.doi.org/10.3390/nu7064875] [PMID: 26091233]

CHAPTER 8

Traditional Cheeses of Eastern Anatolia Region in Turkey

Asli Çelikel[1,*], **Mutlu Buket Akin**[2] and **Semra Gürbüz**[1]

[1] *Gastronomy and Culinary Art, Mardin Artuklu University, Tourism Faculty, Mardin, Turkey*

[2] *Faculty of Engineering, Department of Food Engineering Harran University Sanliurfa Turkey*

Abstract: Cheese which dates back to ancient times is an important part of our daily diet. Cheese constitutes 13% of the registered consumption of milk and dairy products worldwide. Our country has a rich variety in terms of both regional cheeses and industrial cheeses. Many traditional cheeses are produced in the Eastern Anatolia Region. The name of some cheeses produced in the region varies according to the locality. Erzincan Tulum cheese, which is commonly produced in Erzincan, Erzurum, Elazığ, Tunceli and Bingöl, is also called Şavak Cheese. One of the traditional cheeses produced in the region is Kars Kashar Cheese. Also, in recent years, the production of Gravyer Cheese in Kars and Ardahan provinces has become widespread. Various plants grown in the Eastern Anatolia Region are used in the traditional cheese production. The first ones of these cheeses are Herby Cheese and Cacik Cheese. Approximately, 70-80% of Herby Cheese is produced in Van. This cheese is called Van Herby cheese. Another traditional cheese produced in Eastern Anatolia Region is Civil cheese which is widely produced in Erzurum. This cheese plays an important role in the formation of taste and aroma of green mould cheese that develops during ripening. Identification, classification and standardization of the characteristics of traditional cheeses produced in the Eastern Anatolian Region and production technologies play an important role in the introduction of these products. For this purpose, production methods, physical and chemical properties of traditional cheeses produced in Eastern Anatolia Region have been compiled in this section.

Keywords: Cacik Cheese, Civil cheese, Chemical Properties, Eastern Anatolia Region, Gravyer Cheese, Herby Cheese, Kars Kashar Cheese, Microbiological Properties, Physical Properties, Production Methods, Traditional Cheese, Tulum Cheese, Şavak Cheese, Turkey.

INTRODUCTION

Extending across the east of Turkey and covering 21% of the territory of the country, Eastern Anatolia Region is the area surrounded by the inner borders of

* **Corresponding author Asli Celikel:** Gastronomy and Culinary Art, Mardin Artuklu University, Tourism Faculty, Mardin, Turkey; Tel/Fax: +90 482 213 40 02-2010; E-mail: acelikel2@gmail.com

Celile Aylin Oluk & Oya Berkay Karaca (Eds.)

the North Anatolian Mountains in the north and south of the southeastern Taurus Mountains in the south.

There are Ağrı, Ardahan, Bingöl, Bitlis, Elazığ, Erzincan, Erzurum, Hakkari, Iğdır, Kars, Malatya, Muş, Tunceli, and Van provinces in the Eastern Anatolia Region Fig. (**1**). The continental climate prevails across the region and since the altitude of the region is high, summers are cool, winters are very cold and snowy (Çekel & Sert, 2013). The area is mountainous and the average altitude is 2000-2200 m. The lowland and mountainous structure of the Eastern Anatolia Region allow for the rich flora of the region and the widespread use of animal husbandry.

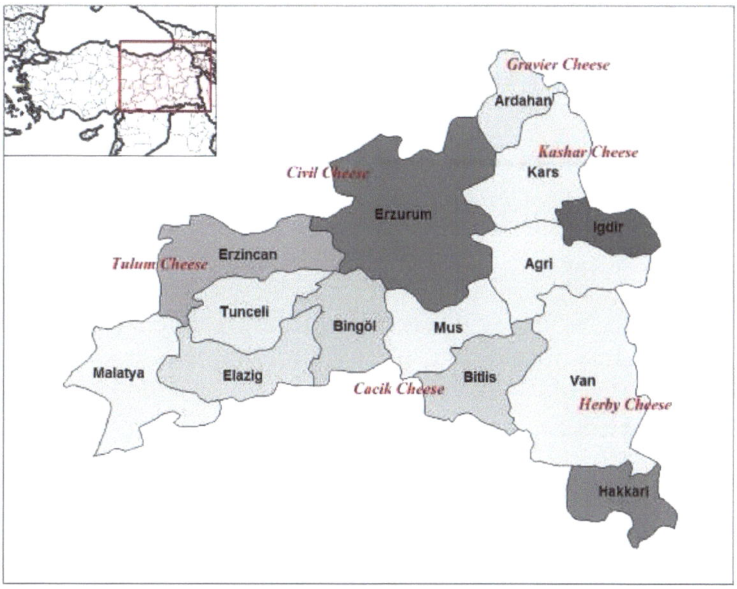

Fig. (1). Eastern Anatolia Region.

According to the Turkish Statistical Institute's 2017 data, about 1.4 million cattle, 9.5 million sheep and 1.5 million goats breeding is carried out in the region. In the same year, the amount of milk produced in the region was 3.5 million tons of cow's milk, 400 thousand tons of sheep milk, 83 thousand tons of goat milk (TUİK, 2017). It is stated that milk production in the North-Eastern Anatolia part of the region is higher than milk production in the Middle-East Anatolia part (Terin & Yavuz, 2015).

The production of milk and dairy products, which is an important part of our daily diet is common in Eastern Anatolia. There are about 19 varieties of cheese produced in the region (Kamber, 2008). Some of these are, Gravyer Cheese, Erzincan Tulum, Kars Kashar Cheese, Van Herby cheese, Civil cheese and Cacık

Cheese. Among these Erzincan Tulum Cheese, Kars Kashar Cheese and Van Herby cheese are among the geographical indicators of the region (Fig. **2**).

Fig. (2). Kars Kashar Cheese, Herby Cheese, Tulum Cheese and Gravyer Cheese.

Van Otlu (Herby) Cheese

Unique flavor and taste of herby cheese are closely related with the species and features of herbs used in production. Herbs, such as *Allium* sp., *Chaerophyllum macropodum Boiss*, *Prangos erulaceae (L.) Lindl.*, *Ferula rigidula DC*, thyme and wild mint are used to produce Herby Cheese (Table **1**). However, it is mandatory to use *Allium schoenoprasum L.* (chives), *Anhriscus nemorosa* (mendo), *Ferula orientalis L.* (ferula), *Mentha spicata* (wild mint), *Thymus migricus* (thyme) and *Ferula sp.* (siyabo) herbs in producing geographically marked Van Herby cheese. It is also stated that *Lepidium sativum L.* (cress), *Ranunculus polyanthemos L.* (buttercup), *Saponaria officinalis* (soapwort), *Anethum graveolens L.* (dill), *Dipsacus laciniatus L.* (teasel), *Menthae pulegium L. püjan* (pennyroyal), *F. Longipedunculata* (pesmen), *Thymus vulgaris L.* (garden thyme), *Ocimum basilium* (basil), *Radix eremuri* (asphodelus), *Allium fuscoviolaceum* (muscari), *Allium scorodoprasum L. subsp. rotundum (L.) stearn* (round headed leek) and *Rumex acetosella L.* (sheep's sorrel) can be used in addition to the above mentioned herbs (Hayaloğlu & Fox, 2008; TPE 2018). The herbs added to the cheese not only give taste and flavor, but also extend the storage duration of product thanks to their antibacterial and antioxidant properties (Durmaz, *et al.*, 2006). It is found in a study for antioxidant properties of some herbs used to produce herb cheese that *Thymus sp.* has the highest antioxidant properties, which is followed by Chaerophyllum sp., Allium sp., Prangos sp., and Ferula sp. (Çelik, *et al.*, 2008).

Table 1. Parts of herbs added to Van Herby Cheese with their Latin names.

Latin Names	Parts Used	References
Allium sp.	Leaf and stem	Çelik, *et al.*, 2008
Allium fuscoviolaceum (Liliaceae)	Leaf and stem	TPE, 2018
Allium scorodoprasum L.Subsp. rotundum (L.) stearn	Leaf and stem	TPE, 2018
Anethum graveolens L.	Leaf and stem	Ocak & Köse, 2015; TPE, 2018
Anthriscus sp.	Leaf and stem	Kavaz, *et al.*, 2013
Anhriscus nemorosa	Leaf and stem	TPE, 2018
Chaerophyllum macropodum Boiss	Leaf and stem	Ocak & Köse, 2015
Dipsacus laciniatus L.	Leaf and stem	TPE, 2018
Ferula Sp.	Leaf	TPE, 2018
Ferula rigidula DC	Leaf and stem	Şenel, *et al.*, 2012
Ferula orientalis L.	Leaf and stem	TPE, 2018
F. Longipedunculata	Flower	TPE, 2018
Lepidium sativum L.	Leaf and stem	TPE, 2018
Mentha spicata	Leaf	Ocak & Köse, 2015; TPE, 2018
Mentha pulegium L. Mint	Leaf	Çakmakçı, 2011; TPE, 2018
Radix eremuri	Leaf and stem	TPE, 2018
Ranunculus polyanthemos L.	Leaf and stem	TPE, 2018
Rumex acetosella L.	Leaf and stem	TPE, 2018
Saponaria officinalis	Leaf	TPE, 2018
Thymus migricus	Leaf	Ocak & Köse, 2015; TPE, 2018
Thymus vulgaris L.	Leaf and stem	TPE, 2018
Prangos ferulaceae (L.) Lindl.	Leaf and stem	Ocak & Köse, 2015
Ocimum basilium	Leaf	TPE, 2018

These herbs can either be used fresh or preserved in brine in order to be used later. Harvested herbs are chopped up (up to 0.5 - 1 cm). After boiling they are taken out without waiting for long, cooled and then put into brine. pH value of this brine, in which herbs are put, must be 3.85% (±8%), acidic value must be 1.06% (±7%) and salt ratio must be 5.8% (±12%) (TPE, 2018). During the production, these herbs are taken out of the brine and washed in order to remove extra salt. Some herbs are preprocessed differently before being preserved in brine (Üçüncü, 2008). Chives (*Allium Schoenoprasum L.*) can be used fresh after being chopped or being preserved in brine that contains salt between 3-8.8%. On the other hand, ferula (*Ferula Orientalis L.)* is kept in boiling water until it is mildly softened after which it is chopped up and then put into brine water (TPE, 2018).

Production

Van Herby Cheese, which is a semi-hard cheese, is similar to white cheese in terms of production method, however, the herbs added to the cheese create a taste and flavor different from white cheese. Van Herby Cheese is produced both through traditional and industrial methods in the region. It can be produced during the lactation period of animals in the houses whereas continuously in industries. Preservation of herbs in brine is a significant factor that facilitates industrial production of Van Herby Cheese. Although sheep's milk is preferred to produce the cheese, due to problems in the supply of sheep's milk in the region, cow's milk or goat's milk can also be mixed with sheep's milk (Yetişmeyen, 1997; Akyüz & Çoşkun, 1996).

Usually raw milk is used in the traditional method, however the milk to be used in the traditional production of cheese, which is produced through salinization method in preparing the brine and which will be consumed fresh, must be pasteurized (TPE, 2018). On the other hand, the milk used for industrial production is pasteurized for 30 minutes at 65ºC (Tunçtürk, *et al.*, 2014). Fermentation temperature of milk is 30ºC (TPE, 2018). Either commercial rennet or the rennet that is prepared locally is used in the fermentation of milk (TPE, 2018). Sheep's stomach, alum, black pepper, ginger, clove, powdered sugar and water are mixed in a cup when producing local rennet, and this mixture is kept for 2 weeks, during which it is stirred once in every 2 or 3 days, and then it is used for fermentation (Üçüncü, 2008; TPE, 2018). No additives or starter culture is used in the traditional production of Van Herby Cheese. However, lyophilized starter culture in a ratio of 1.5% and $CaCI_2$ in a ratio of 0.02% can be added to the milk in industrial production of herby cheese (Tunçtürk, *et al.*, 2014). Fermentation period is 1- hours after adding rennet to the milk (TPE, 2018). Then the coagulum is cut and 1-2% (w/w) herb mixture, which has already been prepared, is added and mixed (Sancak, *et al.*, 2015; Üçüncü, 2008). This ratio must not exceed 2% in the production of geographically marked Van Herby Cheese (TPE, 2018). Traditionally produced cheese is put into cloth sachets and pressed for 3-4 hours before it is shaped in the size of a hand with a thickness of 2-3 cm (TPE, 2018). However, industrially produced cheese is cut into blocks. Cheese is salted either through brine salting or dry salting (Tunçtürk, *et al.*, 2014; TPE, 2018). Brine salting method can be used both in traditional and industrial production. The concentration of brine salt must be between 18-21%, brine temperature must be between 8-16°C and pH value must be between 5.1-4.7 (TPE, 2018). $CaCl_2$ (0.15-0.55%) can also be added to the brine in order to prevent cheese to be softened (TPE, 2018). In brine salting method, cheeses are put into a tin or plastic containers, brine water is added to them and then they are kept in a cool place for 2-3 months. Dry salting method is preferred in traditional production. In dry salting, cheeses are initially kept as salted for a couple of days then they are placed within plastic drums or earthen pots in layers, combined with Cacik cheese, without leaving any gap (Çakmakçı, 2011; Çoşkun, 2005; Üçüncü, 2008; TPE, 2018). The opening of this drum or pot is either covered with a cloth or a wide leaf of grape is placed and then plastered with mud in order to be closed. Later, the cup is turned upside down and buried in a sandy and humid place in order to accelerate dehumidification of herby cheese. The cup is left under the ground for 4-7 months in order to be matured (TPE, 2018). Locals call this method of production of cheese "burial" since it is buried under the ground after dry salting (Çoşkun, 1998; Çoşkun, 2005; Ocak & Köse, 2015).

Physical Properties

It is semi-hard and has a small porous structure (TPE, 2018). Usually, it has chives or thyme taste and smell. The color of cheese may vary from white to yellowish, depending on the type of milk used, and when it is cut it may have a glassy, flat and bright structure (Emirmustafaoğlu & Çoşkun, 2012; Kamber, 2008; TPE, 2018). The appearance of Van Herby cheese is given in Fig. (**2**). The hardness value of Van Herby cheese varies depending on the production method. Tunçtürk, *et al.*, (2014) reported that hardness values of Van Herby cheese, manufactured through the traditional method (10.19 N) is lower than Van Herby cheese, manufactured through the industrial method (18.39 N).

Chemical Properties

It is stated in TS 13205 standard of Turkish Standards Institute that humidity of Herby Cheese must be maximum 60%, the amount of fat in dry matter must be minimum 45%, the amount of salt in dry matter must be maximum 7.5%, and the amount of aflatoxine M1 must be maximum 0.05 µg/kg and it is also mentioned that there must not be any natamycin, vegetable oil and starch in 5 mm. thick cheese, cut from the surface (TSE, 2016a). It is also stated that the amount of dry matter in geographically marked Van Herby cheese must be 43.81-47.78%, the protein must be 20.60-25.52%, the fat must be 16.75-19.21, the ash must be 5.07-7.45%, the salt must be 4.60-6.90%, and the calorie must be between 241-205 kcal/100g (TPE, 2018). Chemical properties of Van Herby cheese vary, depending on the type of milk, the amount of mixture of herbs, salting method and aging level used in the production. It is found in various studies that the amount of dry matter in Van Herby cheese is 55.41-46.29%, fat is 17.29-28.00%, salt is 4.24-9.01%, protein is 17.57-22.17%, ash is 5.25-8.11%, acidic value is 1.05-1.78, pH value is 4.50-5.10, degree of lipolysis (ADV-acid degree value) is

1.66-5.59, water soluble nitrogen (WSN) is 19.59-17.42%, trichloroacetic acid-soluble nitrogen (TCA-SN) is 11.46-15.01%, and phosphotungstic acid-soluble nitrogen (PTA-SN) is between 2.92-5.97% (Table **2**). One of the reasons for high nutritional value of cheese is its rich mineral content. The herbs added during the production of Van Herby cheese enrich the cheese in terms of minerals and Vitamin C (Çakmakçı, 2011). It is reported in the studies that the amount of calcium (Ca) in 100 g of Van Herby cheese is between 678-442.9 mg, phosphorus (P) is 416.00-552.60 mg, potassium (K) is 116.10-180.0 mg, sodium (Na) is 1103-2606 mg, and magnesium (Mg) is 12.70-55.40 mg and there is 11.20-41.79 mg iron (Fe), 15.65-33-99 mg zinc (Zn), 1.25-6.25 mg copper (Cu), and 1.03-2.05 mg manganese (Mn) in 1 kg of Van Herby cheese (Table **3**). It is also reported that amount of calcium (Ca) in geographically marked Van Herby cheese is between 352-394 mg 100 g^{-1}, the amount of phosphorus (P) is 395-433 mg 100 g^{-1}, sodium (Na) is 1057-1149 mg 100 g^{-1}, potassium (K) is 173-189mg 100 g^{-1}, and magnesium (Mg) is 32,10-34.30 mg 100 g^{-1} (TPE, 2018; Tarakçı & Küçüköner, 2008).

Table 2. Chemical properties of Van Herby cheese.

Dry Matter (%)	Fat (%)	Salt (%)	Protein (%)	Ash (%)	Acidity (%la)	pH	ADV	WSN (%)	TCA (%)	PTA (%)	Reference
46.29	22.50	4.56	17.57	5.25	1.18	5.10		19.59	14.48	2.92	Tunçtürk *et al.*, 2014
50.73	28.00	4.24	20.87		1.05	4.52	1.66	17.42	15.01	3.82	Emirmustafaoğlu & Coşkun, 2012
53.00	23.50	-	18.70	7.90	1.78	4.50					Özrenk *et al.*, 2009
52.43	21.65	6.54	20.78	8.11	-	4.52					Tarakçı &Küçüköner, 2008
55.41	24.37	6.64	21.22	-	1.84	4.55	3.92		11.46	5.97	Tarakçı *et al.* 2004
54.3	23.73	9.01	20.51-	-	-	4.6	5.59				Andınç *et al.* 2010.
46.78	17.29	5.73	22.17	6.85	-	-	-	-	-	-	TPE 2018

Microbiological Properties

Herby Cheese of 25 g must not contain *Salmonella* and *Listeria monocytogenes* and the number of coagulase positive Staphylococcus must be less than 10^3 according to the TS 13205 standard of Turkish Standards Institute (TSE, 2016a). It was found in a study, conducted by Şenel *et al.* (2012) that *Enterobacteriaceae* spp. is 2.48 log CFU g^{-1}, Lactic acid bacteria (LAB) is 4.95 log CFU $g^{-1}$and total aerobic mezophlic bacteria (TAMB) is 6.30 log CFU g^{-1}.

Table 3. Mineral contents of Van Herby cheese.

Ca (mg $100g^{-1}$)	P (mg $100g^{-1}$)	K (mg $100g^{-1}$)	Na (mg $100g^{-1}$)	Mg (mg $100g^{-1}$)	Fe (mg kg^{-1})	Zn (mg kg^{-1})	Fr (mg kg^{-1})	Mn (mg kg^{-1})	Reference
678.00	416.00	180.00	1103	33.00	-	-	-	-	Demirci, 1998
313.70	552.60	-	2606	12.70	41.79	33.99	6.25	2.05	Tarakçı & Küçüköner, 2008
442.9	-	116.1	-	55.4	11.20	15.65	1.06	1.03	Ocak & Köse, 2015
378	416	180	1103	33.44	-	-	-	-	TPE, 2018

Cacik Cheese

Cacik Cheese, which is usually produced in the provinces of Van, Hakkari and Bitlis of Eastern Anatolian Region, is also called as "jaji" or herby cottage cheese. It is used in the traditional production of Van Herby cheese and is also consumed fresh during the breakfasts. Cacik cheese is traditionally produced in spring and summer. The cheese that is produced in these seasons is not only consumed fresh but also dried under the sun in the form of pressed balls and then stored to be consumed in winter (Küçüköner & Tarakçı, 1998; Kamber, 2008). Similar to Van Herby cheese, various herbs and some additives are also used in producing Cacik Cheese. Some of these herbs and additives are *Allium spp. Chaerophyllum macropodum, Ferula spp. Prangos spp. Silene vulgaris. Thymus spp. and Mentha spp.* as well as cucumber and green pepper (Küçüköner, *et al.*, 2006).

Production

Although sheep's milk is preferred to produce Cacik Cheese traditionally, it is also possible to use cow's milk. The first step to produce Cacik Cheese is to produce yogurt from milk. The filtered milk is boiled for 20 minutes to ensure that dry matter content is increased. Then the milk is cooled until 43±2°C and one glass of yogurt (1-4%) is added for each 20L of milk for fermentation purposes. Afterwards, the milk must be left for fermentation for 2-3 hours until pH value reaches 4.5. This yogurt is churned and buttermilk is separated. The buttermilk is boiled for 10-15 mins and left to stand. Then, the buttermilk is filtered through cloth bags and kept pressed for 2 days (Çakmakçı, 2011, Kurt & Akyüz, 1984; Çoşkun, 2005; Küçüköner, *et al.*, 2006; Hayaloğlu, 2008; TPE, 2018). The cottage cheese that is produced is removed from its bags and salted and 1-2% herb, which was prepared previously, is added (Üçüncü, 2008). It is stated that the amount of herb must be between 0.84-11.28% in Cacik Cheese, which will be used to produce Van Herby cheese (TPE, 2018). Cacik Cheese is then put into a cloth bag again to be filtered. When the desired thickness is achieved, it is either consumed fresh or dried under the sun to be consumed later and stored in a cool and dry place inside cloth bags after drying (Çakmakçı, 2011; Üçüncü, 2008; Hayaloğlu, 2008).

Physical Properties

Cacik Cheese is white colored and it is thick and spreadable when consumed fresh. Herb mixtures added to the cheese are apparently visible. Its smell and flavor vary, depending on the type and amount of herbs added (Kamber, 2008).

Chemical Properties

Chemical properties of Cacik Cheese, which is produced through the traditional method, vary due to lack of a standardized production method. It is stated that dry matter of Cacik Cheese, which will be used to produce geographically marked Van Herby cheese, must be 22.07%, fat must be 2.69%, protein must be 14.51%, salt must be 1.97%, and acidity must be 1.93% (TPE, 2018). Kücükoner *et al.* (2006) reported that the dry matter must be between 16.45-20.76%, fat must be between 1.45-4.30%, protein must be between 8.14-13.87%, ash must be between 0.96-4.19% and salt must be between 0.25-3.15% whereas acidic value (LA) must be between 0.50-0.89%, pH value must be between 3.08-4.22%, aging index must be between 4.37-11.64%, and non-protein nitrogen must be between 1.61-7.14% within Cacik Cheese. Herbs added to Cacik Cheese affect the mineral substance content, similar to Van Herby cheese. Tarakçı and Küçüköner (2008) reported that the amount of calcium (Ca) is 469.0 mg $100g^{-1}$, phosphorus is 554.5 mg 100 g^{-1}, sodium (Na) is 2830 mg $100g^{-1}$, Magnesium (Mg) is 24.76 mg 100 g^{-1}, iron (Fe) is 52.52 mg kg^{-1}, zinc (Zn) is 34.42 mg kg^{-1}, copper (Cu) is 8.95 mg kg^{-1}, manganese (Mn) is 2.93 mg kg^{-1}, chrome (Cr) is 0.26 mg kg^{-1}, cobalt (Co) is 0.29 mg kg^{-1}, cadmium (Cd) is 0.13 mg kg^{-1}, and nickel is 0.23 mg kg^{-1} within Cacik Cheese.

Microbiological Properties

It is stated in Turkish Food Codex Microbiological Criteria Regulation that the number of coagulase positive Staphylococcus must be less than 10^3 and there must not be any *Salmonella* and *L. monocytogenes* within the samples of Cacik Cheese (TGK, 2011). In a study, in which microbiological properties of Cacik Cheese were researched, it is reported that the total number of total aerobic mezophlic bacteria is 7.44 log CFU g^{-1}, the number of Coliform bacteria is 2.05 log CFU g^{-1}, the number of *E. coli* is 0.14 log CFU g^{-1}, the number of *S. aureus* is 1.35 log CFU g^{-1}, and the number of yeast-mould is 7.37 log CFU g^{-1} (Güneş & Sancak, 2014). Kücükoner *et al.* (2006) found that the total number of total aerobic mesophilic bacteria within Cacik Cheese is 2.63-4.82 log CFU g^{-1}, the number of psychrotrophs is 1.15-3.20 log CFU g^{-1}, the number of lipolitic bacteria is <1-3.20 log CFU g^{-1}, the number of proteolytic organisms is <1-1.54 log CFU g^{-1}, the number of lactobacilli is 2.50-4.26 log CFU g^{-1}, the number of lactococci is 2.14-4.40 log CFU g^{-1}, and the number of mould-yeast is between 2.10-5.40 log CFU g^{-1}.

Gravyer Cheese

Gravyer Cheese is generally considered in the same category as Emmental (Switzerland), Gruyere (France), Fontina (Italy), Gouda and Edam (Netherlands) and Samso (Denmark) (Kamber, 2008). The main origin of Gruyere Cheese is Europe. The development of animal husbandry in Ardahan-Kars Plateau was encouraged in 1850-1855 by the Russian government due to the similarity of the Ardahan-Kars Plateau to the Swiss Alps (Kamber, 2008; Arınç, 2018). The first Gruyere Cheese production facility was established in the region when David Moser, a Swiss cheese producer, found out that the Gruyére Cheese can be produced in the Ardahan-Kars region. There were 32 enterprises producing Gravyer Cheese in Kars and Ardahan in 1905 and they have continued their activities for about 10-15 years. After the Union of Soviet Socialist Republics was founded in 1917, Turkish tribes such as Karapapak (Terekeme), Kıpcak, Komuk and Karakalpak, who did not want to live there and knew how to produce Gravyer Cheese settled in Ardahan-Kars in 1936 and started to produce Gravyer Cheese in this region (Arınç, 2018). Today, this type of cheese is widely produced in the Eastern Anatolia region in Kars and Ardahan.

Production

Cow milk is used in Gravyer Cheese production (Fig. **2**). As raw milk is used in production, some pasteurization process can be applied to milk by some producers as well (TSE, 2016b). The milk to be used in cheese production is standardized in such a way that the amount of fat in dry matter is 45% and acidity is 7-8 SH° (Soxhlet-Henkel). The milk is transferred to a 1 ton capacity container and heated to 37 °C. There should be at least 200-250 L of standardized milk for production. Rennet is added by the proportion of 10 mL /100kg of milk for clotting (Ulutaş, *et al.*, 1993; Kamber, 2008; Arınç, 2018). In this process, some grated Gravyer cheese is added into the milk as a culture (Kamber, 2008). The clotting process takes approximately 45 minutes. The coagulum is cut (0.4-0.8 cm) in the clotting vessel and heated to 55°C with whey. After the curd gains a hard structure, the heating process is terminated and rested for 10-15 minutes (Ulutaş, *et al.*, 1993; Kamber, 2008; Arınç, 2018). Whether the curd has ideal consistency is understood by looking if the curd clings to the palm while squeezing.

A portion of the whey in the curd is filtered through the filter cloths and the clot is molded. The molds are pressed from 500 kg to 3 tons. The pressure process takes about 24 hours, during which the molds are inverted 4-5 times and the cloth is changed.

The outer surface of the cheese removed from the pressure is dry-salted for 4-5 days. The blocks are then stored in ponds containing 20-30% NaCl brine. The blocks are transferred to rooms that have 85-90% relative humidity and 28-32 °C for two weeks and dry-salted for 4-5 days (Ulutaş, 1993; Kamber, 2008; Arınç, 2018). This process takes about 1 month for some producers. Cheese is turned upside down to distribute the heat evenly and to get air. In this process, the cheese begins to mature and pores are beginning to form inside. As a result of the degradation of lactose during ripening, CO_2 accumulating in cheese gives the porous structure to the cheese. Following the gain of the porous structure of the cheese, the cheese blocks are stored for 90-150 days in storages at 10–13 °C (Kamber, 2008; Arınç, 2018).

Physical Properties

Gravyer Cheese has a hard rind and the rind has a light brown color. Cheese blocks range from 40 to 90 cm in diameter and 15 to 45 kg in weight. The color of the cross-sectional surface of the cheese varies from ivory to light yellow. There are pores on the surface of the cross-section, that are 1-2 cm in size and have similar distribution in the cheese structure (Kamber, 2008; TSE, 2016b).

Chemical Properties

According to the Turkish Standards Institute's TS 2174-Gravyer Cheese Standart, the moisture content of the cheese should be at most 40%, the amount of fat in the dry matter should be at least 45% and the amount of salt in the dry matter should be at the most 7.5% (TSE, 2016b).

Kamber *et al.* (2008) found out that the moisture content of Gravyer Cheese as 19.8%, the amount of fat is 32.4%, the protein content is 31.3%, the salt amount is 11.0%, the ash content is 5.4% and acidity value is 1.54% LA. It was reported that the dry matter amount of Kars Gravyer Cheese is 69.88%, the amount of fat is 33.65%, the protein content is 28.89%, the salt amount is 1.03%, the ash content is 4.98% and the pH value is 6.02 (Topuk & Sezer, 2015).

Microbiological Properties

According to the TS 2174 standard of the Turkish Standards Institution the number of coagulase positive staphylococci in 25 gr Gravyer Cheese should be below 10^3 and that *Salmonella* and *L. monocytogenes* should not be present (TSE, 2016b). Topuk and Sezer (2015) reported that the total number of aerobic mesophilic bacteria in Kars Gravyer Cheese was 7.75 log CFU g^{-1}, Staphylococcus-micrococcus 5.48 log CFU g^{-1}, coliform bacteria 1.47 log CFU g^{-1}, Enterobacteriaceae 1.39 log CFU g^{-1}, Lactic Acid Bacteria was 6.67 log CFU g^{-1}, and yeast-moud number was 2.10 log CFU g^{-1}. Turgut *et al.* (2012) reported that the ratio of lactose strains isolated from Kars Gravyer Cheese were 25% *L. brevis* type 1, 18.3% *Pediococcus pentosaceus* type 1, 14.1% *Lactobacillus paracasei ssp paracasei,* 13.3% *Lactobacillus fermentum* type 1, 12.5% *Lactococcus lactis ssp lactis* type 1, 7.5% *Lactobacillus delbrueckil ssp lactis* type 1,5% *Lactobacillus plantarum* type 1 and 4.16% *Lactobacillus rhamnosus*.

Erzincan Şavak Tulum Cheese

The most known among tulum cheeses, which are produced as dry and brined in Turkey, is Erzincan Tulum cheese that is produced as dry type (Figs. **2** & **3**). This cheese, which is produced in Erzincan, Erzurum, Tunceli, Bingöl, and Elazığ provinces of Eastern Anatolia Region, is the most produced cheese in Turkey after White brined cheese and Kashar cheese (Tekinşen & Akar, 2017; Çakmakçı, 2011; Üçüncü, 2008; Kamber, 2008). The word "tulum" in the name of this cheese means goat-skin or sheepskin, which is used for covering the cheese during the production and/or ripening stage.

Fig. (3). Tulum Cheese(sheepskin-bags plastic drum).

Erzincan Tulum cheese is mostly produced by the members of Savak tribe, who deal with sheep and goat breeding and live in their villages in winter and in their large nomad tents on plateaus during the spring and summer months. The cheese is known as Erzincan-Savak Tulum cheese (Akyüz, 1981; Çakmakçı, 2011). The cheese is prepared by fermenting the raw milk of Akkaraman sheep with the home made rennet, which was prepared from calves' stomach, and then adding salt before being ripened in goat-skin or sheep-skin bags and being served for consumption. Sometimes, the milk of goats, which join the herd to lead the sheep, can also be added to the milks used to produce the cheese. The cheese that is produced in high plateaus between May and September is either processed by the animal owners or the production process is completed at dairy farms, which are established at locations close to plateaus and which buy the raw cheese from these owners (Kurt & Öztek; 1984). A Geographical Marking Registration Certificate was obtained for Erzincan Tulum Cheese in 2000, in which the production area was determined as the plateaus of Munzur, Çayırlı, Tercan, Kemah, and Oluk (TPE, 2000). Sensorial, microbiological and chemical properties of Tulum cheese are given in TS 3001 and Cheese Communiqué of Turkish Food Codex (TSE, 2016c; TGK, 2015).

Production

Production method of the cheese is given in Geographical Marking Registration Certificate and in the publications of various researchers (Kurt & Öztek, 1984, Kamber, 2008; Akyüz,1981; Çakmakcı, 2011; Üçüncü, 2008; Tekinşen & Akar, 2017). A starter culture is not usually used in production, but natural microflora is used for fermentation (Öksüztepe, *et al*., 2005).

The milk is taken around 9-10 am in the morning and before the night. The milk that is collected in copper buckets are filtered twice either by using cloth or metal filters, which has started to be used a lot recently, and then poured into boilers that will be used to produce the cheese. The milk's temperature is controlled by the woman, who makes fermentation, and particularly if the milk is hot, it is kept for a while until it reaches to a suitable temperature. It is stated that the fermentation temperature must be between 26-32°C (Kurt & Öztek, 1984; Tekinşen & Akar, 2017).

Homemade rennet is used for coagulation of milk. Rennet is produced by adding salt to calf stomach, which is cleaned and dried, before keeping it within whey. The amount of salt, added to whey, is determined by rule of thumb. Çakmakcı *et al*. (2007) state that this amount is approximately 10%. This mixture is stirred every so often and rennet is broken into pieces during this process. The mixture, the fermentation power of which is controlled, is filtered and used as rennet and this may take some time, ranging from a couple of days to 1-2 weeks.

Then the cheese is put into triangular cloth sachets in order to be filtered from hairs in 1.5 - 2 kg portions and in order to filter coagulation water 1-4 hours after the coagulation (Akyüz, 1981; Kurt & Öztek, 1984; TPE, 2000). The same process is applied to the milk that is taken before the night. Filtering takes approximately 14-16 hours. The curd that is obtained by this way is put into sacks made of gray cotton cloth, which is also known as American cloth, and transported to dairy farms, which are used jointly in plateaus, or to the tents reserved for this process. Filtering of cheese within sacks, which are placed on top of each other and coated to prevent penetration of dust, takes approximately 1 week or 10 days and the sacks are sometimes replaced in order to ensure that all cheeses are filtered. Stones or plastic pallets are placed under the cheese sacks to ensure that the filtered water flows away.

After filtering, the cheese is emptied into wide wooden vessels or, chrome steel vessels and then crushed into small pieces by being trampled by a person, who wears a pair of boots, and then salted with mineral salt or, preferably, Kemah salt (a salt that is obtained through natural methods) in a ratio of 2 - 2.5% (TPE, 2000). The amount of salt is determined by tasting the cheese. The cheese is placed within clean sacks in a manner so as not to leave any air and stitched sacks are piled up and sometimes replaced to ensure that the water is completely filtered from all sacks. After resting for 1 week or 10 days, the cheese is poured again to the vessel, during this processed cheese is tasted and salt can be added again if necessary. The cheese that is cut into pieces smaller than a chickpea is then put into various sizes of goatskin or sheepskin bags, ranging from 2 to 40 kg in weight, or into plastic drums either by hand or by filling machine in dairy farms and then kept in caves at the plateaus or, mostly, in cold storages.

If plastic drums are used, 3-5 holes are opened on the drum by using a packing needle. It is stated in Cheese Communiqué of Turkish Food Codex that bags, made of the skin of sheep, goat, and calf, can be used in producing Tulum Cheese (TGK, 2015). Due to their porous structure, these bags have air and water permeability. The cheese is placed into these bags, which are made of sheepskin and, since it is more durable, goatskin and which are prepared either as a whole bag or divided into two (Hayaloğlu *et al.*, 2007). The cheese that is ripened for a period of three months is then served for consumption (TPE, 2000). The curd that occurs during the production of cheese can be sold to dairy farms either at this stage or later stage and grinding and filling operations are performed by machines (Tekinşen & Akar, 2017).

Physical Properties

Erzincan - Savak Tulum cheese is a semi-hard cheese with rich fat content and it has a buttery flavor, its color is white or light cream, it is not porous, it has a uniquely apparent acidic taste and flavor and it can easily be dissolved in the mouth (Çakmakçı *et al.*, 2008; Hayaloğlu *et al*; 2007; Ücüncü, 2008; Çakmakçı *et al.*, 2011).

Chemical Properties

It is stated in TS 3001 Tulum Cheese standard (TSE, 2016c) that the cheeses will be classified as full fat, semi-fat, low fat and non-fat cheese based on their fat content, which respectively has to contain 45≤ fat, 25≤ fat<45, 10≤ fat <25, 10%> fat (m/m), titrable acidity must be maximum 3% in terms of lactic acid, the amount of salt in dry matter must be maximum 5% and humidity level must be maximum of 45%, they must not contain any vegetable oil and starch, there must not be any natamycin in the cheese, the surface of which is cut by 5 mm and Aflatoxine M1 must be maximum 0.05 μg/kg. It is stated in Cheese Communiqué of Turkish Food Codex (TGK, 2015) that humidity must be maximum 45% and the amount of fat and salt in dry matter must have the same values as required by the applicable standard.

Since Erzincan Tulum-Savak cheese is mostly produced in plateaus and, after a certain stage, in dairy farms, the cheese's properties may vary, depending on the production technique, the person that produces the cheese, quality of the milk and rennet that was used, packaging material, ripening duration as well as storage conditions. The results of studies, which were conducted on chemical quality characteristics of the cheese, are given in Table **4**.

It is found in the studies (Table **4**) that, averagely, the humidity ratio of cheese varies between 37.29-42.59%, dry matter amount varies between 53.3-62.71%, far amount varies between 28.2-34.96%, the fat ratio in dry matter varies between 49.15-52.8%, protein amount varies between 18.5-21.54%, total ash amount varies between 3.37-5.50%, acidity varies between 0.25-1.83, salt amount varies between 2.84-4.66%, the salt amount in dry matter varies between 4.62-8.02, and pH value varies between 4.87-5.153.

Erkan *et al.*, 2018 reported that they found Aflatoxine M1 between 0.64-4.32 $\mu g\ kg^{-1}$ in all samples, taken from 100 Erzincan-Savak Tulum cheese.

Since different covers are used for Erzincan-Savak Tulum cheese, when a study was conducted to research the effect of plastic drums or goatskin bags on proteolysis during the ripening period, it was found that the proteolysis in goatskin bags is higher due to pores in such bags (Çakmakçı, 2011). Similarly, Hayaloğlu, *et al.* (2007), who observed the effect of bags made of skin and plastic drums on the composition of cheese throughout 150 days of ripening period, stated that sharp decrease in the humidity content of cheese due to porous structure of bags made of skin causes a significant difference in the chemical structure, the cheeses that are ripened in plastic drums have more microbiological load, however there is no difference in terms of nitrogenous compounds and total free amino acid amount, flavor patterns are similar and 100 volatile compounds, composed of 11 acids, 16 esters, 12 methyl ketones, 7 aldehydes, 22 alcohols, 7 sulfur compounds, 6 terpenes, and 19 miscellaneous compounds, have been found. While Tarakcı and Durmuş, (2016) who researched the effect of different cheese covers on the composition of cheese, reported that there are differences in the composition of cheese depending on these covers, Aslaner and Bakırcı, (2016) reported that different covers do not have an effect on free fat content.

Table 4. Chemical Properties of Erzincan Tulum-Savak Cheese.

Moisture Content (%)	Dry Matter (%)	Fat (%)	Fat in Dry Matter (%)	Protein (%)	Ash (%)	Acidity (% LA)	Salt (%)	Salt in Dry Matter (%)	pH	References
37.29	62.71	34.96		21.54	5.50	1.66	4.66	7.27	-	Akyüz, 1981
-	54.94	27.35	49.15	20.76	3.37	1.80	3.29	7.19	5.10	Erdem &Patır, 2017
-	53.2	28.2	52.8	18.5	4.7	1.83	3.4	-	-	Kurt *et al.*, 1991
42.59	-	-	-	-	-	0.25	-	-	5.03	Erkan, Demir & Öksüztepe, 2018
-	61.54	-	57.62	21.06	3.92	0.95	2.84	4.62	4.87	*Çakır & Çakmakcı, 2018
42.51	57.49	-	-	-	-	1.09	4.59	8.02	4.92	Patır *et al.*, 2000
-	59.20	31.06	52.47	-	4.96	-	3.95	6.67	5.15	Arslaner & Bakırcı, 2016

*Taken on the 90th day of ripening.

In a study, conducted by Çakır *et al.*, (2016) it is stated that addition of black cumin has positive effects on cheese proteolysis and does increase the volatile compounds in the cheese both quantitatively and qualitatively, 94 volatile compounds that have been identified are from different chemical groups and 31 of them are esters, 14 are alcohols, 12 are terpenes, 10 are carboxylic acid, and 10 are miscellaneous compounds, whereas 8 are methyl ketones, 7 are aldehydes and 2 are sulfur compounds. Çakır and Çakmakcı, (2018) who researched the effects of black cumin on chemical and proteolysis properties of Erzincan Tulum cheese, reported that black cumin causes an increase in the concentrations of dry matter, protein, ash, total nitrogen (TN), water soluble nitrogen (WSN) (% of TN) (ripening index), 5% phosphotungstic acid soluble nitrogen (PTA-SN) (% of TN), and 12% trichloroacetic acid soluble nitrogen (TCASN) (% of TN) and causes a decrease in fat-in-dry matter (FDM), pH, and acidity.

It is reported in a study, conducted by Erdem and Patır (2017) to research ripening levels and histamine amounts, based on some nitrogen fractions, that ripening index was calculated as 24.58% and histamine amount was 70.65 mg 100 g^{-1} in average.

Microbiological Properties

It is stated in Cheese Communiqué of Turkish Food Codex (TGK, 2015) that there must not be any *Salmonella* and *L.monocytogenes* in 25 g of cheese and the number of coagulase positive Staphylococcus must be less than 10^3.

The values found for total aerobic microorganisms, coliform bacteria, staphylococcus-micrococcus, faecal streptococci, moulds and yeast in the studies, conducted by various researchers, are given in Table **5**.

Table 5. Microbiological properties of Erzincan Tulum-Savak Cheese.

Total Aerobic Microorganism	Coliform Bacteria	Staphylococcus-Micrococcus	Fekal Streptococci	Mould and Yeast	References
$2.1x10^9$	$3.2x10^6$	-	-	$1.8x10^6$	**Kurt, 1991
7.1×10^7	9.8×10^6	4.4×10^5	7.9×10^6	2.9×10^5	**Patır *et al.*, 2000
6.85	4.49	3.81	2.19	5.06	*Demir *et al.*,2018
8.89	3.20	-	-	5.18/5.49	**Demir *et al.*, 2017
7.31	2.95	-	-	4.90	**Çakır & Çakmakçı, 2018
-	-	-	-	5.12	*Erkan *et al.*, 2018

*The values are given as log CFU g^{-1} and
** The values are given as CFU g^{-1}

Various researchers have conducted studies to determine the bacteria, which play a role in the fermentation of cheese. Öksüztepe *et al.*, (2005) reported that while lactic streptococci were dominant in the first month of ripening of Erzincan - Savak Tulum cheese, but lactobacillus were dominant in later months. They also reported that *Lactobacillus casei* subsp. *casei*, *Lactobacillus plantarum, Lactococcus lactis* subsp. *cremoris*, *Lactococcus lactis* subsp. *lactis*, and

Leuconostoc mesenteroides subsp. *cremoris* were found among the isolates. In another study, conducted by Patır *et al.*, (2000) the number of lactic streptococci was found as 3.2 x 10^7 g^{-1} and the number of lactobacillus-leuconosto--pediococcus was found as 5.6 x 10^7 g^{-1}, 30.2% of 587 isolated strains were composed of *Streptococcus cremoris*, 21.6% were composed of *Lactobacillus casei* and its subspecies, 9.9% were composed of *Lactobacillus plantarum*, 9.5% were composed of *Streptococcus lactis,* 4.6% were composed of *Lactobacillus curvatus*, 4.9% were composed of species of leuconostoc, and 5.1% were composed of streptococci of enterococcus group and it was concluded that *Streptococcus cremoris, Streptococcus lactis, Lactobacillus casei* and its subspecies, *Lactobacillus plantarum, Lactobacillus curvatus, Leuconostoc cremoris, Enterococcus faecium* and *Streptococcus faecalis* and its subspecies may play an important role in ripening. Çakmakcı *et al.*, (2008) who researched the effect of plastic drums and skin bags, which are used as cheese packages, on the growth of lactic acid bacteria, reported that both types of packages have a similar effect.

Production of cheese from raw milk imposes a significant risk in terms of the existence of food sourced pathogens. In a study, conducted by Azak *et al.*, (2012) to research the existence of *Listeria monocytogenes* in Erzincan-Savak Tulum cheeses, it was found that this bacteria is available in 3% of the analyzed samples.

Demir *et al.*, (2017) reported that vacuumed packaging extends the shelf life of Savak Tulum cheese depending on the amount of potassium sorbate used, but potassium sorbate has an inhibiting effect on *Echerichia coli* and *Staphylococcus aureus.* It is also reported by Tomar *et al.*, (2018) that despite an increase in acidity values in 90 days ripening period of cheese samples, which were produced by adding probiotic bacteria cultures to pasteurized sheep milk, there was a decrease in pH and a_w values, despite an increase in the numbers of TAMB, yeast-mould, lactic acid, lipolytic bacteria, proteolytic bacteria, and *Lactococcus* spp. and *Pseudomonas* spp. type bacteria, a decrease was found in the numbers of total coliforms and Enterobacteriaceae, the number of probiotic cultures, which were added during the production of cheese, has decreased throughout the storage period, and no *E. coli, S. aureus, Salmonella* spp. *Listeria* spp, *Brucella* spp. was found in the cheeses.

Civil Cheese

Civil cheese is a regional cheese variety, that is usually produced from skimmed cow and/or sheep milk (Tekinşen *et al.*, 1996). The geographical boundaries for production of "Civil cheese", which has received Geographical Indication Registration Certificate as "Erzurum Civil cheese" in 2009, were determined as the territory of Erzurum and its districts (TPE, 2009). In addition to Erzurum, Civil cheese is also known, produced and consumed in provinces such as Kars, Muş, Van and Ağrı (Çakmakçı, 2011). This cheese is produced by families, who have their own animals, in the villages or in small sized dairy farms. Only milk, rennet and salt are used to produce this cheese. The cheese is produced from non-pasteurized milk without using any starter culture. There is no production quality standard determined for Civil cheese (Cambaztepe *et al.*, 2009). The cheese is produced by adding rennet after the milk is soured, heating the milk, mixing and kneading the coagulation that was formed, hanging the cheese on slings and forming thin fibers within the mass. The cheese is in the form of four fingers wide molds, its fiber structure is filament like and its fibers do not break (TPE, 2009). Civil cheese is consumed as fresh, brined civil, pressed Civil and pressed civil+ whey (Erzurum Mould Civil cheese or Green Cheese) (Fig. **4**).(TPE, 2009). Green cheese is a regional cheese variety, that is produced after pressing Lor cheese, produced from whey, with Civil cheese into containers and ensuring that they are molded. Civil cheese is a good alternative for the persons, who prefer non-fat foods in their diet (Çakmakçı, 2011).

(a) (b)

Fig. (4). a) Green cheese, **b**) Civil cheese.

Production

There are publications, prepared by various researchers regarding the production method of this cheese (Tekinşen *et al.*, 1996; Çakmakçı, 2011). Furthermore, the production method of this cheese is provided in the document, which forms the basis of Erzurum Civil cheese Geographical Indication Registration Certificate (TPE, 2009).

The milk of cow or sheep or their mixtures, which will be used to produce Civil cheese, are filtered through a separator in order to skim their fat. Whey is added to increase the acidity of skimmed milk or the milk is stored at a cool place in until its acidity is raised. When the milk's acidity reaches 22 Soxhlet-Henkel (^{0}SH), it is considered that suitable acidity level is reached to produce Civil cheese (TPE, 2009). The duration needed to reach above-mentioned acidity level varies, depending on the microorganism load of the milk to be used and hygienic conditions of the person and the equipment used to milk the animal and this duration varies between 12-18 hours. Some cheese producers also determine the sufficiency of the acidity level of milk by tasting it. The milk, acidity of which reaches to the sufficient level, is heated up to 30 °C approximately 1 ml of rennet is added to 10-15 liters of milk (Tekinşen *et al.*, 1996). The milk is continued to be heated and stirred and when the temperature reaches 52-53 °C, coagulation pieces begin to occur. When producing Civil cheese, casein is coagulated as a result the rennet and whey proteins are coagulated due to heat processing (Cakmakçı, 2011). It is ensured by stirring with a scoop that these pieces of coagulation are gathered around the scoop. This process continues until the temperature reaches 65-70 °C (TPE, 2009). The coagulum that is taken out of whey as wrapped around the scoop is kneaded within a container and then hanged on a sling so that it can elongate freely. The elongated parts are then moved up to the sling and this process continues until elongation stops (Tekinşen *et al.*, 1996). Cheeses, which have a fibrous structure and elongation of which has stopped, are removed from the sling, placed on a clean surface and its water content is waited to be flown away for approximately one day. Then they are ripened through dry salting, preserving in brine or vacuumed packaging or by mixing with 20-30% fat free whey cheese (TPE, 2009). Cheeses that were ripened for 3 months or a longer period in the refrigerator or a cooler place are ready for consumption. Moulds, which have naturally spread around the Civil cheese during the ripening process by salting the mixture of Civil cheese and whey cheese within goatskin bags or plastic drums, do reproduce on the surface and inner parts of the cheese and add their own taste to the cheese (Çakmakçı *et al.*, 2012). This cheese, which is called Green cheese as the cheese's color was transformed into a blue-green color, is highly popular among local people.

Chemical Properties

The results of some studies, which were conducted to identify the chemical properties of the cheese, are given in Table **6**. It is found in these studies that the range of dry matter in Civil cheese is 35.19 - 52.49%, the fat amount is 1.79 - 15.64%, fat amount within dry matter is 4.05 - 6.50%, and protein amount is 28.38 - 32.40%.

Table 6. Chemical properties of Civil cheese.

Dry Matter (%)	Fat (%)	Fat in Dry Matter (%)	Protein (%)	Acidity (%LA)	Salt (%)	Salt in Dry Matter (%)	pH	References
35.19	2.30	6.50	32.40	2.39			4.53	Yıldız *et al.*,2010
43.66	1.79	4.05	34.03		4.64	10.71	5.90	Hayaloğlu & Karabulut, 2013
52.49	15.64		28.38					Kamber, 2005
36.06	2.07	5.81		1.25	0.19	0.55		Şengül & Gürses, 2006

It is stated that since Civil cheese was not produced generally under controlled conditions, chemical and biochemical analyses of the samples produce quite different results (Şengül *et al.* 2006). Similarly, in a study conducted by Cambaztepe *et al.* (2009), it is reported that significant differences were found in the chemical compositions of cheeses in chemical analyses made on the 2nd, 30th, 60th, and 90th day of ripening process of Civil cheese, which is made by using aqueous brining, dry salting, combining with Lor (whey) cheese and vacuum packaging methods. It is found in this study that, ripening of the cheese by using different methods is effective on chemical, microbiological and sensorial properties of the cheese. It is also stated in the same study that 90 days ripened cheese samples, which were vacuum packaged and dry salted, are liked by panelists much more than the other samples. Average humidity of 41 moulded Civil cheeses, which were analyzed by (Çakmakcı *et al.*, 2012), was found to be 51.8%, their pH was found to be 6.5, total protein to be 36.95%, fat to be 3.45%, salt to be 5.13%, and fat amount in dry matter was found to be 7.2%. In a study, in which biogenic amine content of Civil cheese was researched, it was reported that putrecine is 0.05 mg $100g^{-1}$, cadaverine is 0.22 mg 100 g^{-1}, histamine is 0.15 mg $100g^{-1}$, and tyramine is 0.88 mg 100 g^{-1} and these values do not exceed toxic dosage limits (Yıldız *et al.*, 2010).

Yazıcı and Dervişoğlu (2003), reported that no negativeness was found, based on storage period, in the chemical, biochemical and sensorial properties of Civil cheeses, which were produced from skimmed milk and their pH values were set as different (5.45, 5.40, 5.35, 5.30 and 5.25) before adding rennet. It is also reported that pH, ash, soluble nitrogen, ripening index and caseine degradation levels of Civil cheeses, which were produced from milks with pH values of 5.45 and 5.40, were high, however Civil cheeses, which were produced from milks with pH values of 5.35 and 5.30, were liked more and caseine degradation level was low during the storage.

Microbiological Properties

Production of cheese from non-pasteurized milk and its fresh consumption without any ripening period causes us to think that there must be standards in terms of health and production quality. Studies conducted to determine the microbiological condition of Civil cheese are given in Table **7**.

Table 7. Microbiological properties of Civil cheese.

Total Aerobic Microorganism	Coliform Bacteria	Staphylococcus-Micrococcus	Mould and Yeast	LAB	References
7.44	-	-	-	6.41	*Yıldız *et al.*,2010
$3.36x10^5$	$3.05x10^4$	-	$1.32x10^5$ / $5.10x10^4$	$2.06x10^4$	**Başkaya *et al.*,2006
7.25	4.35	2.76	6.45	-	*Kamber, 2005

*The values are given as log CFU g^{-1} and
** The values are given as CFU g^{-1}

Moulding of Civil cheese is completely a result of its natural contamination flora. Çakmakcı *et al.* (2012) reported that 165 of 186 mould strains, isolated from 41 moulded Civil cheese samples, were identified as *Penicillium roqueforti* and only 28 of them do not produce penicillin acid, *P. roqueforti* toxin or roquefortin. In a study conducted by Özgören and Şeçkin (2016) on 20 samples of moulded cheese, it is reported that there was not any positive sample in terms of AFM1. In another study conducted by Aydemir Atasever *et al.*, (2010) on the same subject, it is reported that no sample, which exceeds legal limits, was found although 19.4% of 61 Civil cheese samples were found to contain AFM1.

Kars Kashar Cheese

Kashar cheese is one of the most common types of cheese produced in Turkey. There are two different opinions in the literature for the origin of the word “Kashar”. One of these opinions states that the word “Kashar” is related to the Latin word “coerceo”, which means pressing, and the other opinion states that it is related with the Hebrew word “cacher”, which means religiously allowed to be eaten by Jews. The origin of Kashar Cheese goes back to Balkan countries. The literatures state that Kashar cheese was first produced by a Jew girl in Thessaloníki (Ücüncü; 2008; Kamber, 2008; Aydemir 2010; Demir, 2016). This cheese, which is known as Kashar Cheese or Balkan Kashar (Kashkaval), is also called as

"Caciocavallo" in Italy, "Kaskaval" in Bulgaria, "Kackavalj" in Yugoslavia, "Cascaval" in Romania, "Kasseri" in Greece, "Kaskaval" in Hungary, "Kavkazskij" in Russia, and "Rumi" in Egypt (Üçüncü, 2008; Aydemir, 2010).

Although Kashar cheese is produced in all regions of Turkey, it is much wider in Thrace, Central Anatolia and Eastern Anatolia. Particularly "Kars Kashar Cheese", which is produced in Kars province of Eastern Anatolia Region, is one of the traditional geographically marked products of the region.

Kashar cheese was first produced in Kars by four cheese masters, named Bünyamin, Nesim, Nigalay and Nuri and also Fehmi Bey from Plovdiv in Viladikars (Kümbetli) village, which was 5 km far from Kars, in 1926 (Öztek, 1983). Since animal husbandry was widespread in the region, production of Kashar cheese was easily adopted by local people and its production became popular. More than 10 Kashar cheese plants started to operate in Kars between 1927 and 1933 and the production volume reached to approximately 430 tons per year in 1948 (Yazıcı, 2014; TPE, 2015; Demir, 2016). Kars Milk Factory was established in 1969 in Kars with the consultancy and financial assistance of Switzerland and approximately 400 tons of cheese were being produced per year, however, due to lack of milk in the region, the factory was closed later (Demir, 2016). But the establishment of this milk factory has laid the foundations of fabrication of Kars Kashar Cheese, which is currently produced in dairy farms around the region. After the factory was closed, small sized plants, which produce dairy products, particularly Kashar cheese, were opened by private entrepreneurs in Kars (Demir, 2016). From the year 2000 onwards, Food, Industry and Health Ministries have ensured to modernize these production facilities as a result of legal arrangements and incentives and standardization efforts have started regarding the production of Kashar cheese (Demir, 2016). As a result of these efforts, Kars Kashar Cheese was included in the Geographically Marked Products in 2015 following the application no. 2014/013 (TPE, 2015).

Production

Kashar cheese is a pasta-filata type cheese. It is classified as Fresh Kashar Cheese and Mature Kashar Cheese, depending on its ripening period (TSE, 2016d). It is possible to use sheep milk, a mixture of cow milk and sheep milk, and only the cow milk to produce Kashar cheese. The yield of Kashar cheese may vary, depending on the type of milk. 16-18 kg of cheese can be produced from 100 kg of sheep milk, whereas 9-11 kg of cheese can be produced from 100 kg of cow milk (Ücüncü, 2008). Block sizes of Kashar cheese do also vary, depending on the region and ripening condition. Kashar cheeses that are going to be consumed fresh are offered for sale in small-sized and vacuumed packages whereas Kashar cheeses that are going to be left to ripen are produced in larger blocks. While Kashar cheese blocks produced in provinces of Eastern Anatolia weigh approximately 6 kg, those produced in Thrace region weight approximately 11-12 kg (Üçüncü, 2008). It is stated in TS 3272 Kashar Cheese Standard of Turkish Standards Institute that Mature Kashar Cheese is defined as a cheese that is ripened for minimum 120 days under certain conditions in order to have its unique characteristics before it is supplied to the market (TSE, 2016d).

There might be some differences in the production steps of industrial and traditional methods of production of Kashar cheese. In industrial production, the amount of fat in the milk to be used to produce Kashar cheese is standardized and the milk is pasteurized for 30 minutes at 60 °C. After it is cooled down to 34 °C, β-carotene (0.7 mg kg^{-1}), starter culture (9 mL kg^{-1}) and $CaCl_2$ (2 mg kg^{-1}) are added. When the pH value of the milk reaches 6.2-6.3, rennet is added in a manner to generate coagulum within an hour. When the pH value of the coagulum reaches 5.20-5.25, it is cut and boiled at 65-80°C with 4-8% salt. Then the water is filtered and cheese is shaped and the cheese is left for pre-ripening at 15-20°C. Cheeses are transferred to ripening rooms at 8-12°C and 70-75% relative humidity after pre-ripening (Hayaloğlu, 2009; Çakmakçı, 2011; Koca & Metin; 2004). In industrial production of mature Kashar cheese, the cheese's surface can be wiped with potassium sorbate solution in order to prevent growth of mould on the surface (Üçüncü, 2008).

Usually, full-fat or semi-fat cow's milk is used in producing Kars Kashar Cheese and sheep's and goat's milk can also be added to cow's milk in some facilities. After the milk is filtered, it is heated up to 30-35 °C and fermented with commercial cheese yeast. Fermentation period of cheese is approximately 1 hour. When fermentation is completed, the pH value of the cheese is 6.40-6.45. Cheese coagulum is cut in the size of 1 cm^3 and is kept for 5-10 minutes during the coagulum breaking process. Some of the whey in fermentation tank is removed and the coagulum is put into cheese filtering cloth. This cloth is also called "salaspur" in the region (Kamber, 2008). Since the acidity of cheese continues to rise while removing coagulum water, attention is paid to ensure that this process does not last longer than half an hour. Some of the water remaining in cheese filtering cloth is removed by pressing the coagulum and then by using weight. Then the coagulum is removed from the cloth and cut into large pieces and fermented until the pH value of the curd reaches 5.00 - 5.20 under room temperature. Fermentation period might be 3-6 hours, depending on the environmental conditions (Üçüncü,2008; Demir, 2016; TPE, 2016). It can also be understood whether the coagulum is fermented sufficiently or not by looking at whether a piece of cheese is kneaded or not through "spinning" and "leaf opening" experiment in hot water (Üçüncü, 2008). The fermented curd is cut into small pieces (0,3-0,5 mm) and put into a perforated basket and then boiled for 2-3 minutes at 72-78 °C hot salty water, the density of which was set to be 10±2 baume (Demir 2016; TPE 2016). The duration of boiling can be determined when the master pulls the curd up from the basket and check whether it elongates like a gum or not. If the cheese will be dry salted, the density of salt in hot water must be 3 baumes. Boiled curd is kneaded by the cheese master for 3-4 minutes, the boiling water remaining in the curd is removed and the curd is made homogeneous. For this purpose, the entire curd dough is wrapped by a bundle for a couple of times and then pressed by hand, after which the dough is folded in the form of a french roll and cut into blocks. Each cut piece of dough is shaped like a ball and the process called "bloating" is performed. Each piece of dough is twisted by and kept between both hands in bloating and a neck is formed under the block by continuously turning it. Then the dough is held by one hand at its neck line while it is being turned and it is ensured that the dough gets thinner and then is broken from the neck line. Then the dough is placed into blocks in a manner to ensure that broken part does not contact with the circumference. In case the center part is on top or bottom of the block, there might be cracks in the cheese during the ripening (Kamber, 2008, Üçüncü 2008, Demir 2016; TPE 2016). A couple of holes are drilled on the block while Kashar cheese is still hot and it is ensured that the gas moves out (Üçüncü, 2008). Prepared Kashar cheeses are left to cool and rest for 12-24 hours. The blocks are turned upside down for 5-6 times within the first 1-2 hours to ensure that the cheese surface is flattened (Üçüncü, 2008; Kamber, 2008). If Kashar Cheese is going to be ripened, they are taken to pre-ripening room, where the temperature is 15-20°C and relative humidity is 60-70%. On the first day, cheese blocks are stored by pouring salt on top and bottom of them. On the next day, they are stacked in pairs and stored for two days. On the second day, they are stacked in groups of three and stored for four days. This process continues for one month and cheese blocks are always turned upside and down as long as they are stacked. (Üçüncü, 2008; Kamber, 2008, TPE, 2016). After Kashar cheeses, which have completed their pre-ripening process, are wrapped into packaging paper, they are placed into burlap sacks in groups of 5-10, depending on their sizes. (Kamber, 2008: TPE, 2016). After Kashar cheeses are ripened for 30-40 days in pre-resting, they are placed in cold storage rooms, at where the temperature is 8 °C and humidity is 75%, and they are stored therein until they are consumed. (Hayaloğlu, 2009; Çakmakçı, 2011; TPE, 2016). Kars Kashar Cheeses are marketed as "Fresh Kars Kashar Cheese" during pre-ripening period before they are placed into the cold storage room and as "Mature Kars Kashar Cheese" after the 3rd month (TPE, 2016).

Physical Properties

The internal appearance of Kashar Cheese is non-porous and it has a yellowish color. When Kashar Cheese is removed from its mold, it has a white-yellow color, which turns to yellow on the future days of ripening. Furthermore, its color can also change based on the amount of fat within Kashar cheese (Üçüncü 2008; Kamber, 2008). Eroğlu *et al.* (2015) found that L value of Kashar cheese is between 63.90 and 79.03, *a* value is between 1.19 and 1.18, and *b* value is between 7.04 and 21.44. When fresh Kashar cheese is consumed fresh, it has a tasteless taste. The taste is more matured and sharpened thanks to the lipolysis and proteoysis that occur during the ripening (TPE, 2016). The studies showed that flavor compounds of Kashar cheese change throughout the ripening period and butyric acid is the most on the 90th day of storage, which is followed by 2-acety--2-thiazoline, 2-/3-methyl butyric acid, and g-dodecalactone (Yalman *et al.* 2015). Eroğlu *et al.* (2015) reported that Hardness of Kashar cheese is between 34.76 and 112.6 N, external adhesiveness is between -0.310 and -1.176 Nxs, gumminess is between 20.50 and 66.28 N, chewiness is between 0.0127 and 0.0530 N, springiness is between 0.54 and 0.86 mm, internal cohesiveness is from 0.37 to 0.74 and resilience is from 0.11 to 0.32.

Table 8. Chemical properties of Kars Kashar cheese (TPE, 2016).

Ripening	Day 3	Day 30	Day 60	Day 120	Day 180
Dry Matter (%)	54.5-58	57-61	58.5-63	59.5-63	60-63.5
Salt (%)	2.3-3.1	2.7-3.5	2.8-3.5	3.3-3.5	3.4-3.5
Acidity (%LA)	0.83-1.35	1.05-1.61	1.16-1.94	1.42-2.29	1.65-2.29
Protein(%)	24.8-27	25.5-28	26.5-28.5	26.5-28.5	26.5-29
Ash (%)	4.1-4.8	4.5-4.9	4.6-4.9	4.6-5.2	4.7-5.3
Fat in dry matter (%)	43.1-46.86	47.27-50.86	48.1-51.3	48.1-51.3	48.1-52.35
pH*	2.77-3.66	2.99-6.46	3.37-7.95	3.55-8.06	3.59-10.50

*% pH 4.6-Soluble Nitrogen/Total Nitrogen.

Chemical Properties

Kashar cheese is classified as full fat (45≤ milk fat), semi-fat (25 ≤ milk fat <45), low fat, (10 ≤ milk fat <25) and skimmed (10> milk fat) cheese in TS 3272 standard of Turkish Standards Institute. Humidity of non-aged Kashar cheese is reported as 45% and salt amount is 3%, whereas humidity of aged Kashar cheese is 40% and salt amount is 4% (TSE 2016d). Some chemical properties of Kars Kashar cheese, found during the ripening period, are given in Table **8**. Çelik *et al.* (2018) found that dry matter amount of Kashar Cheese is between 48.33 and 51.43%, protein amount is between 29.80 and 26.16%, fat amount in dry matter is between 33.71 and 37.24%, and salt amount in dry matter is between 1.59 and 1.89% throughout the ripening period. Yangılar (2016) found that the dry matter amount of Kashar Cheese, which was ripened for 90 days throughout the traditional method, is 64.48%, fat amount is 31.33%, salt is 4.74%, pH value is 5.31, acidity is 2.82%, protein is 28.49%, water soluble nitrogen is (WSN) 5.34%, trichloroacetic acid-soluble nitrogen is (TCA-SN) 7.25%, pH 4.6-soluble nitrogen is (pH 4.6 SN) %10.59 and ripening index is 18.71%.

Microbiological Properties

25 g of Kashar Cheese must not contain *Salmonella* and *L. monocytogenes* and the number of coagulase positive Staphylococcus must be less than 10^3 according to TS 3272 standard of Turkish Standards Institute (TSE, 2016d). Kamber (2005) found that the number of aerobe mesophilic bacteria of Kashar Cheese is 7,03 log CFU g^{-1}, the number of enterobacteria is 4,30 log CFU g^{-1}, the number of coliform bacteria is 3,91 log CFU g^{-1}, the number of coagulase positive Staphylococci is 2.98 log CFU g^{-1} and the number of yeast and mould is 6.04 log CFU g^{-1}. One of the most important problems encountered in Kashar cheese production is mold development. In recent years, various studies are being carried out to cover cheese edible film in order to solve this problem. These studies have determined that the application of edible film coating limits development mildew growth in cheese (Çelikel, 2017; Yılmaz ve Dağdemir, 2012; Yangılar ve Yıldız, 2016).

CONSENT FOR PUBLICATION

Not applicable.

CONFLICT OF INTEREST

The authors confirm that this chapter contents have no conflict of interest.

ACKNOWLEDGEMENTS

Declared none.

REFERENCES

Arslaner, A & Bakırcı, İ (2016) Effect of milk type, pasteurization and packaging materials on some physicochemical properties and free fatty acid profiles of tulum cheese. *Akademik Gıda,* 14, 98-104. [In Turkish].

Akyüz, N & Çoşkun, H (1996) 'Van Otlu Peynirlerinin üretimi ve Peynire Katılan Otların Peynirin Çeşitli Özellikleri Üzerine Etkileri.*Her Yönüyle Peynir* Hasad Yayıncılık, İstanbul 208-16. In Turkish

Akyüz, N (1981) Erzincan (Şavak) Tulum Peynirinin Yapılışı ve Bileşim. *Atatürk Üniv Ziraat Fak Derg,* 12, 85-112. [In Turkish].

Andınç, S, Genccelep, H & Köse, H (2010) Determination of biogenic amines in herby cheese. *Int J Food Prop,* 13, 1300-14.
[http://dx.doi.org/10.1080/10942910903061869]

Arınç, K (2018) Boğatepe Köyü'nde Gravyer peyniri üretimi ve sürdürülebilir gelişme bakımından önemi (Kars/Türkiye). *Türk Coğrafya Dergisi,* 7-18. In Turkish

Atasever, MA, Adıgüzel, G, Atasever, M & Özturan, K (2010) Determination of aflatoxin m1 levels in some cheese types consumed in erzurum - Turkey. *Kafkas Univ Vet Fak Derg,* 16, 87-91. [In Turkish].

Aydemir, O (2010) *Kars Kaşar Peyniri Karakterizasyonu.* Doktora Tezi, Ondokuz Mayıs Üniversitesi, Fen Bilimleri Enstitüsü, Samsun. In Turkish.

Azak, MG, Kılıç, H, Hızlısoy, H & Abay, S (2012) Erzincan İli Tulum Peynirlerinden Listeria spp. İzolasyonu ve İdentifikasyon. *Erciyes Univ Vet Fak Derg,* 9, 149-56. [In Turkish].

Başkaya, R, Atasever, M, Çakmak, Ö & Yıldız, A (2006) CivilPeynirinin Mikrobiyolojik Nitelikleri. *Istanbul Univ Vet Fak Derg,* 32, 87-94. [In Turkish].

Cambaztepe, F, Çakmakçı, S & Dağdemir, E (2009) Effect of some technological parameters on microbiological, chemical and sensory qualities of Civil cheese during ripening. *Int J Dairy Technol,* 62, 541-8.
[http://dx.doi.org/10.1111/j.1471-0307.2009.00536.x]

Çakır, Y & Çakmakçı, S (2018) Some microbiological, physicochemical and ripening properties of erzincan tulum cheese produced with added black cumin (*Nigella sativa* L.). *J Food Sci Technol,* 55, 1435-43.
[http://dx.doi.org/10.1007/s13197-018-3058-5] [PMID: 29606758]

Çakır, Y, Çakmakçı, S & Hayaloğlu, AA (2016) The Effect of Addition of Black Cumin (Nigella sativa L.) and Ripening Period on Proteolysis, Sensory Properties and Volatile Profiles of Erzincan Tulum (Şavak) Cheese Made From Raw Akkaraman Sheep's Milk. *Small Rumin Res,* 134, 65-75.
[http://dx.doi.org/10.1016/j.smallrumres.2015.12.004]

Çakmakçı, S (2011) Türkiye Peynirleri.*Peynir Biliminin Temelleri* Sidaş Medya Ltd. Şti, İzmir b586-614. In Turkish

Çakmakçı, S, Cantürk, A & Çakır, Y (2017) Peynir üretimi i̇çin sütü pıhtılaştıran enzimlere genel bir bakış ve güncel gelişmeler. *Akademik Gıda,* 15, 396-408. [In Turkish].

Çakmakçı, S, Çetin, B, Gurses, M, Dagdemir, E & Hayaloglu, AA (2012) Morphological, molecular, and mycotoxigenic identification of dominant filamentous fungi from moldy civil cheese. *J Food Prot,* 75, 2045-9.
[http://dx.doi.org/10.4315/0362-028X.JFP-12-107] [PMID: 23127715]

Çakmakçı, S, Dağdemir, A & Hayaloğlu, AA (2008) Influence of ripening container on the lactic acid bacteria population in tulum cheese. *World J Microbiol Biotechnol,* 24, 293-9.
[http://dx.doi.org/10.1007/s11274-007-9470-z]

Çakmakçı, S, Gürses, M & Engin, G (2011) The effect of different packaging materials on proteolysis, sensory scores and gross composition of tulum cheese. *Afr J Biotechnol,* 10, 4444-9.

Çekel, MA & Sert, F (2013) *Doğu Anadolu Bölgesi Süt ve Süt Ürünleri.*Kuzey Anadolu Kalkınma Ajansı Araştırma ve Planlama Birimi.

Çelik, ÖF, Kurt, S, Tufenk, B & Tarakçı, Z (2018) Efficacy of starter culture application using immersion technique on the characteristics of cooked-curd cheeses: kashar cheese sample. *Lebensm Wiss Technol,* 96, 222-7.
[http://dx.doi.org/10.1016/j.lwt.2018.05.031]

Çelik, SE, Özyürek, M, Altun, M, Bektaşoğlu, B, Güçlü, KI, Özgökçe, F & Alpak, R (2008) Antioxidant capacities of herbal plants used in the manufacture of van herby cheese: 'otlu peynir. *Int J Food Prop,* 11, 747-61.
[http://dx.doi.org/10.1080/10942910701594210]

Çelikel, A (2017) *Yenilebilir film ve çeşitli baharat karışımlarının optimizasyonu ve eritme peynirlerinin kaplamasında kullanım olanakları.* Doktora Tezi, Harran Üniversitesi Fen Bilimleri Enstitüsü, Şanlıurfa, Turkiye.

Çoşkun, H (1998) Microbiological and biochemical changes in herby cheese during ripening. *Nahrung,* 42, 309-13.

[http://dx.doi.org/10.1002/(SICI)1521-3803(199810)42:05<309::AID-FOOD309>3.3.CO;2-L]

Çoşkun, H (2005) Otlu Peynir. *Gıda Tek Derneği Yay,* 31, 58.

Demir, M (2016) Coğrafi İşaret Örneği olarak Kars Kaşar Peynirinin Üretimi ve Dağıtımı. *Kafkas Üniversitesi Sosyal Bilimler Enstitüsü Dergisi,* 1, 61-81. [In Turkish].

Demir, P, Erkan, S & Öksüztepe, G (2018) Elazığ'da Satılan Şavak Tulum Peynirlerinin Mikrobiyolojik Kalitesi. *Harran Üniversitesi Veterinelik Dergisi,* 7, 15-20. [In Turkish]. [http://dx.doi.org/10.31196/huvfd.458771]

Demir, P, Öksüztepe, G, İncili, K & İlhak, İ (2017) Vakum Paketli Şavak Tulum Peynirlerinde Potasyum Sorbatın Kullanımı. *Kafkas Univ Vet Fak Derg,* 23, 23-30.

Demirci, M (1998) Trace element levels of our country are some cheese. *Gida,* 13, 17-21.

Durmaz, H, Sağun, E, Tarakçi, Z & Özgökçe, F (2006) Antibacterial Activities of *Allium vineale, Chaerophyllum macropodum* and *Prangos ferulacea. Afr J Biotechnol,* 5, 1795-8.

Emirmustafaoğlu, A & Çoşkun, H (2012) Keçi Sütü, İnek Sütü ve Bu Sütlerin Karışımından Yapılan Otlu Peynirlerde Olgunlaşma Boyunca Meydana Gelen Değişimler. *Gida,* 37, 211-8.

Erdem, G & Patır, E (2017) Elazığ'da Tüketime Sunulan Tulum Peynirlerinde Histamin Düzeyleri ile Bazı Kimyasal Kalite Parametreleri Üzerine Araştırmalar. *Fırat Üniversitesi Sağlık Bilimleri ve Veterinerlik Dergisi,* 31, 235-41. [In Turkish].

Erkan, S, Demir, P & Öksüztepe, G (2018) Elazığ'da Satışa Sunulan Şavak Tulum Peynirlerinin Aflatoksin M1 (AFM1) ve Bazı Kimyasal Parametreler Bakımından İncelenmesi. *Fırat Üniversitesi Sağlık Bilimleri ve Veterinerlik Dergisi,* 32, 45-51. [In Turkish].

Eroğlu, A, Doğan, M, Toker, ÖS & Yılmaz, MT (2015) classification of kashar cheeses based on their chemical, color and instrumental textural characteristics using principal component and hierarchical cluster analysis. *Int J Food Prop,* 18, 909-21. [http://dx.doi.org/10.1080/10942912.2013.864673]

Güneş, RM & Sancak, YC (2014) Van Kahvaltı Salonlarında Tüketime Sunulan Cacıkların Mikrobiyolojik Kalitesi. *YYU Veteriner Fakultesi Dergisi,* 25, 31-5.

Hayaloğlu, AA & Fox, PF Cheeses of Turkey:3. Varieties Containing Herbs or Spices. *Dairy Sci Technol,* 88, 245-56. (2008) [http://dx.doi.org/10.1051/dst:2007015]

Hayaloğlu, A A *Türkiye'nin Peynirleri - Genel Bir Perspektif* Erzurum, Türkiye 10. Gıda Kongresi (2008) 729-32. In Turkish

Hayaloğlu, AA (2009) Volatile Composition and Proteolysis in Traditional Produced Mature Kashar Cheese. *Int J Food Sci Technol,* 44, 1388-94. [http://dx.doi.org/10.1111/j.1365-2621.2009.01968.x]

Hayaloğlu, AA, Çakmakçı, S, Brechany, EY, Deegan, KC & McSweeney, PL (2007) Microbiology, biochemistry, and volatile composition of Tulum cheese ripened in goat's skin or plastic bags. *J Dairy Sci,* 90, 1102-21. [http://dx.doi.org/10.3168/jds.S0022-0302(07)71597-7] [PMID: 17297085]

Hayaloğu, AA & Karabulut, I (2013) Primary and secondary proteolysis in eleven turkish cheese varieties. *Int J Food Prop,* 16, 1663-75. [http://dx.doi.org/10.1080/10942912.2011.604890]

Kamber, U (2005) Kars'ta Satışa Sunulan Kaşar ve Çeçil (Civil) Peynirlerinin Bazı Mikrobiyolojik ve Kimyasal Kalite Nitelikleri. *Kafkas Univ Vet Fak Derg,* 11, 33-8. [In Turkish].

Kamber, U (2008) The Traditional Cheeses of Turkey: Eastern Anatolia Region. *Food Rev Int,* 24, 148-74. [http://dx.doi.org/10.1080/87559120701764589]

Kamber, U, Elmalı, M & Yaman, H (2008) Microbiological and chemical properties of gravyer cheeses.

Atatürk Üniv Vet Bilim Derg, 3, 1-7.

Kavaz, A, Bakırcı, İ & Kaban, G (2013) Some physico-chemical properties and organic acid profiles of herby cheeses. *Kafkas Univ Vet Fak Derg,* 19, 89-95.

Koca, N & Metin, N (2004) Texturel, melting and sensory properties of low-fat kashar cheese produced by using fat replacers. *Int Dairy J,* 14, 365-73. [http://dx.doi.org/10.1016/j.idairyj.2003.08.006]

Kurt, A & Akyüz, N (1984) Van Otlu Peynirinin Yapilışı ve Mikrobiyolojik, Fiziksel ve Kimyasal Nitelikleri. *Gida,* 9, 141-6. [In Turkish].

Kurt, A & Öztek, L (1984) Investigations on Production Techniques of The Şavak Tulum Cheese. *Atatürk Üniv Ziraat Fak Derg,* 15, 65-77.

Kurt, A, Çakmakçı, S & Çağlar, A (1991) Microbiological Studies on the Erzincan Tulum (Savak) Cheese. *Turk J Vet Anim Sci,* 15, 41-50.

Küçüköner, E & Tarakçı, Z (1998) Van ve Yöresinde Üretilen Cacık (Otlu Çökelek) Bazı özelliklerinin Araştırılması. *Geleneksel Süt Ürünleri V Süt ve Süt Ürunleri Sempozyumu* MPM Yayinlari Mert Matbaasi, Tekirdağ 175-784. In Turkish.

Küçüköner, E, Tarakçı, Z & Sağdıç, O (2006) Physicochemical and Microbiological Characteristics and Mineral Content of Herby Cacik, a Traditional Turkish Dairy Product. *J Sci Food Agric,* 86, 333-8. [http://dx.doi.org/10.1002/jsfa.2265]

Ocak, E & Köse, Ş (2015) Van Otlu Peynirinin Üretimi ve Mineral Madde İçeriği. *Gida,* 40, 1-6. [In Turkish].

Öksüztepe, G, Patır, B & Çalıcıoğlu, M (2005) Identification and Distribution of Lactic Acid Bacteria During the Ripening of Şavak Tulum Cheese. *Turk J Vet Anim Sci,* 29, 873-9.

Özgören, E & Seçkin, AK (2016) Aflatoxin M1 Contaminations in Mouldy Cheese. *Mljekarstvo,* 66, 154-9.

Özrenk, E & Köse, Ş Ekin & M. (2009) *Van Piyasasında Satılan Otlu Peynirlerin Bazı Kimyasal Özellileri ve Enerji Değerleri.* Van, Türkiye, 2. Geleneksel Gıdalar Sempozyumu 24-26 Mayıs. In Turkish.

Öztek, L (1983) *Kars İlinde Yapılan Kaşar Peynirlerinin Yapılışları, Bileşimleri ve Olgunlaşmaları Üzerinde Araştırmalarla Bunların Diğer Peynir Çeşitleri ile Kıyaslanmaları.*Erzurum Atatürk Üniversitesi Basımevi. [In Turkish].

Patır, B, Ateş, G, Dinçoğlu, AH & Kök, F (2000) Elazığ'da Tüketime Sunulan Tulum Peynirlerinin Mşkrobiyolojik ve Kimyasal Kalitesi ile Laktik Asit Bakterileri Üzerine Araştırmalar. *Fırat Üniversitesi Sağlık Bilimleri Dergisi,* 14, 75-83. [In Turkish].

Sancak, YC, Sancak, H, İşleyici, Ö & Durmaz, H (2015) Presence of *Escherichia coli* O157 and O157:H7 in Raw Milk and Van Herby Cheese. *Bull Vet Inst Pulawy,* 59, 511-4. [http://dx.doi.org/10.1515/bvip-2015-0076]

Şenel, E, Yıldız, F, Yetişemiyen, A & Durlu-Ozkaya, F (2012) Evaluation of the biogenic amine content and some chemical and microbiological properties of urfa and van herby cheeses. *Kafkas Univ Vet Fak Derg,* 18, 537-44.

Şengül, M, Gürses, M, Dervişoğlu, M & Yazıcı, F (2009) A survey on the some chemical and biochemical properties of civil cheese, a traditional turkish Cheese. *Int J Food Prop,* 9, 791-801. [http://dx.doi.org/10.1080/10942910600596415]

Tarakçı, Z & Durmuş, Y (2016) Effects of packaging materials on some ripening characteristics of Tulum cheese. *Mljekarstvo,* 66, 293-303. [http://dx.doi.org/10.15567/mljekarstvo.2016.0405]

Tarakçı, Z & Küçüköner, E (2008) Comparison of basic nutrients, mineral and heavy metal contents of herby dairy products. *Int J Food Sci Technol,* 43, 216-9. [http://dx.doi.org/10.1111/j.1365-2621.2006.01411.x]

Tarakçı, Z, Çoşkun, H & Tunçtürk, Y (2004) Some properties of fresh and ripened herby cheese, a traditional variety produced in turkey. *Food Technol Biotechnol,* 41, 47-50.

Tekinşen, OC, Atasever, M & Keleş, A (1996) Civil Peynirinin Kimyasal ve Organoleptik Özellikleri. *Veteriner Bilimleri Dergisi,* 12, 65-71. [In Turkish].

Tekinşen, KK & Akar, D (2017) Erzincan Tulum Peyniri. *Atatürk Üniv Vet Bilim Derg,* 12, 218-26. [In Turkish].
[http://dx.doi.org/10.17094/ataunivbd.347980]

Terin, M & Yavuz, F (2015) Türkiye Bölgeler Arası Optimum Süt ve Ürünleri Akışı:Spatial Denge Modeli. *Ege Univ Ziraat Fak Derg,* 52, 207-17. [In Turkish].
[http://dx.doi.org/10.20289/euzfd.27975]

TGK (2011) Türk Gıda Kodeksi Mikrobiyolojik Kriterler Yönetmeliği, Resmi Gazete Sayı:28457. Ankara.

TGK (2015) Türk Gıda Kodeksi Peynir Tebliği, Resmi Gazete:29261. Ankara.

Tomar, O, Akarca, G, Beykaya, M & Çağlar, A (2018) Some characteristics of erzincan tulum cheese produced using different probiotic cultures and packaging material. *Kafkas Univ Vet Fak Derg,* 24, 647-54.

Topuk, Ş & Sezer, Ç (2015) Some quality characteristics of kars gravyer cheese. *Gida,* 40, 69-75.

TPE (2000) *Tulum peyniri Coğrafi İşaret.* Türk Patent Ensititüsü, Ankara.

TPE (2009) *CivilPeyniri Coğrafi İşaret.* Türk Patent Ensititüsü, Ankara.

TPE (2015) *Kars Kaşarı Coğrafi İşaret Tescil.* http://www.turkpatent.gov.tr

TPE (2018) *Van Otlu Peyniri Coğrafi İşaret.* Türk Patent ve Marka Kurumu, Ankara.

TSE (2016a) *TS 13205 Otlu Peynir.* Türk Standardları Enstitüsü, Ankara.

TSE, (2016b). TS 2174 Gravyer Peyniri. basım yeri bilinmiyor: Türk Standartları Enstitüsü

TSE (2016c) *TS 3001Tulum Peyniri.* Türk Standartları Ensititüsü, Ankara.

TSE (2016d) *TS 3272 Kaşar Peyniri.* Türk Standartları Ensititüsü, Ankara.

TUİK (2017) *Türkiye İstatistik Kurumu.* http://www.tuik.gov.tr

Tunçtürk, Y, Ocak, E & Köse, Ş (2014) Farklı Süt Türlerinden Üretilen Van Otlu Peynirlerinin Fiziksel ve Kimyasal Özellikleri ile Proteoliz Profillerinde Olgunlaşma Sürecinde Meydana Gelen Değişimler. *Gida,* 36, 163-70. [In Turkish].

Turgut, T, Erdoğan, A & Atasever, M (2012) Identification of nonstarter indigenous làctobacilli from kars gravyer cheese. *Kafkas Univ Vet Fak Derg,* 18, 781-4.
[http://dx.doi.org/10.9775/kvfd.2012.6386]

Ulutaş, Z, Çağlar, A & Kurt, A (1993) Kar Gravyer Peynirinin Yapılışı, Duyusal, Fiziksel ve Kimyasak Özellikleri Üzerine Bir Araştırma. *Gida,* 18, 197-202. [In Turkish].

Üçüncü, M (2008) *A'dan Z'ye Peynir Teknolojisi-2* izmir: Meta Basım Matbaacılık Hizmetleri

Yalman, M, Güneşer, O & Karagül Yüceer, Y (2017) Evaluation of some physical, chemical and sensory properties of kasar cheese and its processed and analogue types. *J Agric Sci,* 23, 63-75.

Yangılar, F (2016) Effects of natamycin edible films fortified with essential oils on the safety and quality parameters of kashar cheese. *J Food Saf,* 37, 1-10.

Yangılar, F & Oğuzhan Yıldız, P (2016) Casein/natamycin edible films efficiency for controlling mould growth and on microbiological, chemical and sensory properties during the ripening of Kashar cheese. *J Sci Food Agric,* 96, 2328-36.
[http://dx.doi.org/10.1002/jsfa.7348] [PMID: 26198464]

Yazıcı, F & Dervişoğlu, M (2003) Effect of pH adjustment on some chemical, biochemical, and sensory properties of civil cheese during storage. *J Food Eng,* 56, 361-9.

[http://dx.doi.org/10.1016/S0260-8774(02)00208-X]

Yazıcı, S (2014) *Dünden Bugüne Peynirin Kars Serüveni "Alplerden Kafkaslara Kars Peynirciliğinin 150 Yıllık Tarihi.*Boğatepe Çevre ve Yaşam Derneği-Tarih Vakfı Yayını, Ankara. In Turkish.

Yetişmeyen, A (1997) Otlu Peynir Üretim Tekniğinin ve Kalite Özelliklerinin Geliştirilmesi Üzerine Bir Araştırma. *Turk J Agric For,* 21, 237-47. [In Turkish].

Yıldız, F, Atilla Yetişemiyen, A, Şenel, E, Durul Özkaya, F, Özteki, Ş & Şanlı, E (2010) Some properties of civil cheese: A type of traditional turkish cheese. *Int J Dairy Technol,* 63, 575-80. [http://dx.doi.org/10.1111/j.1471-0307.2010.00612.x]

Yılmaz, F & Dagdemir, E (2012) The effects of beeswax coating on quality of kashar cheese during ripening. *Int J Food Sci Technol,* 47, 2582-9. [http://dx.doi.org/10.1111/j.1365-2621.2012.03137.x]

CHAPTER 9

Traditional Cheeses in Southeastern Anatolia Region in Turkey

Mutlu Buket Akin* and **Musa Serdar Akin**

Faculty of Engineering, Department of Food Engineering, Harran University, Sanliurfa, Turkey

Abstract: In Southeastern Anatolia region, various cheese varieties such as Urfa Cheese, Antep Cheese, Diyarbakır Örgü Cheese, Adıyaman Şuji Cheese and Siirt Otlu Cheese are produced from ewe's and goat's milk.

Urfa cheese is a semi-hard, white-brined cheese with generally a spherical shape and the diameter of around 5-7 centimetres. Antep cheese is another semi-hard cheese type which is manufactured from ewe's or goat's milk or the mixtures of them and stored in brine. This cheese is usually 3-4 cm long and 2 cm thick, rectangular or cube shaped. Diyarbakır Örgü cheese is a hard pasta-filata type cheese variety which is stored in brine. This cheese is made from ewe's or goat's milk or the mixtures of them and is like a hair braid is shape. Adıyaman Şuji cheese is a semi-hard, white-brined cheese variety which is made from cow's milk or sheep's milk or goat's milk or the mixtures of them. This cheese is in the form of an accordion-like circle with a diameter of about 7-10 centimetres. Although Otlu cheese is mainly produced in Eastern Anatolia region, it is also produced in Diyarbakır and Siirt villages in the Southeastern Anatolia region. The main difference of these cheeses is Van Otlu cheese is ripened in brine, but Siirt Otlu cheese is ripened without brine. Siirt Otlu cheese is manufactured from ewe's or goat's milk and stored in a plastic container or sheep's skin bag by burying it underground.

Keywords: Adıyaman Şuji cheese, Antep cheese, Diyarbakır Örgü cheese, Siirt Otlu cheese, Urfa cheese.

INTRODUCTION

Southeastern Anatolia Region (Fig. **1**) is situated in southern Turkey (Mesopotamia), where goat and sheep breeding (Fig. **2**) is widely carried out.

There are 17.166.194 head cattle, 172.181 head buffaloes, 33.677.636 head sheep and 44.312.308 head goats in Southeastern Anatolia Region. Approximately 8.4%, 18.3% and 25.1% of the existing cattle, sheep and goats, respectively, in

* **Corresponding author Mutlu Buket Akın:** Faculty of Engineering, Department of Food Engineering Harran University Sanliurfa Turkey; Tel/Fax: +90414 318 37 25; E-mail:mutluakin@harran.edu.tr

Celile Aylin Oluk & Oya Berkay Karaca (Eds.)

Turkey, are in this region. 1.442.766 tonnes cow's milk, 281.988 tonnes sheep's milk and 149.463 tonnes goat's milk are produced in this region (TUİK, 2017).

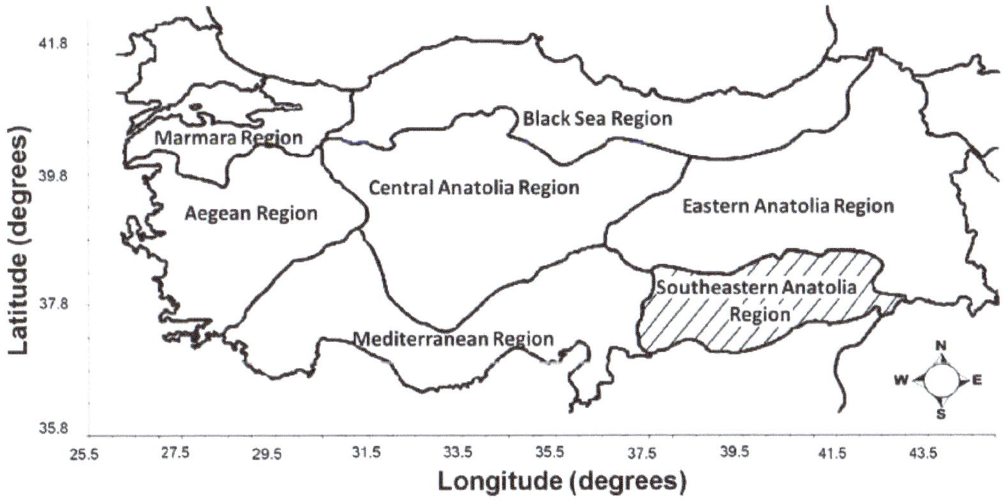

Fig. (1). Southeastern Anatolia Region (map prepared by M. Ali Çullu).

Fig. (2). Sheep breeding in Southeastern Anatolia Region (Photographed by M. Serdar Akin).

In this region, various cheese types such as Urfa Cheese, Antep Cheese, Diyarbakır Örgü Cheese, Adıyaman Şuji Cheese and Otlu Cheese are produced from alone or a mixture of cow's, sheep's and goat's milk. Among these cheeses, Antep cheese (with 356 certificate number) and Diyarbakir Orgu cheese (with 170 certificate number) received Protected Designation of Origin (PDO) and Protected Geographical Indication (PGI) from the Turkish Patent Institute, respectively. Traditional production methods and the properties of these cheese varieties are given below:

URFA CHEESE

Description

Urfa cheese is a semi-hard, white-brined cheese, the shape of which is generally spherical and the diameter is around 5-7 centimetres. Its characteristic features are lustrous white colour, no rind, closed texture and a very salty acidic taste. This cheese is generally manufactured from raw sheep's or goat's milk or the mixture of them between February and July. After May, when the quantity of sheep's and goat's milk is reduced, it is manufactured from cow's milk also (Kamber & Terzi, 2008; Akin *et al.*, 2015a; Bintsis *et al.*, 2018).

Production Method

A large amount of Urfa cheese is still produced in villages according to ancestral methods in primitive and unhygienic conditions. On the other hand, a lot of studies have been carried out on standardization and modernization of traditional Urfa cheese (Akin & Sahan, 1998; Atasoy *et al.*, 2008; 2013; Atasoy & Türkoğlu, 2009; Kirmaci *et al.*, 2014).

Urfa cheese is traditionally produced from raw sheep's milk without using a starter culture. Production steps of Urfa cheese are given in Fig. **3**

After milking (Fig. **4a**), milk is filtered by using a cloth and coagulated with animal rennet at 32°C (Fig. **4b-c**). The coagulation process lasts about 60 min (Fig. **4d**). After coagulation, the curd (without cutting) is poured into muslin bags called parzin that has a conical shape, and approximately, 0.5 kg capacity (Fig. **4e-f**). Parzıns are hung up for whey separation and drained by its own weight (Fig. **4g-h**). This process is achieved in a cold room (about 4-8°C) between 6-18 hours. These cheese blocks are called deleme (Fig. **4i**). Delemes are dry-salted overnight by turning them at 4-6 hours intervals in order to make each side of them salted. Then they are scalded in their own whey for 1–3 min (Fig. **4j-k**). Afterwards, the scalded cheeses are filled into tins or drums with brine (16-20% NaCl) (Fig. **4l**) and matured until autumn in cold storage (less than 10°C, for more than 3 months).

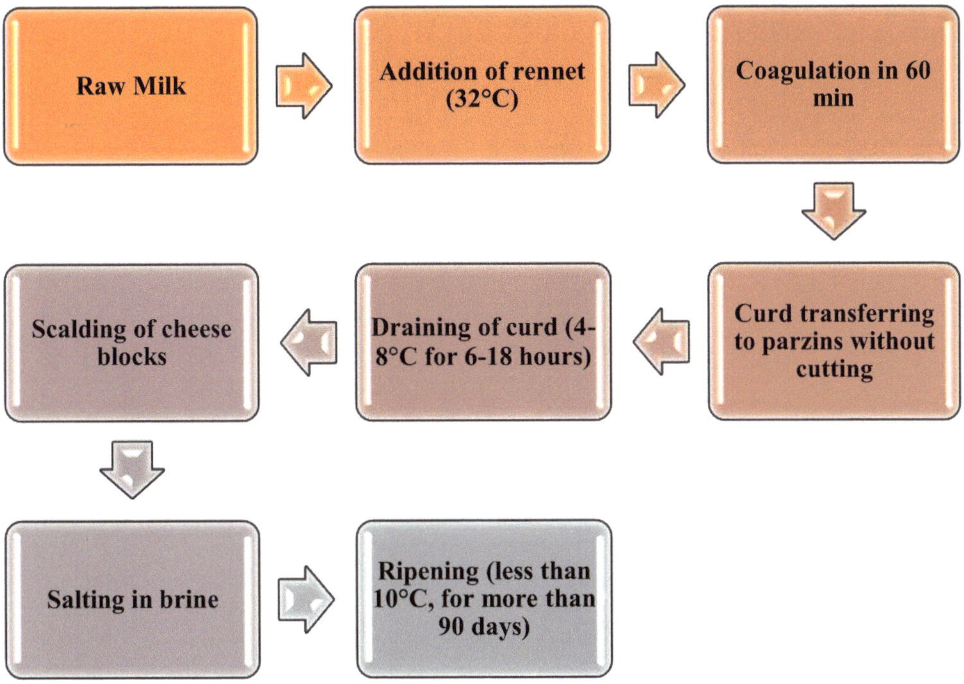

Fig. (3). Production steps of traditional Urfa cheese.

Chemical Properties

According to the Turkish Food Codex Cheese Communique (2015), maximum moisture and salt contents of cheeses matured in brine must be 60% and 4%, respectively. The mean chemical composition of Urfa cheese is within the following averages: titratable acidity 1.20%, dry matter 44.25%, fat 21.42%, protein 17.11%, ash 1.74% and salt 3.5% (Akin *et al.*, 2016).

Microbiological Properties

Yalcın *et al.* (2007) determined $2.3x10^8$ CFU g^{-1} total aerobic mesophilic bacteria, $1.2x10^5$ CFU g^{-1} coliform bacteria, $3.1x10^4$ CFU g^{-1} yeast and mould, $5.4x10^2$ CFU g^{-1}*Staphylococcus aureus*, and $2.0x10^7$ CFU g^{-1}*Lactobacillus* in Urfa cheese sold in Sanlıurfa markets.

Relevant Researches

Chemical and microbiological properties of Urfa cheese sold in Sanlıurfa were studied by Akin and Sahan (1998), Sahan *et al.* (1998a, 1998b), Yetismeyen and Yildiz (2003), and Yalcin *et al.* (2007). Biochemical changes in Urfa cheese during ripening were studied by Atasoy (1999), Ozer *et al.* (2002, 2003), Atasoy and Turkoglu (2009), Atasoy *et al.* (2008, 2013) and Kirmaci *et al.* (2014).

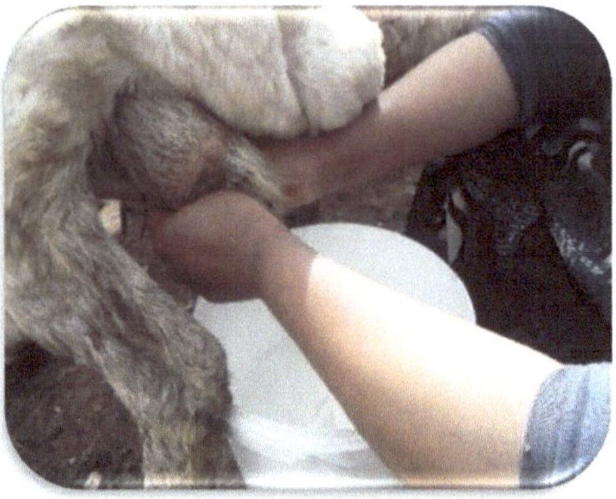

(a)

(b)

(c)

(d)

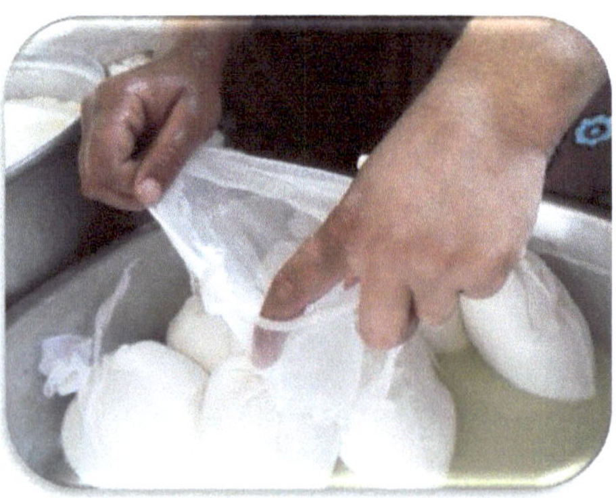

(e)

(f)

Fig. (4) cont.....

Fig. (4). Production steps of Urfa cheese: **(a)** Milking of sheep, **(b)** Filtering of milk **(c)** Renneting of milk, **(d)** Coagulated milk **(e, f)** Transferring coagulated milk into parzın, **(g, h)** Draining of curds **(i)** Delemes, **(j, k)** Scalding cheese blocks, **(l)** Urfa cheese in brine (Photographed by M. Serdar Akin).

Ozer *et al.* (2003) found that the cheese manufactured from concentrated milk by using UF technique had a more compact structure and fragile body than the cheese produced from the unconcentrated milk. In addition, they suggested that very dense brine concentration, which is used for reducing microbiota of the cheese, could be reduced by scalding of cheese blocks.

Atasoy (2004) determined that the type of starter cultures does not affect fat hydrolysis in Urfa cheeses. He also reported that the cheeses manufactured from pasteurised milk had noticeably lower amounts of volatile fatty acids (2.90 mL NaOH 100 g^{-1} cheese) than the cheeses made from raw milk (14.55 mL NaOH 100 g^{-1} cheese).

Atasoy and Turkoglu (2008) showed that heat treatments adversely affected the lipolysis and sensory properties of Urfa cheeses. They found that cheese made from pasteurized milk and inoculated with thermophilic or mesophilic lactic starters milk had a lower level of lipolysis than the cheeses made from raw milk.

Kav *et al.* (2011) detected the presence of Aflatoxin M1 (AFM1) in 127 samples of Urfa cheese. They found that 28.3% of cheese samples were contaminated with AFM1 between 70.61 and 770.97 ng kg^{-1}. It was determined that 10.2% of these samples exceeded the legal limits of AFM1 (250 ng kg^{-1}).

Hayaloglu & Karabulut (2013a, b) determined the free fatty acid profile and volatile compounds of Urfa cheeses and classified this cheese as mild lipolysed based on its free fatty acids profile.

70 volatile compounds were recovered from Urfa cheese made from sheep's milk by Kırmacı *et al.* (2015). Of these, 20 were aldehydes and ketones, 11 were ester, 10 were alcohol, 10 were acids, 6 were terpenes and 13 were other compounds.

ANTEP CHEESE

Description

Antep cheese is another semi-hard cheese type which is manufactured from ewe's or goat's milk or the mixtures of them and stored in brine. This cheese is usually 3-4 cm long and 2 cm thick, rectangular or cube shaped. It is characterised by greyish white colour, no rind, closed texture and a taste of salty acid (Kamber & Terzi, 2008; Akin *et al.*, 2015a; Bintsis *et al.*, 2018).

Production Method

Traditional Antep cheese is manufactured from raw milk without using a starter culture. Production steps of Antep cheese are given in Fig. (**5**).

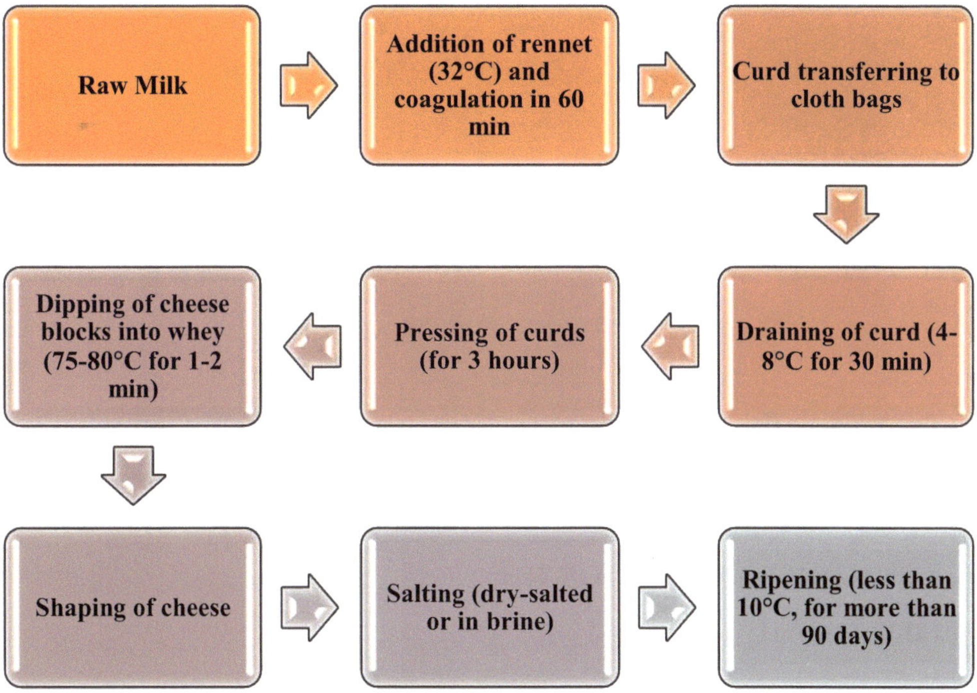

Fig. (5). Production steps of traditional Antep cheese.

The raw milk (at body temperature) is coagulated with the commercial rennet in sufficient amount to coagulate milk within 1–2 hours. Milk was used to be coagulated with fig juice before, but it is no longer practised these days. After coagulation, the curds, which are not cut into pieces, are transferred into cloth bags being spherical in shaped and they are placed on an inclined surface for about 30 min to drain off the whey. Then they are pressed to drainage all the whey by placing stones on them. Once the curd hardens, spoon-sized cheeses are put into small bags which called *sesbezi or cibinlik.* After slightly bound, the bags are dipped into their own whey (about 75-80°C) for 1-2 min. The cheeses are then allowed to cool slightly and are shaped in the palm of a hand. Shaped cheeses are transferred into another container, sprinkled with salt and sold in the market (Fig. **6**). For maturation of cheese, they are tranferred into cans of brine and kept in cold storage (less than 10°C, for more than 90 days) (Kaya, 1995; Kahyaoglu *et al.*, 2005; Kamber & Terzi, 2008).

Fig. (6). Antep cheese (Photographed by Mutlu B. Akin).

Chemical Properties

According to the Turkish Food Codex Cheese Communique (2015), maximum moisture and salt contents of cheeses matured in brine must be 60% and 4%, respectively. The mean chemical composition of Antep cheese is within the following averages: titratable acidity 1.71%, dry matter 53.00%, fat 23.28%, protein 20.25% and salt 3.24% (Caglar *et al.*, 1998).

Relevant Research

Chemical properties of Antep cheese sold in Gaziantep were studied by Kaya (1995) and Caglar *et al.* (1998). Biochemical changes of Antep cheese depending on the salt concentration of brine were studied by Kaya & Oner (1996) and Kaya *et al.* (1999).

The effect of salt concentration on Gaziantep cheese's hardness and whiteness values during short-term brining was investigated by Kaya (2002). It was found that the hardness values of the samples in low salt concentrations were the lowest while the L-value was the highest.

Kahyaoglu ve Kaya (2003) showed that reduction in the fat content of Gaziantep cheeses caused significant increases in moisture and protein contents. They also reported that the elastic shear modulus (G′) and melting temperature increased as the fat content decreased, but the meltability decreased.

Kahyaoglu *et al.* (2005) found that fat reduction decreased viscoelasticity *L* value and *b* value, but increased hardness, cohesiveness, gumminess and springiness values of Gaziantep cheese. They also determined that curd dipping process increased hardness, cohesiveness, gumminess and springiness of cheeses.

Karatop (2010) determined the volatile compounds of Antep cheese sold in Gaziantep markets. A total of 65 volatile substances were found in the cheeses including 48 from neutral/basic fractions (2 ketones, 19 esters, 3 alcohols, 4 lactones, 2 aldehydes, 8 hydrocarbons, 9 terpens) and 17 from acidic fractions (4 branched chain, 12 alifatic, 1 aromatic).

DIYARBAKIR ORGU CHEESE

Description

Diyarbakır Orgu cheese is a hard pasta-filata type cheese variety which is stored in brine. This cheese is made from ewe's or goat's milk or the mixtures of them and is shaped like a hair braid. The features of this cheese are light yellow colour, homogenous and elastic consistency and the distinctive taste and odour (Kamber and Terzi, 2008; Akin *et al.*, 2015 a, b).

Production Method

Although Diyarbakir Orgu cheese is generally produced in households, it has started to be manufactured in small dairies and modern dairy plants due to increasing consumer demand (Hatipoglu, 2014). Although sheep milk is preferred in Diyarbakır Orgu cheese production, sheep's milk or cow's milk or mixtures thereof can be used to produce this cheese due to limited lactation period of sheep (between April and June). Production steps of Diyarbakir Orgu cheese are given in Fig. (**7**).

Animal rennet is added to raw milk in such a quantity to coagulate milk within 60-90 min at 28-35°C (Fig. **8a**). Then the curd is cut into small pieces (1-3 cm^3 parts), separated from the whey and pressed for 5-6 hours (Fig. **8b**). Afterwards, the curd is left to fermentation at room temperature for 80-180 min.

A piece of curd is immersed in hot water (70-75°C), then kneaded and strecthed like a string. If it does not break, curd ripening is sufficient. The ripened curd is cut into small thin slices (Fig. **8c**) and immersed in water at a temperature of 70-75°C for 2-7 minutes. The curd is kneaded and then a portion of curd is strecthed and formed to the shape of a hair braid (Fig. **8d-e**). The cheeses are put into tins with NaCl brine (12Be') (Fig. **8f**) and ripened at least 30 days (Kamber & Terzi, 2008; Hatipoğlu, 2014).

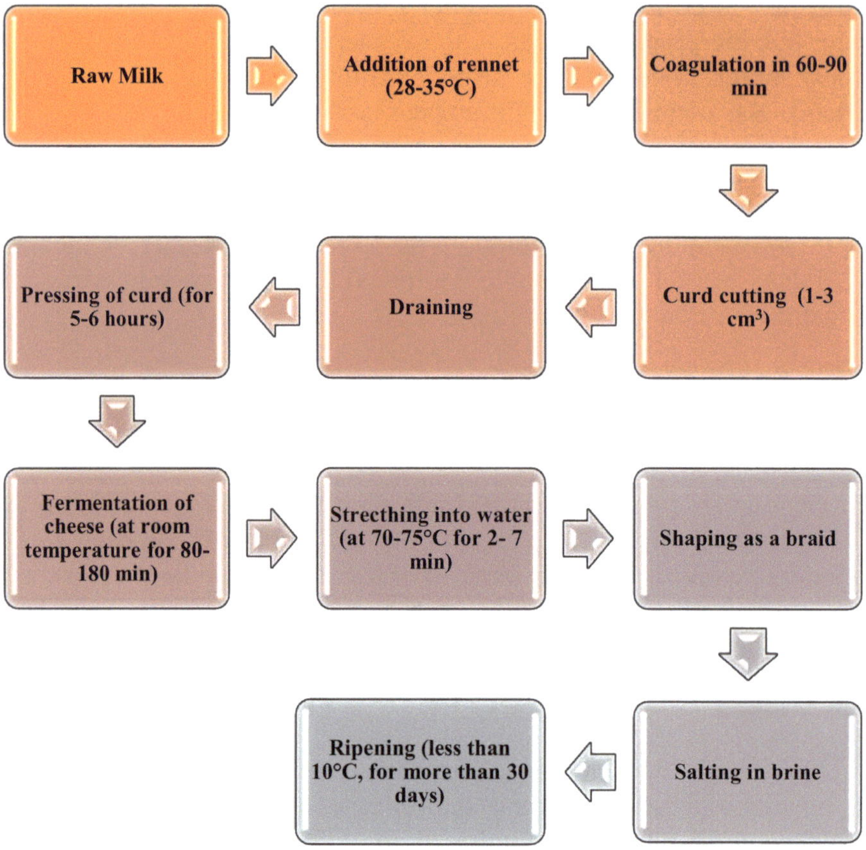

Fig. (7). Production steps of traditional Diyarbakir Orgu cheese.

Chemical Properties

According to the Turkish Food Codex Cheese Communique (2015), maximum moisture and salt contents of cheeses, which are made by the process in which curd is immersed in hot water and kneaded, must be 45% and 4%, respectively. The mean chemical composition of Diyarbakir Orgu cheese is within the following averages: dry matter 54.43%, fat 20.57%, protein 25.20%, ash 8.77% and salt 7.67% (Hatipoglu, 2014).

Microbiological Properties

Hatipoglu (2014) reported that the mean counts of *Lactococcus* sp. 7.57 log CFU g^{-1}, *Lactobacillus* sp. 7.26 log CFU g^{-1}, coliform bacteria 3.39 log CFU g^{-1}, and coliform bacteria with fecal originated 2.72 *E.coli,* 1.76 log CFU g^{-1} yeast and mould 2.63 log CFU g^{-1}, in Diyarbakir Orgu cheese.

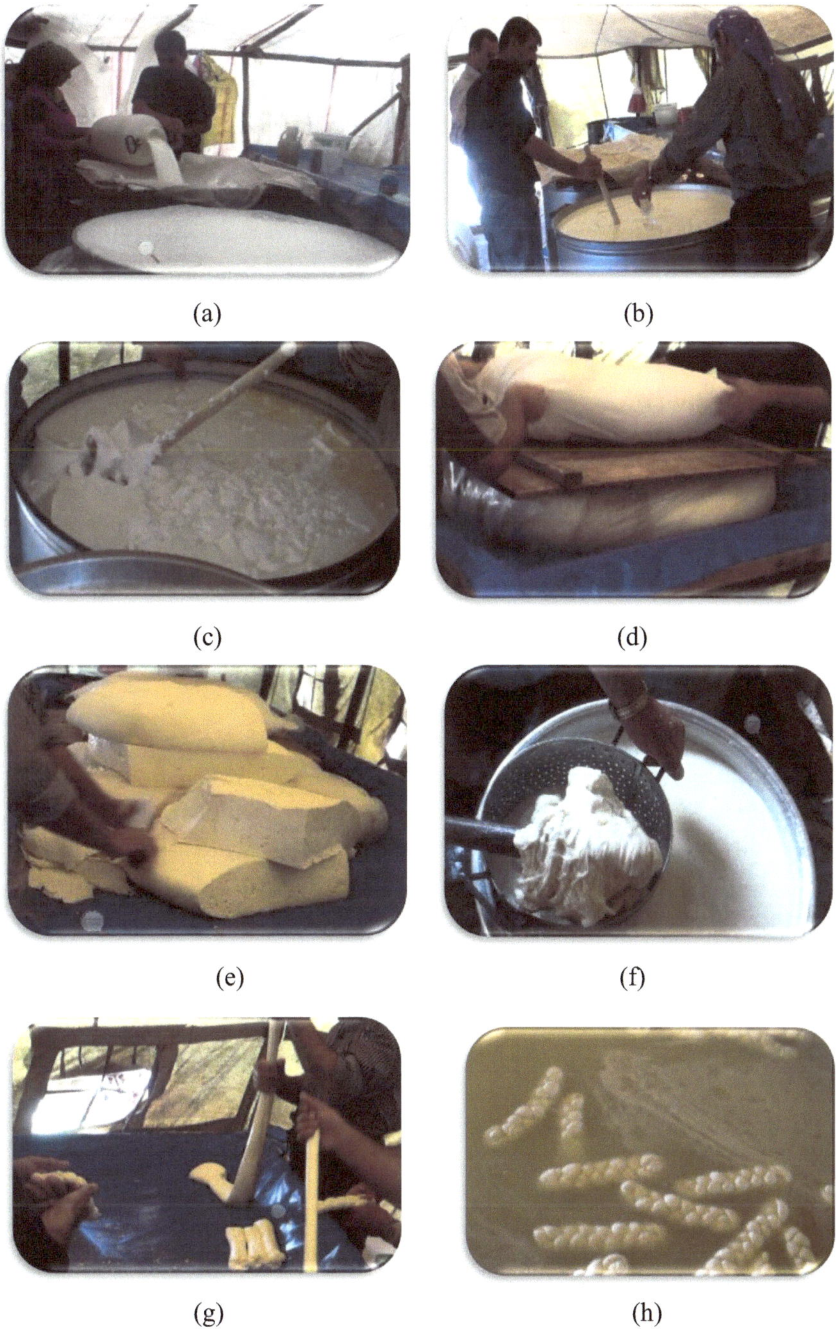

Fig. (8). Production steps of Diyarbakir Orgu cheese **(a)** Filtering of milk, **(b)** Addition of rennet, **(c)** Coagulation of milk, **(d)** Pressing the curd, **(e)** Slicing the curd, **(f)** Kneading and strecthing of the curd, **(g)** Shaping the curd as a braid, **(h)** Putting the cheese into the brine (Photographed by Abdulkerim Hatipoglu).

Relevant Researches

Chemical and microbiological properties of Diyarbakir Orgu cheese sold in markets were studied by Ozdemir *et al.* (1998), Aksu *et al.* (1999), Anar *et al.* (2000), Turkoglu *et al.* (2003), Vural *et al.* (2010) and Akin & Akin (2015b).

Çelik *et al.* (2005) reported that the microbiota of Diyarbakir Orgu cheese consisted of *Lactobacillus* spp. (40%) and *Enterococcus* spp. (38.87%). Researchers have isolated lactic acid bacteria from the cheeses and defined them as *Lb. cruvatus, Lb. parabuchneri, Lb. paracasei, Lb. plantarum, E. durans, E. faecalis, and E. faecium* ve *E. hirae*. They also determined *Lactococcus lactis* and *Pediococcus acetilactici.*

Çelik *et al.* (2006) investigated the biochemical changes in Orgu cheese containing different fat levels during 90 days ripening period. Whilst the dry matter of cheeses remained unchanged, the fat and protein contents of them decreased obviously. In addition, the titratable acidity decreased as the fat content of cheese increased. During storage, the lipolysis and proteolysis of cheeses increased regularly.

Some properties of Örgü cheeses manufactured from the raw and pasteurized cow milk were examined by Celik & Turkoglu (2007) during a 3 months ripening period. They found that salt and salt in total solid were lower, but total solids, protein, water-soluble nitrogen, trichloro-acetic acid soluble nitrogen and acid contents of cheeses made from pasteurized milk were lower than that made from raw milk. In addition, in both kind of cheeses, the total solid and protein contents were lower, but salt, salt in total solid, water-soluble nitrogen, trichloro-acetic acid soluble nitrogen and acid degree decreased during ripening.

Erkan *et al.* (2009) evaluated the presence of Aflatoxin M1 and Verotoxin 2 in Diyarbakir Orgu cheese produced in Diyarbakir. In the examined samples, 46.67% of samples of Aflatoxin M1 were detected, but no verotoxin was detected. The amount of Aflatoxin M1 in the Orgu cheese samples (14.44%) was found above the values which were permitted by the Turkish Food Codex.

Çelebi (2011) studied the effects of coagulating enzyme type (calf rennet, microbial enzymes, recombinant chymosin) on the properties of Orgu cheese during ripening and concluded no differences among the chemical and sensory properties of cheeses depending on the enzyme types.

Hayaloglu & Karabulut (2013a, b) investigated free fatty acid profile and volatile compounds of Orgu cheeses and classified this cheese as mild lipolysed based on its free fatty acids profile. They reported that the levels of volatile compounds of

Orgu cheese were probably low because of the scalding or cooking process in the manufacture of these cheeses.

Hatipoğlu (2014) determined the characteristic properties of the traditional Diyarbakir Orgu cheeses, collected from different dairy plants and ripened for 120 days, and evaluated them in terms of food safety. According to the result, the fresh cheese had a potential infectious risk in terms of consumers' health. However, the risk was minimized for ripened cheese after 60th days storage.

ADIYAMAN ŞUJI CHEESE

Description

Adıyaman Şuji cheese is a semi-hard, white-brined cheese variety which is made either from cow's milk, sheep's milk or goat's milk or the mixtures of them. This cheese is in the form of an accordion-like circle with a diameter of about 7-10 centimetre. It is characterized by the yellowish-white colour, no rind, closed texture and a taste of salty acid (Akin *et al.*, 2010).

Production Method

Adıyaman Şuji cheese is still produced in villages according to ancestral methods in primitive and unhygienic conditions. Adıyaman Şuji cheese took its name from the process applied after the cheese was produced (Akin *et al.*, 2010). Traditional Adıyaman Şuji cheese is manufactured from raw milk and no starter culture is used in traditional production. Production steps of this cheese are given in Fig. (**9**).

The raw milk is coagulated immediately after milking within 60-90 min at 32-36°C. After coagulation, curd is cut into 1-3 cm^3 and transferred into cloth bags called parzik (Fig. **10a**). Then they are put on the pallet consisting of reed rods to separate whey (Fig. **10b**) and a container is placed underneath the reed rods to collect the whey. To accelerate the whey separation from the curd, another pallet consisting of reed rods is put on the curd and stones are placed top of the pallet. After the whey is released completely from teleme, the curd is removed from parzik. On the other hand, the whey is boiled and the cheese blocks are put into the boiling whey (Fig. **10c**). They are left inside this whey until cooled (about 40°C). This process is referred to as the Şuji in the region. The cheeses are placed between the reed again and drained to release the whey. The reeds give the shape of a unique wavy cheese. The fresh cheeses are salted by dry salt and filled into canisters (Fig. **10d**). The cheeses released their water and can be stored into its own water for a year.

Chemical Properties

According to the Turkish Food Codex Cheese Communique (2015), maximum moisture and salt contents of cheeses matured in brine must be 45% and 4%, respectively. The mean chemical composition of Adiyaman Şuji cheese is within the following averages: titratable acidity 0.316% (as L.A.), dry matter 46.94%, fat 24.8%, fat in dry matter 52.77%, salt 3.72%, salt in dry matter 7.67%. (Akin *et al.*, 2010).

Microbiological Properties

According to microbiological analysis of Adiyaman Şuji cheese sold in Adiyaman markets by Akin *et al.* (2010); the mean count of coliform bacteria is $3.2x10^3$ CFU g^{-1} and mould and yeast counts $8.8x10^3$ CFU g^{-1}.

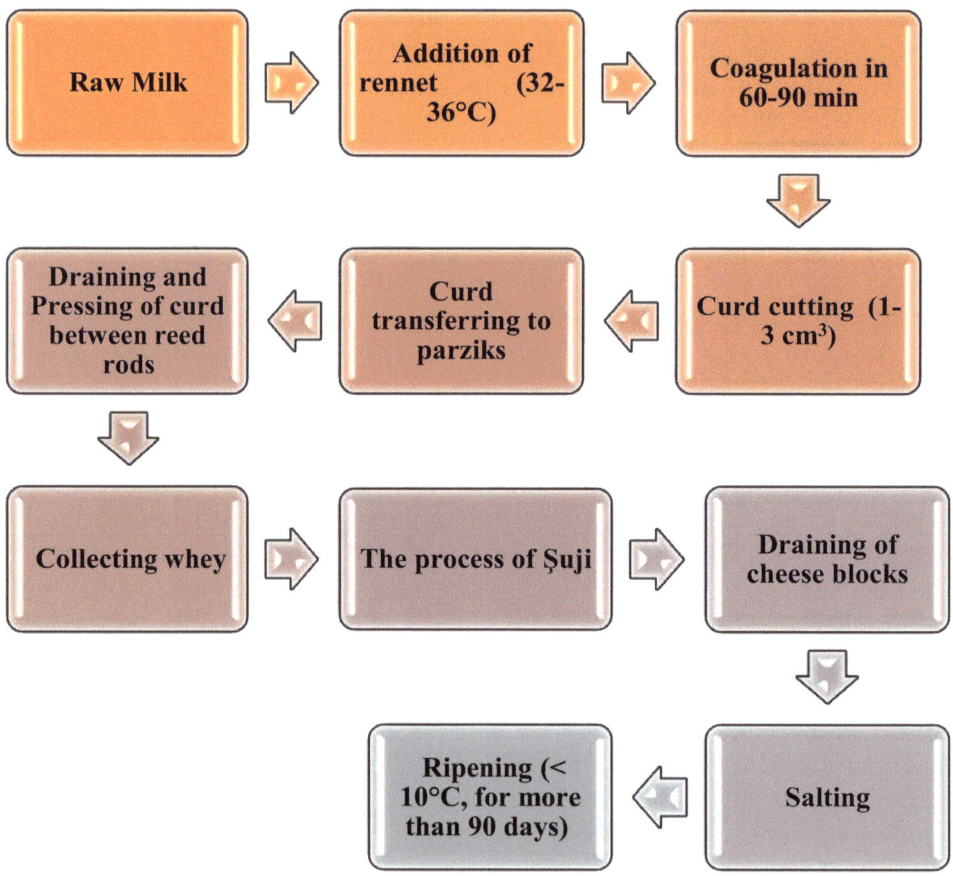

Fig. (9). Production steps of traditional Adıyaman Şuji cheese.

Fig. (10). Production steps of Adiyaman Şuji cheese (a) Coagulated milk, **(b)** Transferring coagulated milk into parzık, **(c)** Reed pallet, **(d)** Draining and pressing of curd between reed pallet, **(e)** Şuji process, **(f)** Adiyaman Şuji cheese (Photographed by Pelin Dolek and M. Serdar Akin).

Relevant Researches

Chemical and microbiological properties of Adiyaman Şuji cheese sold in Adiyaman were studied by Akin *et al.* (2010).

SIIRT OTLU (HERBY) CHEESE

Description

Although Otlu cheese is mainly produced in Eastern Anatolia region, it is also produced in Diyarbakır and Siirt villages in the Southeastern Anatolia region. The main difference in this cheese is that Van Otlu cheese is ripened in brine, but Siirt Otlu cheese is ripened without brine. Siirt Otlu cheese is manufactured from ewe's or goat's milk and stored in a plastic container or sheep's skin bag by burying underground. Its characteristic features are white or yellow colour, densely or sparsely herbed appearance, medium-hard or hard texture and salty taste.

Production Method

Raw sheep's milk is used for making of Otlu (Herby) cheese. In case sheep's milk is insufficient, goat's milk or mixture thereof is used for the cheese production. Production steps of this cheese are given in Fig. (**11**).

Milk is coagulated by using animal rennet within 1 or 2 hours. After coagulation, the curd is cut into 2 cm, transferred to cloth bags and pressed about 5-6 hours by using stones for whey drainage. On the other hand, herbs are prepared to add into the cheese. Generally, the herbs consisting of *Allium schoenoprasum, Allium aucheri Boiss* and *Allium paniculatum L.* subsp. *paniculatum* are used for cheese production. These herbs are collected in May, dried and salted. The amount of herbs added is up to the consumers' preference. The curd released from whey is crumbled by hand and mixed with the herbs and salt. The step of adding herbs differs depending on the manufacturers. Some of them add the herbs before coagulation, some of them add them after cutting curd and some of them add after pressing curd. Then, the cheese is filled into the plastic container or sheep's skin bag. Afterwards, the container is covered with vine leaves, then sealed with mud and is buried underground to ripe for at least 5-6 months (Fig. **12**) (Dogan, 2011; Dogan, 2012).

Chemical Properties

According to the Turkish Food Codex Cheese Communique (2015), maximum moisture and salt contents of cheeses, which are dry salted and similar to Tulum cheese, must be 45% and 4%, respectively. The mean chemical composition of Siirt Otlu cheese is within the following averages: titratable acidity 2.44±0.62% (lactic acid), dry matter 53.18±2.68%, fat 25.01±3.89%, protein 21.65±1.69%, ash 7.07±1.15% and salt 6.93±1.26% (Dogan, 2011).

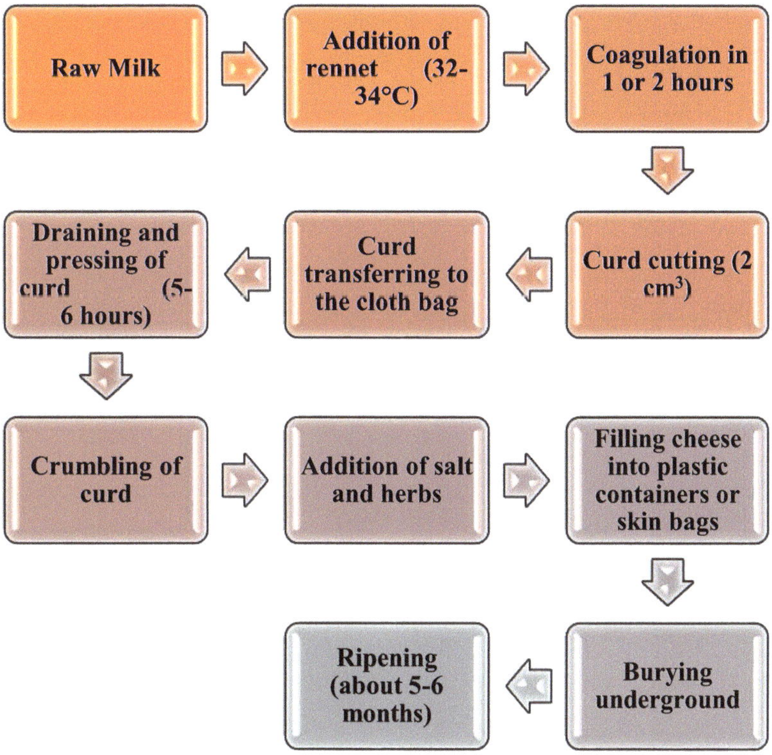

Fig. (11). Production steps of traditional Siirt Otlu cheese.

Fig. (12). Siirt Otlu cheese (Photographed by Nurcan Dogan).

Microbiological Properties

Microbiological properties of Siirt Otlu cheese are as follows: Total aerobic mesophilic bacteria 9.19 log CFU g^{-1}, Enterococcus count 3.57 log CFU g^{-1}, Lactococcus count 7.04 log CFU g^{-1}, Lactobacillus count 6.69 log CFU g^{-1}, yeast-mould count 6.37 log CFU g^{-1}. *E. coli* is under detectable amount (Dogan, 2012).

Relevant Researches

Chemical and microbiological properties of Otlu cheese sold in Eastern and Southeastern Anatolian region were studied by Tekinsen (2004) and Vural *et al.* (2007).

Tarakci & Akyuz (2009) studied the effects of packaging materials and filling methods on some properties of Otlu (Herby) cheese. They used different containers and different filling methods for the preparation of Otlu cheeses. They reported that while containers influenced acidity, pH, dry matter, fat contents and water-soluble nitrogen, proteolysis was higher in the cheeses cut in pieces in the pot compared to others and the acceptability of cheeses in block form was higher.

Dogan (2011) examined the traditional production process of Siirt herby cheese by interviewing personally with the producers and making surveys on the traditional production process of that cheese which is produced in the centre of Siirt, its towns like Eruh, Pervari, Kurtalan, the villages of these towns. In addition, he determined the chemical composition of 30 Siirt herby cheese collected from different micro-domestic establishments.

Dogan (2012) studied microbiological, chemical, biochemical and sensory characteristics of 20 Otlu cheese samples produced according to traditional methods in Siirt Province. Ripening index with respect to WSN and NPN was found as 20.66±4.95% and 15.72±1.98%, respectively. According to the results Total free fatty acids (TFFAs) amount of cheese samples was 32 324 mg kg^{-1} cheese. It was reported that the average 20% of the total amount of TFFAs was short-chain, 56% medium length, and 18% of the long-chain fatty acids.

LIST OF ABBREVIATIONS

AFM1 Aflatoxin M1

Be Baume degree

CFU Colony forming unit

L.A. Lactic acid

NaCl Sodium chlorid

NPN Non protein nitrogen

PDO Protected Designation of Origin

PGI Protected Geographical Indication

TFFA Total free fatty acid

WSN Water soluble nitrogen

CONSENT FOR PUBLICATION

Not applicable.

CONFLICT OF INTEREST

The authors confirm that this chapter contents have no conflict of interest.

ACKNOWLEDGEMENTS

Declared none.

REFERENCES

Akin, MS, Celikel, A, Unver, N & Senocak-Soran, G (2016) Some properties of urfa cheeses sold in şanlıurfa province *International Congress on Food of Animal Origin,* Industry Partnership 13.

Akin, M & Sahan, N (1998) Şanlıurfa'da Üretilen Taze Urfa Peynirlerinin Kimyasal ve Duyusal Özelliklerinin Belirlenmesi Üzerine Bir Araştırma*V Süt ve Süt Ürünleri Sempozyumu (Geleneksel Süt Ürünleri)* Tekirdağ, Turkey. 282-296.

Akin, MB, Akin, MS, Karaca, OB & Kırdar, SS The Traditional Cheeses Produced from Ewe's Milk in Southeast Anatolia Region of Turkey (TF3-P368). The 3rd International Symposium on Traditional Foods from Adriatic to Caucasus. (2015) Sarajevo/ Bosnia and Herzegovina. 01-04.

Akin, MS Physicochemical and microbiological properties of orgu cheeses sold in sanliurfa markets. The 3rd International Symposium on Traditional Foods from Adriatic to Caucasus (2015) Sarajevo–Bosnia And Herzegovina. 192.

Akin, MS, Guler-Akin, MB & Dolek, P (2010) Adıyaman Şuji Peynirlerinin Kimyasal Ve Bazı Mikrobiyolojik Özelliklerinin Belirlenmesi Üzerine Bir Araştırma. *The First International symposium on Traditional Foods from Adriatic to Caucasus,* 268-70.

Aksu, H, Colak, H, Vural, A & Erkan, ME (1999) Diyarbakır Bölgesinde Üretilen Örgü Peynirlerde Mikrobiyolojik ve Kimyasal Özellikler Üzerine Bir Araştırma. *YYÜ Veteriner Fakültesi Dergisi,* 10, 8-11. [In Turkish].

Anar, S, Soyutemiz, E & Cetinkaya, F (2000) Örgü Peynirin Üretim Aşamalarında Görülen Bazı Mikrobiyolojik ve Kimyasal Değişimler. *Veteriner Fakültesi Dergisi,* 19, 81-5.

Atasoy, AF & Turkoglu, H (2008) Changes of composition and free fatty acid contents of Urfa cheese (a white-brined Turkish cheese) during ripening: effects of heat treatments and starter cultures. *Food Chem,* 110, 598-604.
[http://dx.doi.org/10.1016/j.foodchem.2008.02.046]

Atasoy, AF & Turkoglu, H (2009) 'Lipolysis in Urfa cheese produced from raw and pasteurized gaots' and cows' milk with mesophilic or thermophilic cultures during ripening'. *Food Chem,* 115, 71-8.
[http://dx.doi.org/10.1016/j.foodchem.2008.11.061]

Atasoy, AF (1999) A study on the determination of level of proteolysis in Urfa cheese sold in local markets', MSc Thesis, Harran University, Institute of Natural and Applied Sciences, Sanliurfa.

Atasoy, AF (2004) Farklı Tür Sütlerden Yapılan Urfa Peynirlerinin Nitelikleri Üzerine Değişik Pastörizasyon Normlarının Ve Starter Kültürlerinin Etkileri', PhD. Thesis, Ankara Üniversitesi, Ankara. [In Turkish].

Atasoy, AF, Hayaloglu, AA, Kirmaci, H, Levent, O & Turkoglu, H (2013) Effects of partial substitution of caprine for ovine milk on the volatile compounds of fresh and mature Urfa cheeses. *Small Rumin Res,* 115, 113-23.
[http://dx.doi.org/10.1016/j.smallrumres.2013.09.002]

Atasoy, AF, Yetişmeyen, A, Turkoğlu, H & Ozer, BH (2008) Effects of heat treatment and starter culture on the properties of traditional Urfa cheeses (a white-brined Turkish cheese) produced from bovine milk. *Food Control,* 19, 278-85.
[http://dx.doi.org/10.1016/j.foodcont.2007.04.004]

Bintsis, T, Alichanidis, E, Uzunsoy, I, Ozer, B, Papademas, P, Radulovic, Z & Miocinovic, J (2018) *Global Cheesemaking Technology 'White-Brined Cheeses.* John Wiley & Sons, Ltd., UK.

Celebi, M (2011) *'Farklı Pıhtılaştırıcı Enzimlerin Olgunlaşma Süresince Örgü Peynirinin Özellikleri Üzerine Etkilerinin Belirlenmesi'.* MSc Thesis, Süleyman Demirel Üniversitesi, Isparta. [In Turkish].

Celik, S & Turkoglu, H (2007) Ripening of Örgü Cheese Manufactured with Raw or Pasteurized Milk: Composition and Biochemical Properties. *Int J Dairy Technol,* 60, 253-8.
[http://dx.doi.org/10.1111/j.1471-0307.2007.00344.x]

Celik, S, Erdogan, A & Gurses, M (2005) *Geleneksel Taze Örgü Peynirinden İzole Edilen Laktik Asit Bakterilerinin Tanımlanması* GAP IV. Tarım Kongresi. 21-23 Eylül, Şanlıurfa 634-637.

Celik, S, Turkoglu, H & Erdogan, A (2006) *Farklı Oranlarda Yağ İçeren Pastörize Süt ile Yapılan Geleneksel Örgü Peynirinin Olgunlaşma Periyodu Boyunca Bileşim ve Bazı Biyokimyasal Özelliklerinin Değişimi 2,* Türkiye 9. Gıda Kongresi, 24-26 Mayıs, Bolu, Turkey 867-868.

Dogan, C (2011) *Siirt Otlu Peynirinin Geleneksel Üretim Yöntemi ve Bileşimi.* MSc Thesis, Kahramanmaraş Sütçü İmam Üniversitesi, Kahramanmaras, Turkey [In Turkish].

Dogan, N (2012) 'Siirt İlinde Üretilen ''Siirt Otlu Peynirinin''Bazı Özelliklerinin Belirlenmesi', MSc Thesis, Harran Üniversitesi, Sanliurfa.

Erkan, ME, Vural, A & Guran, HŞ (2009) Diyarbakır Örgü Peynirinde Aflatoksin M1 ile Verotoksin 1 ve 2 Varlığının Araştırılması'. *Dicle Üniversitesi Veteriner Fakultesi Dergisi,* 1, 19-25. [In Turkish].

Hatipoglu, A (2014) Geleneksel Diyarbakır Orgu Peynirinin Karakteristik Özelliklerinin Tespit Edilmesi ve Gıda Güvenliği Açısından Değerlendirilmesi', PhD Thesis, Harran Üniversitesi, Sanliurfa, Turkey. [In Turkish].

Hayaloglu, AA & Karabulut, I (2013) SPME/GC-MS characterization and comparison of volatiles of eleven varieties of Turkish cheeses. *Int J Food Prop,* 16, 1630-53. b
[http://dx.doi.org/10.1080/10942912.2011.587625]

Hayaloglu, AA & Karabulut, I (2013) Characterization and comparison of free fatty acid profiles of eleven varieties Of Turkish cheeses. *Int J Food Prop,* 16, 1407-16. a
[http://dx.doi.org/10.1080/10942912.2011.587626]

Kahyaoglu, T & Kaya, S (2003) Effects of heat treatment and fat reduction on the rheological and functional properties of Gaziantep cheese. *Int Dairy J,* 13, 867-75.
[http://dx.doi.org/10.1016/S0958-6946(03)00113-4]

Kahyaoglu, T, Kaya, S & Kaya, A (2005) Effects of fat reduction and curd dipping temperature on viscoelasticity, texture and appearance of gaziantep cheese. *Food Sci Technol Int,* 11, 191-8.
[http://dx.doi.org/10.1177/1082013205055002]

Kamber, U & Terzi, G (2008) The traditional cheeses of Turkey: Southeast anatolia region. *Food Rev Int,* 24,

62-73.
[http://dx.doi.org/10.1080/87559120701762237]

Kav, K, Col, R & Tekinsen, KK (2011) Detection of aflatoxin M1 levels by ELISA in white-brined Urfa cheese consumed in Turkey. *Food Control,* 22, 1883-6.
[http://dx.doi.org/10.1016/j.foodcont.2011.04.030]

Kaya, S & Oner, MD (1996) Water activity and moisture sorption isotherms of Gaziantep cheese. *J Food Qual,* 19, 121-32.
[http://dx.doi.org/10.1111/j.1745-4557.1996.tb00409.x]

Kaya, S (2002) Effect of salt on hardness and whiteness of Gaziantep cheese duringshort-term brining. *J Food Eng,* 52, 155-9.
[http://dx.doi.org/10.1016/S0260-8774(01)00098-X]

Kaya, S, Kaya, A & Oner, MD (1999) The effect of salt concentration on rancidity in Gaziantep cheese. *J Sci Food Agric,* 79, 213-9.
[http://dx.doi.org/10.1002/(SICI)1097-0010(199902)79:2<213::AID-JSFA170>3.0.CO;2-H]

Kirmaci, HA, Hayaloğlu, A, Ozer, B, Atasoy, F & Turkoğlu, H (2014) Evolution of proteolysis in Urfa cheese made from ewe's milk by wild-type starter culture. *Small Rumin Res,* 119, 120-9.
[http://dx.doi.org/10.1016/j.smallrumres.2014.01.006]

Ozdemir, S, Celik, S, Ozdemir, C & Sert, S (1998) Diyarbakır'ın Karacadağ Yöresinde Mahalli Olarak Yapılan Örgü Peynirinin Mikrobiyolojik Ve Kimyasal Özellikleri', In: V. Süt ve Süt Ürünleri Sempozyumu. 21-22 Mayıs, Tekirdağ,154-166. [In Turkish].

Ozer, BH, Atasoy, AF & Akin, MS (2002) Some properties of Urfa cheese (a traditional white-brined Turkish cheese) made from bovine's and ovine's milk. *Int J Dairy Technol,* 55, 94-9.
[http://dx.doi.org/10.1046/j.1471-0307.2002.00040.x]

Ozer, BH, Robinson, RK & Grandison, AS (2003) Textural and microstructural properties of Urfa cheese (A white-brined Turkish cheese). *Int J Dairy Technol,* 56, 171-6.
[http://dx.doi.org/10.1046/j.1471-0307.2003.00100.x]

Sahan, N, Var, I & Akin, MS (1998) Taze Urfa Peynirlerinin Mikrobiyolojik Özellikleri ve Bazı Patojen Bakterilerin Aranması', In: V. Süt ve Süt Ürünleri Sempozyumu (Geleneksel Süt Ürünleri), 21-22 Mayıs 1998, Tekirdağ, Turkey, 315-327. In Turkish.

Sahan, N, Var, I & Akin, MS (1998) 'Olgunlaştırılmış Urfa Peynirlerinde Mikrobiyolojik Bir Çalışma', In: Gıda Mühendisliği Kongresi, 16-18 Eylül 1998, Gaziantep, Turkey, 337-346. In Turkish.

Tarakci, Z & Akyuz, N (2009) Effects of packaging materials and filling methods on selected characteristics of otlu (Herby) Cheese. *Int J Food Prop,* 12, 496-511.
[http://dx.doi.org/10.1080/10942910701813941]

Tekinsen, KK (2004) Hakkari ve Çevresinde Üretilen Otlu Peynirlerin Mikrobiyolojik ve Kimyasal Kalitesi. *Veteriner Bilimleri Dergisi,* 20, 79-85. [In Turkish].

TUIK (2017), 'Turkish Statistical Institute. Animal husbandary data by regions in Turkey', Food, Agriculture and Animal Ministry of Turkey. https://www.tarimorman.gov.tr/sgb/Belgeler/SagMenuVeriler/HAYGEM.pdf

Turkish Food Codex Cheese Communique (2015), NO: 2015/6. resmigazete.gov.tr/eskiler/2015/02/20150208-16.htm

Turkoglu, H, Ceylan, ZG & Dayisoylu, KS (2003) The microbiological and chemical quality of örgü cheese produced in Turkey. *Pak J Nutr,* 2, 92-4.
[http://dx.doi.org/10.3923/pjn.2003.92.94]

Vural, A, Erkan, ME & Guran, HŞ (2010) The examination of the microbiologic quality in örgü cheese (braided cheese) samples. *Kafkas Univ Vet Fak Derg,* 16 (Suppl. A), 53-8.

Vural, A, Narin, I, Erkan, ME & Soylak, M (2008) Trace metal levels and some chemical parameters in herby cheese collected from south eastern Anatolia-Turkey. *Environ Monit Assess,* 139, 27-33. [http://dx.doi.org/10.1007/s10661-007-9806-z] [PMID: 17562196]

Yalcın, S, Ardıc, M & Nizamoglu, M (2007) Urfa Peynirinin Bazı Kalite Nitelikleri. *Atatürk Üniv Vet Bilim Derg,* 2, 90-5. [In Turkish].

Yetismeyen, A & Yıldız, F (2003) Urfa Peynirlerinin Mikrobiyolojik, Kimyasal ve Duyusal Niteliklerinin Saptanması. *Gida,* 28, 287-94. [In Turkish].

CHAPTER 10

Traditional Cheeses of the Central Anatolia Region in Turkey

Ebru Şenel* and **Tuba Şanli**

Faculty of Agriculture, Department of Dairy Technology, Ankara University, Ankara, Turkey

Abstract: In recent years, methods of production and composition of many traditional cheeses have been determined. There are more than a hundred different types of cheese in Turkey, which are generally classified according to the origin, production methods and ripening conditions. In this paper, 15 different types of cheese produced in Central Anatolia Region of Turkey have been introduced. The most common ones are Konya Küflü (Moldy) cheese, Divle Tulum (animal skin bag) cheese, Gölbaşı Tulum cheese, Çepni Tulum cheese, Ereğli Cloth cheese, Karaman Tulum cheese, Yozgat Çanak (Pot) cheese, Çömlek cheese, Kayseri Çömlek cheese, Niğde Küp (Gödelek) cheese, Sivas Küp cheese, Çankırı Küpecik cheese, and Ayaş Küpecik (Basma) cheese. Two of them, Küp and Tulum cheeses, are more frequently encountered and generally produced from sheep's milk and goat's milk. Most of these traditional cheeses have been produced using different manufacturing techniques depending on the geographical location and habits.

Key words: Ayas Basma, Central anatolia region, Cömlek cheese, Divle tulum, Konya küflü, Küp cheese, Traditional cheeses.

INTRODUCTION

Cheese has been produced since ancient times by processing milk into a more resistant product. Different methods of productions have generated different flavors and textural properties. Cheese may be classified according to the types of raw material, mechanism of clotting, lipid content, dry matter content, the method of production, ripening level, method of salinization, the region of manufacturing *etc.* (İşleyici *et al.*, 2018). Central Anatolia has been the crossroad of civilizations during thousands of years. Geographical conditions, agriculture and animal husbandry play a pivotal role in the development of cheese varieties (Kamber 2015).

* **Corresponding author Ebru Senel:** Faculty of Agriculture, Department of Dairy Technology, Ankara University, Ankara, Turkey; Tel: +90312 596 1300; Fax: + 90 312 318 2219; E-mail senel@agri.ankara.edu.tr

Celile Aylin Oluk & Oya Berkay Karaca (Eds.)

According to the Turkish Statistical Institute, 20.699,894 tons of milk were processed into 687.206 tons of cheese in Turkey (Anonymous 2017). Turkey has an important potential in terms of traditional cheese varieties. Recently, the interest of consumers in taste and aroma of traditional cheeses have been increased. In Turkey, more than 100 varieties of cheese are produced (Hayaloğlu and Karabulut 2013a, Kamber 2015, İşleyici *et al.,*2018). However, most of traditional cheeses are known as different or same names in different regions (Kamber 2015). Therefore, the number of cheese varieties is not certain. Many of these cheeses are still produced according to ancient practices and traditional methods in restricted geographical areas and consumed locally in Turkey.

Central Anatolia Region covers approximately 21 percent of the land of Turkey, with a surface area of 151,000 km^2. It is the second largest state by surface of the country where more than 15 types of cheeses are produced in different cities which are shown in Fig. (**1**). The majority of these cheeses are Konya Küflü (Moldy) cheese, Divle Tulum cheese, Gölbaşı Tulum cheese, Çepni Tulum cheese, Ereğli Cloth cheese, Karaman Tulum cheese, Yozgat Çanak (Pot) cheese, Çömlek cheese, Kayseri Çömlek cheese, Niğde Küp (Güdelek) cheese, Sivas Küp cheese, Çankırı Küpecik cheese, and Ayaş Küpecik (Basma) cheese.

Fig. (1). Central Anatolia region of Turkey and its provinces.

In this chapter, some chemical, physicochemical, sensorial and microbiological properties of 15 varieties of cheese produced from Central Anatolia Region and its production techniques are introduced. Central Anatolian cheeses can be classified according to manufacturing method and area as follows:

- Cheeses ripened in animal skin (Tulum) with/without mold
- Cheeses ripened in a pot/jar/pan (Çömlek/Küp/Çanak)

CHEESES RIPENED IN ANIMAL SKIN BAG (TULUM)

Some cheeses are ripened in an earthenware pot/jar/pan or in animal skins (Tulum) under soil or sand or caves to extend the shelf life. Some of these cheeses are ripened without moldy in Tulum, such as Çepni Tulum cheese, Gölbaşı Tulum cheese, Ereğli Tulum cheese, and Karaman Tulum cheese, but some of them are ripened with moldy species, such as Konya Küflü (Moldy) cheese and Divle Tulum cheese.

Küflü cheese, a mold-ripened variety, produced from raw milk has a complex microflora. The growth of molds during the ripening of cheese contributes to proteolysis and lipolysis, resulted in dramatic changes in cheese since they have greater enzymatic potential than bacteria (Hayaloğlu *et al.,* 2008). These changes in ripening period with and without mold are summarized into three groups.

a. Lactic acid produced from lactose by the metabolic activity of lactic acid bacteria, is metabolized by molds, resulting in a pH increases of above 6.0 at the end of the cheese ripening.
b. Mold activity increases lipolysis. More than 5% of total fatty acids are hydrolyzed to free fatty acids due to the extracellular lipases of molds.
c. Proteolysis level in mold-ripened cheese is very extensive due to fungal and non –fungal (plasmin, nonstarter organism, *etc.*) proteases activity and casein is degraded to peptides and amino acids. (Hayaloğlu *et al.* 2008).

All of these changes contribute to textural properties, aroma, and flavor of the cheese.

KONYA KÜFLÜ (MOLDY) CHEESE

Konya Küflü cheese is a mold-ripened variety, produced from some local manufacturers in Konya province. This cheese is well known and consumed in the area by local people. Konya Küflü cheese has semi-hard texture, rectangular shape, and weights between 300 and 500 g. The color of this cheese is generally white with blue or green grainy, besides blue or green color dominates after more growth of molds on the cheese surface (Fig. **2**). This cheese is also known as "Gök cheese" (Blue cheese) because of its blue-green color (Tekinşen *et al.,* 2009). Konya Küflü cheese has intense aroma and flavor due to its complex microflora and use of home-made calf rennet in manufacturing and long ripening period (Hayaloğlu *et al.,* 2008). Since Konya Küflü cheese is produced under uncontrolled manufacturing conditions such as the use of undefined starter culture and use of raw milk, it has a potential risk to human health. However, there have not been reported microbial health problems associated with the consumption of Konya Küflü cheese (Güley *et al.,* 2014).

Method of Production: Konya Küflü cheese is produced using skimmed or partially skimmed raw sheep's milk without using starter culture (Güley *et al.,* 2014). It is kept in cylindrical plastic bags weighing 55-60 kg during 40-45 days at cold storage. Then, cheese is cut into blocks and covered with a paper. It is ripened at room temperature until blue-green molds are spontaneously grown on the surface of the cheese. The ripened temperature may vary between 18 and 30 °C according to producers. Some of them prefer between 18°C and 23 °C temperatures, while some prefer between 25°C and 30 °C temperatures.

In the other way of production, fresh cheese is salted and placed into cloth bags and kept for 3 days until serum drains. Then, the cheese is transferred into wooden troughs where it is broken up to tiny bits pounding by hand. Slivered cheese is pressed with a wooden stick into bags where a nylon bag has also been placed inside. These bags are left at the cold storage for 2-3 months for ripening. After this period, the cheese is taken off by cutting the cloth bags. The cheese is cut into smaller size portions (5-6 cm wide) and subsequently it is placed into bags again. Therefore, the cheese is preserved at warmer temperatures for a second ripening period. This cheese has been ripened until the molds have grown on the surface. If any black or brown mold is observed on the cheese surface, these cheeses are not consumed.

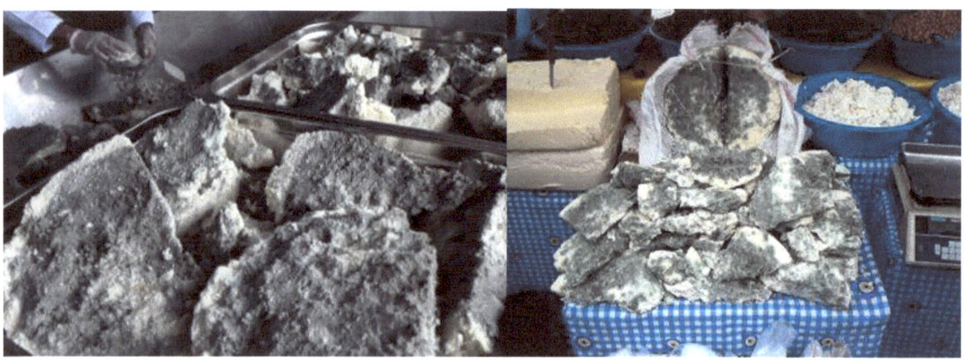

Fig. (2). Konya Küflü cheese.

Properties of Cheese: Until today, chemical composition, biochemistry, volatile compounds and microbiota of Konya Küflü cheese have been investigated (Hayaloğlu *et al.,* 2008, Yılmaz *et al.,* 2005, Güley *et al.,* 2014). Dry matter content, fat content, salt content, pH and acidity of Konya Küflü cheese was found as 44.84-53.86%, 27.25- 28.25%, 1.76-5.62%, 4.94-7.45 pH, 0.27-1,53%LA (12-68 °SH), respectively. Hayaloğlu *et al.* (2008) referred that ripening index of cheese was of a very high level due to the breakdown of casein into peptides and free amino acids by the activity of molds proteases. In this study, free amino acids amounts, pH-4.6-soluble nitrogen, and TCA-soluble nitrogen were found as 9.75-36.55 (mg Leu/g), 12.39-28.44 (%TN), 10.30-27.70 (%TN), respectively. Also, they referred that many volatile compounds which consist of 11 acids, 21 esters, 20 ketones, 6 aldehydes, 32 alcohols, 8 sulfur compounds, 10 terpenes, and 30 miscellaneous compounds, were resulting from complex proteolysis. In addition, it was determined that the principal volatile compounds of Küflü cheese were ketones and alcohols. In another study, the level of AFM1 was determined in some moldy cheeses. AFM1 was found in 5 of the 20 Konya Küflü cheese samples. The concentration of AFM1 positive samples was determined as 41.4 ±10.9 ng/kg (Özgören and Seçkin 2016). Another research was conducted on the identification of lactic acid bacteria (LAB) in mature Konya Küflü cheese. LAB microbiota of this cheese mainly consists of species belong to two genera; Lactobacillus and Enterococcus. Predominantly determined species of two genera in Konya Küflü cheese are as follows; *Lactobacillus brevis, Lactobacillus paracasei, Lactobacillus casei, Lactobacillus plantarum, Enterococcus faecium, Enterococcus faecalis* (Güley *et al.*, 2014).

A survey was conducted on the presence of *Listeria monocytogenes* for some traditional semi-hard cheeses. *Listeria monocytogenes* was detected in 7 of the 30 Konya Küflü cheese samples (%23.3) (Güner and Telli 2011). Researchers recommended that it is important to create standardization and to take measures recontamination with L. monocytogenes associated with possible listeriosis outbreaks in all manufacturing stages and storage of Küflü cheese (Güner and Telli 2011).

DIVLE TULUM CHEESE

Divle Tulum cheese is a traditional cheese with the geographical indication label on December 8, 2017, by the Turkish Patent and Trademark Office. Divle Tulum cheese also called as "Divle Obruk cheese" is produced in Üçharman, a village located on the north slopes of Taurus mountains overlooking Central Anatolia in Ayrancı district of Karaman province.

Divle Tulum cheese is taken the name of the cave where the cheese is ripened in. It is claimed that this cheese has been produced for 700 years and the mold of Roquefort cheese which is known all over the world were taken from this cave (Kamber and Terzi 2008). In recent years, Divle Tulum cheese and its production have become the focus of attention due to its characteristic flavor and aroma. The internal color of Divle Tulum cheese changes from white to creamy (Fig. **3**) but the external color of the cheese in contact with the animal skin is dirty grey and pale (İşleyici *et al.,* 2018, Kamber and Terzi 2008). Divle Tulum cheese is produced between April and June and put into the cave for ripening. Ripened cheeses are taken out at the end of October.

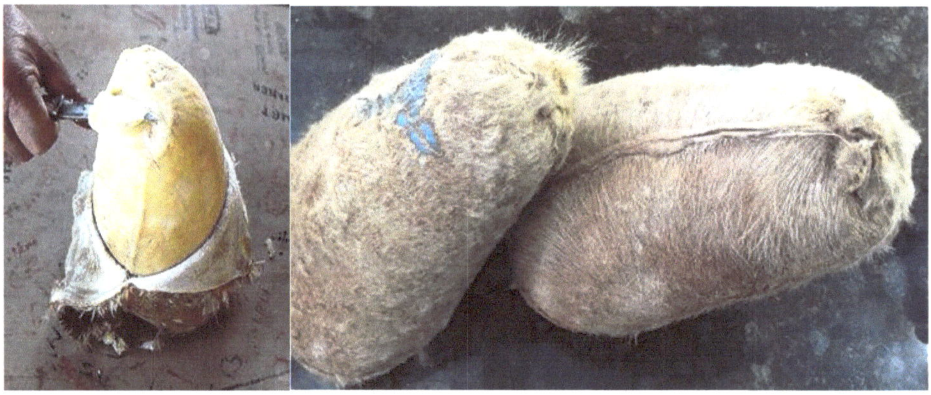

Fig. (3). Divle Tulum cheese.

Method of Production: Divle Tulum cheese is mainly produced from sheep's milk. When sheep milk is unavailable, it is mixed up with primarily goat milk and/or cow milk. Since traditional cheeses are generally produced by families, there are small differences between the cheeses (İşleyici *et al.,*2018).

After milking, raw milk is filtered through cheesecloth to remove coarse dirt. Milk is not pasteurized or heated. It is coagulated at own natural temperature. Approximately, a half a cup (50 mL) of natural rennet diluted with cold water is added to per a hundred liter of milk for coagulation. The milk is clotted within 1.5-2 hours. The cloth is pressed with a finger to check whether the clotting has been completed. If the finger does not adhere to the clot or if the clot surface does not break or if green whey is separated on the surface, the clotting process is completed. Then, the coagulum is cut into small cubes and slightly heated so that whey is drained easier. The curd is transferred into a cotton bag and hanged up to a high and cool place. When excessive whey is completely drained out, the bag is put on a wooden surface and pressed under a weight for reducing whey amount. Pressing is gradually increased weights on curd takes between 12-18 hours. Pressing continues up to 24 hours once the coagulum is not heated. After pressing, the fresh cheese is placed into a suitable container and cold water is added on the top. The water is changed once it is cloudy and it is repeated for 2-3 days until the water is clear. This stage is crucial for the removal of sour taste and yellowish whey. Fresh raw cheese is cut into slices, crumbled by hand and salted with rock salt. The amount of salt is adjusted according to the expected taste of cheese. When the cheese is excessively salted, ripening does not occur. These cheeses taste bitter and are locally called "Keşleşmiş" (Kamber and Terzi 2008, İşleyici *et al.,* 2018). Salted and crumbled fresh cheese is put into an animal skin bag called "Tulum" which is made of goat, sheep or lambskin. The fresh cheese is compressed into Tulum by hand or with a wooden stick (locally named "küsküç") in order to evacuate the air. If the cheese amount is enough to fill a Tulum, the surface of cheese is sealed with salt. If it is not enough, Tulum is kept and used for the second part of fresh cheese production. Then, the mouth of Tulum is closed with salt and sewed completely. For faster drying, Tulum is pierced in 15-20 holes with a large needle. All prepared Tulums are placed on sand or gravel in a cool place of house for a week. When skin starts to get drier, the Tulum is cleaned, marked and placed into cave for main ripening (Kamber and Terzi 2008, İşleyici *et al.,* 2018). The cave is located under a 37-m hill in 1.5-2 km southeast from Divle, in a sinkhole called "Obruk" (Fig. **4**). The depth and length of the cave are around 50 m from surface and 257 m, respectively. The temperature of the cave changes between 2 and 5°C and its humidity is around 85-90%. Although the total storage capacity of the cave is around 100 tons, only 40% is currently used (Gönç 1974, Kamber 2005, Kamber and Terzi 2008, İşleyici *et al.* 2018).

Fig. (4). Divle Obruk (Cave).

Tulums are used for ripening of Divle tulum cheeses, are prepared from skins of sheep, goat or lamb. The goatskin is generally preferred since it is sturdy. The procedure for the prepartion of Tulum is described as follows: Meat and fat on the surface of animal skin are taken away from. Then, the animal skin is salted and kept in a dry and cool place for 3-4 days. Before use, the dried animal skin is firstly immersed underwater for 1-2 days, and then the hairs on the surface are shaved. The inner side of the skin is completely rubbed with salt and flour. The skins tanned are used directly if they are small in size, or cutting down to necessary size if they are larger. After they are stitched to the Tulum shape, they are used. (İşleyici *et al.*2018).

Properties of The Cheese: There are many studies conducted on chemical, microbiological properties of Divle Tulum cheese and its production (Gönç 1974, Keleş and Atasever 1996, Morul and İşleyici 2012, Hayaloğlu and Karabulut 2013, İşleyici *et al.,* 2017, İşleyici *et al.,* 2018). First research on Divle Tulum cheese was performed by Gönç (1974). In this research, stages of manufacturing of the cheese and some chemical properties of 28 Divle Tulum cheese samples were determined. Also, Keleş and Atasever (1996) investigated on fat-free dry matter content, moisture, fat content, total ash, salt, pH and titratable acidity of Divle Tulum cheese. In a study conducted on primary and secondary proteolysis in eleven Turkish cheese varieties, some chemical and biochemical properties of Divle Tulum cheese were determined by Hayaloğlu and Karabulut (2013a). Some chemical properties of Divle Tulum cheese are provided in Table **1**.

Coliform bacteria, fecal streptococcus, yeast and fungi counts in Divle Tulum cheese were detected 1.64×10^6 cfu/g 5.58×10^7 cfu/g, 3.5×10^6 cfu/g, respectively (Keleş and Atasever 1996). In a study investigated microbial composition of 50 Divle Tulum cheese samples, anaerobic mesophilic count of 6.78x10 log cfu/g, *E.coli* count of 3.61 logcfu/g in 9 samples, coliform count of 3.04 log cfu/g in 20 samples, *S.aureus* count of 5.04 logcfu/g in 40 samples, coagulase (+) *S. aureus* of 4.82 log cfu/g in 25 samples, *Enterococcus* of 6.69 logc fu/g in 48 samples, *Enterobacteriacea* count of 2.90 log cfu/g in 40 samples, psychrophilic count of 4.29 log cfu/g in 25 samples and yeast/mold count of 6.63 log cfu/g in all samples were found (Morul and İşleyici 2012). Researchers explained that Divle Tulum cheese has potentially a risk for public health regarding the level of investigated specific organisms.

In another study inspected the presence of AFM1 in 55 Divle Tulum cheeses, 18.18% of samples had between 5.15 and 26.44 ng/kg, an average level of 10.84 ng/kg AFM1, 41.82% had less than 5 ng/kg AFM1 was found, but AFM1was not detected in 40% of samples. The results showed that the level of AFM_1 in analyzed Divle Tulum cheeses had less than the maximum limit (500 ng/kg) of specified in Turkish Food Codex and there was no risk for public health in related with AFM1 level (İşleyici *et al.,* 2011).

Water-soluble nitrogen and TCA-soluble nitrogen and total free amino acids level of cheeses were found 21.47%, 12.29%, 4.19 mg Leu/100 g, respectively in Divle Tulum cheese. The concentrations of total free amino acids were determined to be highest in Divle Tulum cheese among the investigated cheeses (Hayaloğlu and Karabulut 2013a). In another study conducted by Hayaloğlu and Karabulut (2013b), free fatty acids content of Divle Tulum cheeses was found as C_4: 13.9, C_6: 10.7, C_8:14.9, C_{10}:19.5, C_{12}:12.5, C_{14}:32.8, C_{16}:75.8, C_{18}:22.5, $C_{18:1}$:87.2, $C_{18:2}$:7.9 mg/100g cheese. These results indicate that lipolysis occurs at high levels in Divle Tulum cheese and this cheese has a strong characteristic aroma due to the high content of short-chain free fatty acids (> 40 mg / 100 g).

The heavy metal and mineral contents in 20 ripened Divle Tulum cheese samples were investigated by İşleyici *et al.* (2017). The mean level of lead (Pb) of 0.026 mg/kg (n=11), Cadmium (Cd) of 0.020 mg/kg (n=), Arsenic (As) of 0.068mg/kg (n=16), Mercury (Hg) of 0.009 mg/kg (n=8), Copper (Cu) of 0.201mg/kg (n=2), Iron (Fe) 10.892 mg/kg (n=14), Zinc (Zn) of 21.699mg/kg (n=16) were found. Heavy metal level of Divle Tulum cheese did not exceed limits defined in national and international standards. The mean level of calcium (Ca), phosphorus (P), magnesium (Mg), aluminum (Al), potassium (K), tin (Sn) and nickel (Ni) were found 6333.96 (n=20), 4802.26 (n=20), 203.76 (n=20), 0.513 (n=3), 1153.23 (n=20), 0.061 (n=10) and 0.167 mg/kg (n=7), respectively. The consumption of Divle Tulum cheese contributes significantly to human nutrition depending on mineral substances.

ÇEPNI (AKÇABELEN) TULUM CHEESE

Çepni Tulum cheese is produced traditionally by nomadic people in the Akçabelen (Çepni) tableland in the Beyşehir district of Konya province. This cheese made from full-fat goat's milk is ripened in Tulum.

Method of Production: Rennet is added to fresh raw goat milk after milking. After 1-2 hours, the curd is formed and put into a cloth bag and then hang a high spot. It is left for 7-8 hours for whey draining. The drained curd is crumbled by hand and salted. Then, it is pressed into a Tulum. After 10-15 days, the cheese is removed from the Tulum. It is crumbled again and pressed into another Tulum. Thus, prepared cheeses are ripened in caves located in that area for 3-4 mounts.

GÖLBAŞI TULUM CHEESE

Gölbaşı Tulum Cheese is produced in Gölbaşı district of Ankara province. This cheese is made from skimmed or semi-skimmed milk at the end of the spring or summer (between May and July). Gölbaşı Tulum cheese has very small granules. It has a slightly salty and ripened flavor. The color of the cheese is bright white

(Fig. **5**) (Kamber and Terzi 2008). The cheese is ripened in Tulum which are obtained from lamb. Lambskin's hairs are shaved. Then, it is kept in salty water for a day and is dried under the sun. The skin is kept for a few hours in warm water to soften prior to use.

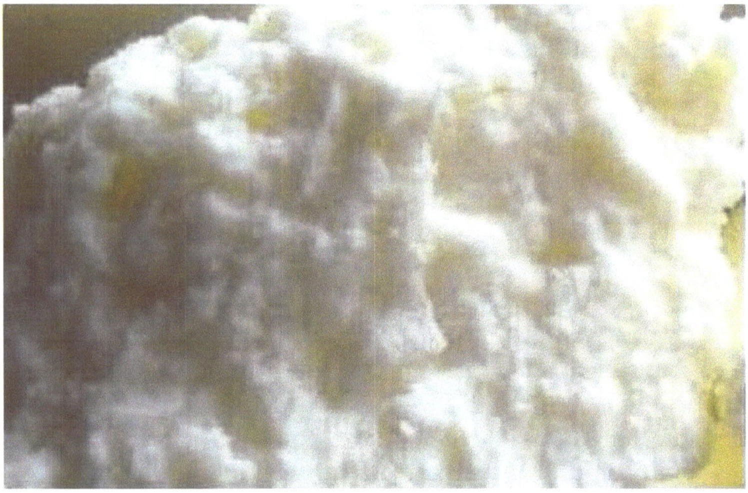

Fig. (5). Gölbaşı Tulum Cheese.

Method of Production: The raw milk is heated to 31-34 C° and added rennet. Then the milk is left for 2-3 hours to coagulation. After curd formation, the curd is filled into clot bag and kept for 10-12 hours to whey drainage on a wooden surface. During the draining, the pressure is applied to the curd. The cheese is cut into large pieces and salted. After a day, the large curd pieces are put into the brine. It is kept for 2-3 mounts in the brine. The cheeses are removed from the brine in autumn and are grated. Finally, the cheeses are filled into Tulum by pressing firmly. Tulum's mouths are closed tight and placed on soil or sand surface. Then weights are placed on it. Tulum is wiped with a dry cloth every day and removed whey. In this process, the soil or sand placed under the Tulum is changed every 2 days. After a month, the cheese is poured into a large basin and then some amount of cream is added. The cheese mixed with cream is filled into another Tulum and ripened for 2-2.5 months in a cool environment. Mild molded cheese is preferred by the consumer.

EREĞLI CLOTH TULUM CHEESE

Ereğli Cloth Tulum cheese is made of cow milk, sheep milk or mixture in Ereğli district of Konya province. This cheese has a mild rancid aroma. It has a bright creamy color and smooth texture (Fig. **6**).

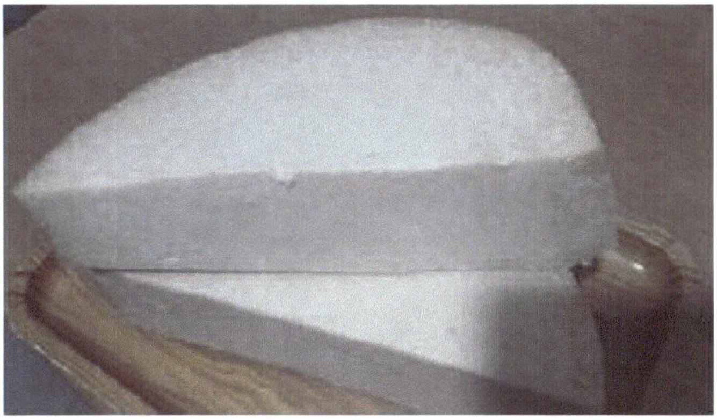

Fig. (6). Ereğli Cloth Tulum Cheese.

Method of Production: Ereğli Cloth Tulum cheese is currently produced from pasteurized milk in industrial scale, but few people make this cheese by the traditional way in villages at the household scale. After milking, raw milk is heated to warm temperature and rennet is added for coagulation. When the coagulation is completed within 4 hours. Curd is placed in a cloth bag. The cheese is pressed until to obtain dry structure. Then, the cheese is crumbled into small pieces and salted with rock salt. Salted cheese is boiled in the hot water. While melted cheese is hot, it is pressed into again cloth Tulum. The mouth of Tulum is closed when the cheese is cool down (Fig.**7**). Then, it is placed in a cool area of house and ripened for one or two months (Kamber and Terzi 2008, Anonymous 2018a).

Fig. (7). Ereğli Cloth Tulum Cheeses in cloth bags.

KARAMAN TULUM CHEESE

Karaman Tulum cheese is made in Karaman province. This cheese is similar to

Ereğli Cloth Tulum cheese and Konya Molded cheese. Karaman Tulum cheese is made from whole sheep milk or goat milk. After coagulation, the cheese is pressed into a cloth bag or an animal skin. The cheeses are ripened for two or three months (Kamber and Terzi, 2018).

Ermenek Tulum cheese of Karaman is also popular known cheese in Ivriz area. Karaman- Ermenek Tulum cheese is made from whole sheep milk. It has a less salty taste and good flavor. Raw milk is coagulated with rennet. When the coagulation is completed, the coagulum is cut into nodule shaped. The curd is heated at 60-80 °C for 10 min. After 30 minutes, it is placed on wood and kept under press for a day. Then curd is crumbled and salted and pressed into Tulum. Sometimes, black sesame is also added to the cheese (Ünsal, 2000).

CHEESES RIPENED IN POT, JAR, AND PAN

These types of cheeses are produced in Kayseri, Ankara, Niğde, Nevşehir, and Sivas provinces. These cheeses are ripened in Jar called as "Küp or Küpecik", Pan "Çanak" and Pot "Çömlek". They are similar to Tulum cheese, but their flavor-aroma and characteristics are very different.

ÇÖMLEK (POT) CHEESE

Çömlek cheese is made in Kırşehir, Çankırı and Nevşehir provinces in Turkey. It is usually made in villages at household scale and it is named after the province such as Kırşehir Çömlek cheese, Çankırı Çömlek cheese, Nevşehir Çömlek cheese. The making of these cheeses is similar to Niğde Küp cheese and Yozgat Çanak cheese. The main difference from Niğde Küp cheese is that black sesame added to Kırşehir Çömlek cheese when the cheese is crumbled. The other difference is that Nevşehir Çömlek cheese is not buried underground or sand for ripening, it is ripened on the ground within the pot that is hermetically sealed (Fig. **8**). The mouth of the pot or jar is sealed with dough instead of sludge. The color of the cheese is generally white or dirty white. The flavor and aroma of the cheese are mild. It has medium salty, fatty, and ripen taste. Its texture consists of small granules (Kamber and Terzi, 2008, Ünsal, 2000).

Method of Production: Raw milk is used for manufacturing. After milking, rennet is added to the milk at its own temperature for coagulation. After the coagulation, the curd is warmed up to 50-55 °C without being cut. Then it is poured into a cloth bag and hung in a high place for draining. After removing a large amount of whey from the cheese till the next day, it is pressed among two heavy stones for one more day. Then fresh cheese is crumbled with hand and salted. Then, it is placed into the cloth bag and the remaining whey is removed until a dry form is obtained. The cheese is removed from cloth bag and crumbled again. Also black sesame is added for Kırşehir Çömlek cheese. Then, the cheese is pressed into the pot. While the mouth of pot is sealed with dough in Nevşehir Çömlek cheese making, a leaf is placed in the mouth of pot and it is closed with a clean cloth in Kırşehir Çömlek cheese making. Nevşehir Çömlek cheese is ripened in the pot for 2 -3 months in cold place of house or storeroom. But Kırşehir Çömlek cheese is buried upside down into sand in a cold place of house and ripened for 3 months (knowledge obtained from Latife Yeni by personal communication, Kırşehir, 2018).

Fig. (8). Nevşehir Çömlek Cheese.

Properties of the Cheese: Chemical composition of Çömlek cheese is non-fat dry matter 26.9%, fat 40.9%, ash 6.5%, salt 5.0%. The acidity of the cheese is 0.4% lactic acid (LA) (Şahin, 1997). Total aerobic microorganisms, lactic streptococci bacteria, proteolytic bacteria, lactobacillus, coliform, yeast and mold of Çömlek cheese are found as $9.9x10^9$cfu/g, $3.0xx10^7$ cfu/g, $6.0x10^6$cfu/g, $3.0x10^7$cfu/g, $3.3x10^4$cfu/g, $2.4x10^6$cfu/g, respectively (Şahin, 1997). The prevalence of *Listeria* spp. of Çömlek cheese obtained from Kırşehir and Nevşehir province was investigated in 70 samples (Kaptan, 2016). *Listeria* spp. were found in 6 (18.8%) of 32 samples obtained from Kırşehir. *L. monocytogenes* was determined in 5 of these samples, *L. grayi* was determined in one. *Listeria* spp were found in 9

(23.7%) of 38 samples obtained from Nevşehir. *L. monocytogenes*was determined in 7 of these samples *L. grayi* was determined in one and *L. ivonavii* in the other one (Kaptan, 2016).

KAYSERI ÇÖMLEK (POT) CHEESE

This cheese differs from the other pot cheeses produced in Central Anatolia (Fig. **9**). In the production of Kayseri Çömlek cheese, the curd is washed. In addition, the property of this cheese is related to the rennet used in clotting. The home-made rennet is traditionally prepared using herbs collected from the mountains. In spring, 8 different herbs, such as thyme, spindle leaves, pear leaves have been used for the preparation of rennet.

Method of Production: Fresh cheese is kept in cold water for washing. Then the curd is boiled at 60-65 °C for a short time. It is crumbled by hand. Salt and black susame are added at the same time. Prepared cheese is placed into a clear pot with a capacity of 1-2 kg. The filled pot is completely closed with a 2-3 cm thick melted animal fat to prevent the entry of air. Then, the pot is buried upside down into river sand and left for ripening for 3-4 months (Kamber and Terzi, 2008, Yalman *et al.,* 2016).

Fig. (9). Kayseri Çömlek (Pot) Cheese.

There are very limited studies conducted on Kayseri Çömlek cheese. The traditional manufacturing method of Kayseri Çömlek cheese was introduced in a research (Kamber and Terzi, 2008), the other one was conducted on mineral and trace compounds of Kayseri Çömlek cheese. In this study, concentrations of some elements in Kayseri Çömlek cheese were analyzed by Mendil (2006). The average levels of Fe, Mn, Zn, Cu, Pb, Cr, Ni, Na, K, Ca and Mg are 8.9, 1.1, 10.1, 0.19, 1.2, 0.02, 0.20, 5037, 311,3473 and 28.9 µg/g, respectively.

NIĞDE KÜP (JAR) CHEESE (GÖDELEK CHEESE)

Niğde Küp cheese also known as Gödelek cheese, is a local Turkish cheese (Fig. **10**). The name "Gödelek" refers to the jar, which is of smaller size in Niğde province Prior to using in cheese manufacturing, these jars are filled with salty water or whey and left for 24 hours. Then, they are dried under the sun (Ünsal, 2000). Niğde küp cheese is mainly produced from raw sheep milk. However, goat milk or cow milk or goat/cow mixture can also be used. Its production method is similar to that of Yozgat Çanak cheese, but there are minor differences (Kamber and Terzi, 2008).

Fig. (10). Niğde Küp (Jar) Cheese.

Method of Production: Rennet is added to the freshly milked milk. After 1 hour, the milk is converted to the curd. Then, the curd is mixed with rolling pin and is allowed to rest for a short time. The curd is transferred into a cloth bag for whey drainage. A heavyweight is placed on the clot bag and it is left for one day. Then the cheese is removed from the cloth bag and put into a large copper basin called "strike" and broken into hand size pieces and salted. Then cheese pieces are again filled into cloth bag and left to drainage for 6 hours using a heavyweight above it. After drainage, the cheese is again placed in copper basin and is cut into small thin pieces. Then, it is placed into a cloth bag and pressed again under pressure. At the end of the draining, the cheese is poured into copper basin and crushed by hand. If necessary, salt may be added at this stage. Also salt is put into the bottom of Gödelek before using, then the cheese is tightly filled into it. The pot is closed with a cheesecloth. Finally, Gödelek is buried upside down in the river sand renewed every year. Gödelek cheese is ripened in cold rooms for 4-5 months. Niğde Küp cheese is produced in June and consumed in October and November (Ünsal, 2000, Kamber and Terzi, 2008).

SIVAS KÜP (JAR) CHEESE

Sivas Küp cheese ripened underground in the jar, is produced in Sivas province of Central Anatolia (Fig. **11**) (Kamber and Terzi, 2008; Pekel and Korukluoğlu, 2009). The name of "Küp" refers to clay pot, in which the curd is filled and ripened. Nowadays, plastic containers are also used for the ripening of Küp cheese. This cheese is similar to the Çanak (Pan) and Gödelek cheeses which are produced in Yozgat and Niğde provinces. There are little differences in the production processes.

Fig. (11). Sivas Küp (Jar) Cheese.

The production of Küp cheese is generally made from skim milk or semi skim milk in June and July. The cheeses made within this period are ripened for 2-3 months and are consumed in October.

Only one study has been conducted regarding Küp cheese until today in the literature. The chemical composition of Küp cheese was investigated and 49.51%

dry matter, 9.44% fat, 4.53% salt, 24.78% protein and 5.82 pH were found (Pekel and Korukluoğlu, 2009). In addition, these researchers reported that coliform bacteria were not found in the microflora of 25 Küp cheese samples. However, strains of Aspergillus and Penicillium were determined in the Küp cheese samples. Yeast and mold counts were determined as 7.57 log cfu g^{-1} in these samples.

Method of Production: This cheese is traditionally made from skimmed milk. After clotting, the curd is cut into large pieces and heated at 55-80 ° C for 10-15 minutes. The heated curd is filled into a cloth bag and placed on a wooden floor to drain for 10-15 hours. Simultaneously, a heavyweight pressure is applied to the curd (Fig. **12**). Then, the curd is divided into medium pieces by hand and firmly pressed into a jar or plastic bag with a capacity of 10 kg. During this process, salt (100 g salt for 2 kg of cheese) and sometimes cream are added between each layer of cheese. Once it is filled completely, the jar is closed with a cloth and placed upside down for 2 - 3 days. Then, it is buried in the sand for ripening for 2-3 months in a cool environment (Kamber and Terzi, 2008; Pekel and Korukluoğlu, 2009).

Fig. (12). Draining stage of cheese within clot bag in Sivas Küp cheese making.

AYAŞ KÜPECIK (JAR) CHEESE

There are two ways to make Küpecik cheese in Ayaş district of Ankara province: "Basma" and "Ovma" methods. The pot used in both manufacturing is called "küpecik" and is also utilized for cooking kebabs by making a hole in the bottom that allows the air to escape. Küpecik cheese is no longer produced or rarely produced due to changes in living conditions in rural areas. (Ünsal 2000, knowledge obtained from Sedat Ceylan by personal communication, Ankara, 2018).

AYAŞ BASMA CHEESE

Ayaş Basma cheese is two times ripened. The name of Basma refers to the pressure of the cheese in the pot. Generally, Ayaş Basma cheese is consumed with a traditional bread called "Bazlama". The production method is as follows: The fresh cheese or curd made from low-fat milk is drained out of a cloth bag and then is round-shaped and salted. Then, it is left for drying on of the rooftop of the houses for a month. During the pre-ripening period, the cheese gets harder. Once it becomes stiff, the cheese is taken off and often used by wetting. Then it is sliced and the solid surface is removed. Small pieces are boiled in hot water. Simultaneously, the cheese is stretched to obtain flexible structure similar to Kashar cheese. The cheese is pressed into the pot and ripened for 2 months. After ripening, the pot is broken and cheese is removed out.

AYAŞ OVMA CHEESE

Ayaş Ovma cheese is consumed with a traditional soup called "Tarhana". The name of this cheese refers to the term "rub" with hand. The production method is as follows: The raw fatty sheep milk is coagulated with rennet called "Şirden". Coagulum is cut into small pieces. Curd is crumbled by the way of rubbing. Then it is mixed with cream and fatty cheese. Salt and black sesame are simultaneously added to the mixture. Finally, it is pressed into a pot called "Küpecik". The mouth of the pot and the hole on bottom of pot are covered with tallow or grape leaves. After then, the pot is placed in an absorber soil called "bebe toprağı" in a cold environment for 3 months.

ÇANKIRI KÜPECIK CHEESE

Çankırı Küpecik cheese is made in Çankırı province. This cheese is named after Küpecik which is short and smaller, wide-mouthed earthenware pots. Characteristic of this pot can be able to release the serum of cheese from inside but not to leak fat of cheese. Actually, this cheese is similar to other Jar cheeses produced in this area. Heated milk is used for manufacturing of this cheese. Also curd is not heated.

Method of Production: Fresh raw milk is left for a while and the formed cream layer is removed. Semi-skimmed milk is heated and cooled to renneting temperature around 28-35 °C. Then rennet is added to milk. Rennet is prepared traditionally by cleaning and drying of abomasum. Commercial rennet is not used for the production of cheese. Today, a few people produce their own rennet. After coagulation, curd is placed into clot bag and hung to a high pot for draining. Drained cheese is removed from bag, crumbled with hand and salted using rock salt. Then if it is necessary, it is put into the clot bag again, and placed under a heavy stone and kept for about more day. Finally, the cheese is pressed into Küpecik (pot). Water and air inside the pot should not be released during the pressing. Then, the pot is covered with a cloth and placed reversibly on the sand in a cold place of house. Küpecik cheese is generally ripened for 2 or 3 months (knowledge obtained from Halim Hıcıplı by personal communication, Çankırı, 2018).

YOZGAT ÇANAK (PAN) CHEESE

Çanak cheese which can be classified as a semi-hard cheese is slightly salty, ripened and white color and has a homogeneous granulated structure (Kamber and Terzi, 2008; Kırdar and Kursun, 2011). Yozgat Çanak Cheese is registered by geographical indication on December 18, 2017 (Anonymous 2017b). Yozgat Çanak Cheese is described as mild ripened and low-fat cheese. This cheese is produced from milk of sheep, goat, cows feeding in natural vegetation and water sources of Yozgat and its vicinity. This cheese gets its name from the pot in which it is put into the process of ripening called "Çanak" (Fig. **13**). Similar cheeses are also produced with different names in neighbor provinces, such as Küp Cheese in Sivas, Küpecik Cheese in Çankırı, Gödelek Cheese in Niğde and Ovma Cheese in Ankara Ayaş (Kamber and Terzi, 2008).

Fig. (13). Yozgat Çanak (Pan) cheese.

Çanak cheese, a kind of local cheese, is produced traditionally with a small scale for self-consumption. Due to the lack of a standard production technique, the chemical composition of Çanak cheese is very variable. It can be produced from whole or skimmed milk depending on the producer. There is very limited research regarding Çanak cheese. Akyüz and Gülümser (1984) reported that dry-matter fat,

fat in dry-matter, protein, salt in dry-matter, the acidity of Çanak cheese were as 51.49%, 13.5, 25.54%, 26.6%, 11.92% and 1.62% lactic acid, respectively. Kırdar *et al.*, (2013) investigated mineral content of 40 Çanak cheese samples. In this study, the mean calcium, magnesium and potassium concentrations of samples were found to be 250.01-3487.96 mg / kg, 10.01-171.40 mg/ kg and 225.62-296.74 mg/kg, respectively. In another study, coliform bacteria, *Staphylococcus*, *Micrococcus* spp., *Enterococci* and *Enterobactericeae* were determined in 35 Çanak cheese samples at an average rate of 2.62, 3.97, 3.34 and 3.34 log cfu g^{-1} respectively (Kırdar and Kursun 2011). Same authors reported that yeast and mold counts were between 2.00-6.08 log cfu g^{-1} and 2.00-4.78 log cfu g^{-1} respectively in Çanak cheese samples. The high number of microorganisms content of the Çanak cheese was related to the use of raw milk, non-hygienic production and storage condition. Kaptan (2016) reported that the incidence of *Listeria* spp., *L. monocytogenes*, *L. grayi*, *L.innocua*, and *L. innovai* in 42 Yozgat cheese samples was 21.4%, 9.5%, 4.8%, 4.8%, and 2.4%, respectively.

Method of Production: The raw milk is heated to the coagulation temperature (32-35 °C) and added home-made rennet (100 ml rennet per 10 kg milk). While the rennet is slowly added, the milk is stirred continuously at the same time. Clotting time varies between 1-1.5 hours.

After the curd is formed, it is allowed to rest for 15-20 minutes by dividing into 4 equal parts. The broken curd is transferred to the cloth bags and cloth bags are hanged in a high place and left to drain for 30-120 minutes (until the whey flow stops). After draning, the pressure is applied to the curd (3-5 kilograms of weight for each kilogram of cheese) for 10-12 hours. Then, the curd is placed into a large tray and cut into large pieces with a knife and salted (75 g salt for 1 kg of cheese). The salted curd is placed into the last cloth bag. Second time pressure (10 kilograms of weight for each kilogram of cheese) is applied to the cloth bags in a cool environment (12-16 °C). This second pressure process that lasting 2-7 days is called "Tercan", "Tecin", "Tecen" and "Bastırık". After pressing, the cheese is again poured into a large tray and is crumbled by hand. Salt is added again at this stage (75 g salt for 1 kg of cheese). The cheese is placed into the pan (Çanak) and it should be pressed firmly without gaps. After the pan is filled, salt is added to the top of cheese and the mouth of the pan is covered with a clean cloth. After 2-3 days, the pan is turned upside down. During this stage, the cloth in the mouth of the pan is changed as it gets wet. The cheese is left until the cloth is completely dry (3-5 days). Finally, the cloth covering the pan is replaced with a clean one and is covered with a thick piece of dough. Then, the pan is buried upside down in the sand (Fig. **14**). When the number of pan is limited, the pans are placed into a box filled with sand. The ripening is achieved in a cool and slightly humid environment for 3-4 months. Generally, Çanak cheese is produced in summer and buried until the ripening in autumn (between September and November), then it is consumed in winter (Akyüz and Gülümser, 1984, Kamber and Terzi, 2008).

Fig. (14). Ripening upside in sand in Yozgat Çanak Cheese.

CONSENT FOR PUBLICATION

Not applicable.

CONFLICT OF INTEREST

The authors confirm that this chapter contents have no conflict of interest.

ACKNOWLEDGEMENTS

Declared none.

REFERENCES

Anonymous, (2017a) *Turkish Statistical Institute* ,Turkey.

Anonymous, (2017b) https://www.turkpatent.gov.tr/TURKPATENT/resources/temp/FE34F670-2762-40-8-87DB-940CB13D6376.pdf.

Anonymous, (2018) http://www.eregli.bel.tr/tr-407946456-a407207559/Accessed.

Akyuz, N & Gulumser, S (1984) Production, composition and ripening of yozgat çanak cheese. *J Food,* 9, 231-8.

Güley, Z, Uysal, HR & Kilic, S (2014) Lactic acid bacteria flora of konya küflü cheese: a traditional cheese from konya province in Turkey. *J Microbiol Biotechnol Food Sci,* 04, 238-42. [http://dx.doi.org/10.15414/jmbfs.2014-15.4.3.238-242]

Güner, A & Telli, N (2011) a survey on the presence of *Listeria monocytogenes* in various semi-hard cheeses from different regions of Turkey. *J Anim Vet Adv,* 10, 1890-4.
[http://dx.doi.org/10.3923/javaa.2011.1890.1894]

Gönç, S (1974) Studies on divle tulum cheese technology and its constituents. *J Ege Univ Agric Fac Seri A,* 11, 515-33.

Hayaloglu, AA & Karabulut, İ (2013) Primary and secondary proteolysis in eleven turkish cheese varieties. *Int J Food Prop,* 16, 1663-75. a
[http://dx.doi.org/10.1080/10942912.2011.604890]

Hayaloglu, AA & Karabulut, İ (2013) Characterization and Comparison of Free Fatty Acid Profiles of Eleven Varieties of Turkish Cheeses. *Int J Food Prop,* 16, 1407-16. b
[http://dx.doi.org/10.1080/10942912.2011.587626]

Hayaloglu, AA, Brechany, EY, Deegan, KC & McSweeney, PLH (2008) Characterization of the chemistry, biochemistry and volatile profile of küflü cheese a mould-ripened variety. *Swiss Soc of Food Sci and Technol,* 41, 1323-34.
[http://dx.doi.org/10.1016/j.lwt.2007.08.020]

Isleyici, O, Sancak, YC, Tuncay, RM & Yucel, UM (2017) Determination of mineral substance and heavy metal levels in divle tulum cheese. *Van Veteri J,* 28, 151-6.

Isleyici, O, Sancak, YC & Tuncay, RM (2018) Divle Tulum Cheese. *Van Veteri J,* 29, 119-24.

Kamber, U (2015) inary Traditional turkey cheeses and their classification. *Van Veteri J,* 26, 161-71.

Kamber, U & Terzi, G (2008) The traditional cheeses of Turkey: Central anatolian region. *Food Rev Int,* 24, 74-94.
[http://dx.doi.org/10.1080/87559120701764548]

Kaptan, B (2016) Prevalence of *Listeria spp and L. monocytogenes* in Home made Pottery Cheese. *Tekirdag Ziraat Fak Derg,* 13

Kirdar, SS & Kursun, O (2011) Microflora and pathogen bacteria in traditional anatolian canak (Pan) cheese. *J Anim Vet Adv,* 10, 2511-5.
[http://dx.doi.org/10.3923/javaa.2011.2511.2515]

Kirdar, SS, Ocak, E, Kose, S & Ozer, E (2013) Seasonal changes in mineral and trace elements of çanak cheese. *Asian J Chem,* 25, 6177-80.
[http://dx.doi.org/10.14233/ajchem.2013.14304]

Mendil, D (2006) ' Mineral and trace metal levels in some cheese collected from Turkey'. *Food Chem,* 96, 532-7.
[http://dx.doi.org/10.1016/j.foodchem.2005.03.006]

Morul, F & Isleyici, O (2012) Chemical and microbiological properties of divle tulum cheese. *Van Veterinary Journal,* 23, 71-6.

Özgören, E & Seckin, AK (2016) Aflatoxin M1 contaminations in mouldy cheese. *Mljekarstvo,* 66, 154-9.

Pekel, M & Korukluoglu, M (2009) A study on microbiological, chemical qualities and moulds flora of kup (crock) cheese produced in sıvas province *Anadolu Journal Agricultural Sciences,* 24, 1-7.

Şahin, E *The Effects of manufacturing Çömlek cheese from raw and pasteurized milk and ripening in different storage conditions PhD*Thesis, No.58656 (1997) Selçuk University, Institute of Medical Sci, Konya Turkey.

Tekinsen, KK, Cebirbay, MA & Elmalı, M (2009) Moldy Cheese of Konya. *Acad Food J,* 7, 31-5.

Ünsal, A (2000) *Süt uyuyunca Türkiye Peynirleri Yapı Kredi Yayınları, No1339, 2* Baskı, İstanbul 221.

Yalman, M, Tepeli, SO & Zorba, NND (2016) 'Mold Flora of Traditional Cheeses Produced in Turkey', Turkish Journal of Agriculture –. *Food Sci Technol (Campinas),* 4, 926-33.

Yilmaz, G, Ayar, A & Akin, N (2005) The Effect of Microbial Lipase on the Lipolysis During The Ripening of Tulum Cheese. *J Food Eng,* 69, 269-74. [http://dx.doi.org/10.1016/j.jfoodeng.2004.08.017]

İşleyici, O, Sancak, YC & Morul, F (2011) A study on the level Aflatoxin M1 in Divle Tulum Cheese. *Van Veter J,* 22, 105-10.

CHAPTER 11

Traditional Cheeses of the Aegean Region in Turkey

Cem Karagözlü* and **İlyas Erdem Tonguç**

Faculty of Agriculture, Department of Dairy Technology, Ege University, Bornova – İzmir, Turkey

Abstract: Cheese production has much developed in the Aegean Region, where the Anatolian lands meet the sea. The milk used to make cheese is obtained from the animals that give birth in February and March. In mountainous areas and highlands, this period also covers April and May, and, therefore, goat milk becomes important. Especially, fatty milk is collected in the spring season, after the calves and lambs are born. During this period, ewe's milk is very fatty and used in combination with cow's milk. Therefore, ewe's milk is preferred for cheese production during the spring season. Small ruminants are of importance in the Inner Western Anatolia. Today, technological applications in cheese production, as is the case in many areas, have diversified the cheese production with the modernization of traditionalism. As in the Aegean Region, in many countries, there are numerous cheese types for which traditional production is still preferred and offered to the world market. The factors that affect the taste of cheese range from the properties of grass growing at a different flora where the animal from which the milk was obtained grazed, to production technique, sanitation and hygiene. Unfortunately, it has become harder to find local cheeses in markets with their genuine traditional tastes. At this point, the necessity of reflecting the local tastes obtained through traditional production methods should not be overlooked while making a transition in production techniques, from traditional to modern. Locally the most popular cheese types produced in the Aegean Region are İzmir Tulum Cheese, Sepet Cheese, Kirlihanım Cheese, Kopanesti Cheese, Armola Cheese, Lor Cheese, Tire Çamur Cheese, Aydın Çörekotlu Cheese, Çoban Cheese, Köy Cheese, and Peppered Lor.

Keywords: Aegean region, Armola, İzmir Tulum cheese, Kopanesti, Turkish local cheeses.

INTRODUCTION

Aegean Region, with its unique beauty where the Anatolian lands meet the sea, covers 11% of Turkey with a surface area of approximately 85,000 km^2 (Fig. **1**). It

* **Corresponding author Cem Karagözlü:** Faculty of Agriculture, Department of Dairy Technology Ege University Bornova – İzmir Turkey; Tel: +902323112902 Fax: +902323881864; E-mail: cem.karagozlu@ege.edu.tr

Celile Aylin Oluk & Oya Berkay Karaca (Eds.)

is the 5th largest region in Turkey. The characteristics of the mountainous areas in the Aegean Region are that they lie in the form of mountain lines in the east-west direction. These mountain lines in the east-west direction are separated with valleys lying in the same direction. In the Edremit, Bakırçay, Gediz, Küçük Menderes and Büyük Menderes basins, which form the grabens in the east-west direction in the Aegean Region, the plains other than the deltas fall into the alluvial-filled sedimentary plains group. Also, the plains in the Inner Western Anatolia fall into this group (Kütahya, Simav, Afyon, Sandıklı, Şuhut plains). The Aegean Sea coast is the longest coast of Anatolia, mostly divided into bays and coves. The climate effective until the inner parts of the region is the Mediterranean climate. Summers are hot and arid while winters are warm and rainy. The average temperature of the warmest month is 27-28 °C, while the average temperature for the coldest month is 8-10 °C. The average yearly temperature is 17-18 °C. Snowfall and frost are very rare. The highest precipitation is in winter whereas the lowest is in summer. The yearly precipitation varies depending on the altitude. It ranges from 600 to 1000 mm. The vegetation consists of scrub plants, called maquis. Maquis consists of plants including myrtle, laurel, arbutus, olive, oleander, locust, kermes oak, *etc*. Higher places are covered with forests (red pine forests). The Mediterranean climate in the region loses its dominance in the northern parts and, with the effect of elevation in the eastern parts. Towards the inner parts, the continental climate becomes dominant due to the increase in elevation and distance from the sea effect. Therefore, winters are cold and snowy in the inner parts. The vegetation is steppe. Higher altitudes are partly covered with oak forests. Cattle breeding is common in the larger cities and vicinity whereas sheep breeding is common and developed in the inner parts. Goats are bred in the coastal and mountainous areas (Kanat, 2016).

On the other hand, the Aegean Region has been involved in an intensive commercial activity, dating back to the ancient era, covering the Aegean and the Mediterranean. Beyond exchanging the goods, trade is a phenomenon where different societies come together and interact with each other in a cultural sense. The coastal cities such as Efes, Bergama, Foça, Erythrai, Milet, Bodrum were places where trade and culture were traded for years. Especially after the 11th Century, Venetians become dominant in the trade. In the course of this trade history, the records are uncommon for cheese. David Jacoby, in his study entitled "Cretan Cheese: A neglected Aspect of Venetian Medieval Trade" states that cheese was produced commonly in the Mediterranean in the Medieval Age. It was the food of the poor who can rarely eat meat. In addition, cheese had an important place in the foods consumed by the Byzantine monks. Monks did not consume meat. Also, Jacoby states that cheese has a revitalizing effect on those who travel for long periods on the land and the sea. In the expeditions of the armies, the

Byzantine, Arabic, Turkish and Jewish sources state that cheese was transported as a trade product (Swan, 2009).

Fig. (1). Some cheeses of the Aegean region in Turkey.

With almost ten-thousand-year-old cuisine culture and tradition, the most popular local cheese types in the Aegean Region are, İzmir Tulum Cheese, Bergama Tulum Cheese, Ayvalık Sepet Cheese, Ayvalık Kirlihanım Cheese, Kopanesti, Karaburun Goat's Sepet Cheese, Seferihisar Armola Cheese, Tire Çamur Cheese, Aydın Çörekotlu Cheese, Denizli Derecik Cheese, Çoban Cheese, Köy Cheese, Milas Kazıklı, Milas Kırktokmak, Bodrum Testi, Peppered Lor and Yörük Cheese (Fig. **1**). To make the cheese more durable and to extend its life, the tradition of using skin-bags, cubes, kegs, bags and baskets was adopted from Central Asia to the Balkans. These choices are also seen in the traditional cheese of the Aegean Region. Most of the cheeses that we will discuss in this section are produced by small producers in the villages by traditional methods and sold in local markets. There are no academic researches on most of these cheese types while some limited scientific data about some of these cheeses have been published.

İZMIR TULUM CHEESE

Although Tulum cheese is named after the packaging material, the majority of Tulum cheeses available on the market today are in a tin or plastic packing. İzmir Tulum Cheese (Brined Tulum Cheese), in TS 11966 standard, is defined as a product produced from pasteurized cow's, ewe's, buffalo's and goat's raw milk or mixtures of them with the proper processing, and using additives during production, obtained at the end of a certain ripening process (Anonymous, 1996).

İzmir Tulum cheese is different from the similar cheese produced in Anatolia in terms of production method. This difference is also reflected in its taste and structure. Today, the production of İzmir Tulum in the skin bags continue only in very few dairy plants. In many dairy plants, skin bag is replaced with tin containers and ewe's milk is replaced with ewe's or goat's milk mixed with cow's milk. İzmir Tulum cheese is produced from a mixture of high-quality ewe's and cow's milk.

In traditional Tulum cheeses, milk is not heated, milk is filtered through the cheesecloth and heated up to the fermentation temperature 28-35°C. Following the rennet addition, renneting procedure lasts between 45 minutes and 1 hour 15 minutes. The curd is broken down to the size of a chickpea using wooden sticks. The curd is rested for 15 minutes and water at 40 - 50 °C is poured onto the curd. The curd is separated from whey using a cheesecloth. Two different methods are adopted to separate the curd from the whey. In the first one, a suitable weight is placed on the curd, while in the second one, cheesecloth is hanged to drain the whey. In this spontaneous process, the resulting cheese is harder. The skin-bah in which the cheese will be placed in, preserved and ripened is usually made of goatskin. Goat skin-bag is softer and more elastic. The hair of the skin is first cut using scissors and cleaned using a brush with water and soap. The possible holes are closed. Unripe cheese is taken from the cheesecloth, cut in large pieces and placed in skin-bags by dabbling coarse-grained salt between the cheese pieces. The salinity level is kept at 7-9%. The opening of the skin-bag is closed using a rope. The skin-bag should be airtight to prevent deterioration. The skin-bag is kept in the dairy plant for a week and then sent to cold storage for ripening. Skin-bags are checked at least once a month. The cheese in the cover absorbs the brine in the ripening process, releasing some of it from the skin-bag. The ripening should last for at least three months for a good Tulum Cheese. The most important difference is that, in traditional Tulum Cheese production, cheese is ripened on the non-hairy side of the skin bag whereas, in the Tulum Cheese produced in İzmir and vicinity, the cheese ripened in the shaved hairy side of the skin-bag (Unsal 1997; Karagozlu & Atilla, 2018).

As mentioned above, İzmir Tulum cheese is now widely ripened in the tin containers. Therefore, it is also called İzmir Teneke Tulum Cheese (teneke means tin in Turkish) or Brined Tulum Cheese. For the production, the milk that will be used in cheese production is usually not pasteurized according to the producer's preference or heated to 65°C for 10 min. Following the heating to the renneting temperature, renneting process is carried out. Renneting temperature is 32–37°C and renneting duration is 45–60 minutes. The curd formed is broken using 1.5-meter long sticks with smaller sticks attached to it. This process continues until the curd is broken down to the size of a chickpea. Following this procedure, the curd and the whey is left to rest approximately for 15 minutes. In some dairy plants, the curd is rested after it is broken, water is poured into the vat at 40–50°C' and thoroughly mixed. The rested curd is gathered at the bottom of the cheese vat. The curd is taken in a mesh cloth in the vat. Pressure applied along the boat to release the whey in the cloth, or, in small dairy plants, the mesh cloth is removed from the vat by connecting and folding its edges. To drain the whey, in some dairy plants, the curd taken out of the vat in the mesh cloth is hung on the hooks at the ceiling. The cheese produced from the remaining curd following the draining is called the "Askı Tulumu (referring to the hanging procedure)" by dairy farmers. In some dairy plants, the curd broken in the vat is taken on the mesh cloth spread on the counter and pressure applied by placing weights on it. This type of cheese, from which the whey is separated by this method is called the "Baskı Tulumu (referring to the pressure application procedure)". The curd obtained following the whey separation is cut in large pieces. The pieces are kept in brine prepared using rock salt and then placed in tin containers. Salt is dabbled between the cheese pieces. The tin containers are covered and left to rest for 1-2 days. Then the tin containers are either salted and closed or the cheese pieces are taken in vat and left to fermentation after dabbling some salt on the pieces. This period is approximately 1-3 days. During this time, cheeses are sometimes turned upside down for pore formation. Then, the cheese pieces are placed in tin containers and rested for a couple of days. In the meantime, the water collected in the tin container is removed. Then the tin containers are filled with brine, sealed shut and left to ripen at 4–6 °C for 3.5–4 months. The yield in Brined Tulum Cheese depends especially on the dry matter content of the milk used in production. In general, 4.5-5.0 L ewe's milk, 5.5-6.0 L goat's milk and 8-9 L cow's milk are needed to obtain 1 kg Tulum Cheese (Unsal 1997, Karagozlu & Atilla, 2018).

While traditional Tulum cheeses in Anatolia are a single mass, Tulum cheeses in İzmir region are in large pieces. İzmir Tulum is harder and saltier than the White Cheese (Fig. **2**). Uniformly spread pores in the size of a pinhead on the cheese indicate a good quality in White Cheese. However, if the pores are larger, and if there are even slits on the cheese, this is regarded as faulty. Tulum has a longer shelf-life than the fresh cheeses and it is very tasty due to the ripening procedure (Karagozlu & Atilla, 2018).

Fig. (2). İzmir Tulum Cheese (Photographs by Cem Karagözlü).

Different studies have reported the chemical composition of İzmir Tulum Cheese. Eralp (1967), in their study in İzmir, have reported the average dry matter as 61.00%, fat as 26.60%, salinity as 5.96%, protein as 27.44, water-soluble Nitrogen as 0,739% and acidity as 115.6 SH. Yaygın (1971) collected İzmir Tulum Cheese samples from different cities (İzmir, Aydın, Balıkesir and Manisa) and reported the dry matter as 57.13%, fat as 20.70%, fat-free dry matter as 28.19%, protein as 21.28%, salinity in the dry matter as 10.26% and acidity as 68.4 SH. Kılıç and Gönç (1990), in their study on İzmir Tulum Cheese, reported the average dry matter as 51.87%, fat as 22.91%, fat in dry matter as 44.13%, salinity as 4.94%, salinity in dry matter as 9.53%, protein as 23.37%, protein in dry matter as 45.14%, water-soluble Nitrogen as 2.17%, ripening index as 58.44 and acidity as 71.07 SH. Koca (1996), for İzmir Tulum Cheese, has reported the average dry matter as 54.30 - 54.30%, fat as 28.42 - 31.17%, fat in dry matter as 51.68-54.98%, salinity as 2.42-2.56%, salinity in dry matter as 4.26-4.58%, Nitrogen as 3.49-3.67%, water-soluble Nitrogen as 0.308 - 0.451%, ripening index as 8.79 - 12.62%, free fatty acids as 0.77 - 0.895 KOH/g, acidity as 0.793 - 0.929%, and pH as 4.64 - 4.93. Kılıç *et al.* (1998), in their study on İzmir Tulum Cheese production, have reported the dry matter content as 43.68%, fat as 18.58%, fat in dry matter as 42.29%, salinity as 5.25%, salinity in dry matter as 11.15%, protein as 15.82%, pH as 5.18 and acidity as 2.48%.

Karagözlü *et al.* (2016), have reported the average dry matter, fat, protein, salinity, milk acid, pH and cholesterol levels in İzmir Tulum Cheese as 541.1 g kg^{-1}, 260.0 g kg^{-1}, 258.1 g kg^{-1}, 35.7 g kg^{-1}, 5.54%, 4.87 and 115.33 mg 100 g^{-1}, respectively.

A total of 37 İzmir Tulum cheese samples, 21 produced from cow's milk, 16 produced from the mixture of cow's, sheep's and goat's milk were analyzed. The pH, titratable acidity, dry matter contents, fat and fat-in-dry matter, salt and salt-in-dry matter, total nitrogen, protein, Water Soluble Nitrogen (WSN), Soluble Nitrogen in Trichloroacetic acid (TCA-N), Protease-Peptone Nitrogen (PPN) values of 21 İzmir Tulum cheese samples produced from cow's milk varied between 4.11-4.92, 0.522-1.408%, 49.920-68.648%, 17.50-30.50%, 35.06-51.98%, 2.340-4.563%, 3.75-8.66%, 3.29-4.29%, 20.99-27.37%, 0.47-1.21%, 0.32-0.91%, 0.12-0.46%, respectively. In addition, characteristics stated above for a total of 16 İzmir Tulum cheese samples produced from cow's, sheep's and goat's milk varied between 4.01-4.75%, 0.86-1.447%, 54.587-71.854%, 24.00-32.00%, 33.59-52.71%, 1.872-5.265%, 3.02-9.65%, 3.37-4.31%, 21.50-27.50%, 0.56-1.29%, 0.34-1.03%, 0.12-0.60%, and 14.51-33.78, respectively (Akpınar *et al.*, 2017)

In terms of the microflora of Salamura Tulum, the following microorganisms have been found to play a role in the ripening of the cheese: *S.lactis, S.cremoris, S.faecalis, S.faecalis spp., S.faecium, S.durans, S.liquefaciens, Lactobacillus casei, L.plantarum, L.helveticus, L.citvorum, L.fermenti, L.bulgaricus, L.leischmanii, Leuconostoc dextranicum, and L.brevis.*

Gökovalı (1980), during the ripening period of İzmir Teneke Tulum Cheese, identified 128 *Streptococcus* spp., and identified 74 of them as *E. durans*, 31 as *E.faecium*, 16 as *E. faecalis* and 7 as *E. faecalis* subsp. *liquefaciens.* In the study, it was determined that the dominant microorganisms were *Streptococcus* and *Lactobacillus* species and, among the streptococci, *E. durans* had the highest ratio with 58%. At the end of the research, after the milk is pasteurized in the production of Brined Tulum Cheese, it has been proposed to use *E. faecalis, E. durans, Lb. casei, Leu. dextranicum* and *Leu. citrovorum* as a pure culture. The researcher has reported *E. durans*, *E. faecalis* had a role during the ripening period of Brined Tulum Cheese while *Lb. casei*, *Lb. plantarum*, *Leu. dextranicum* ile *Leu. citrovorum* played a role during the further stages of the ripening period.

Kılıç & Gönç (1992), found that mostly *Lactobacillus, Enterococcus* and *Lactococcus* in their study about isolation from İzmir Tulum cheese. Öner *et al.* (2004) stated that *Lactobacillus* was the dominant flora in their research about Tulum cheese. They found that the ratio of *Lactococcus* was 15.8%.

As for the lactic bacteria, which plays a significant role in maturity, it is reported to be 9.15x10^5 cfu g^{-1} in traditional methods, while it is 8.49x10^5 cfu g^{-1} in cheese samples produced with a starter culture. Streptococcal bacteria exist in the natural flora of raw milk. So, it can be encountered in every phase of maturity of cheeses. The mean of Streptococcal bacteria is 4.23×10^6 cfu g^{-1} in cheese samples that are produced without starter culture, while it is 6.17×10^6 cfu g^{-1} when culture is used (Kamber, 2008). Büyükyörük *et al.* (2010), studied with 90 traditional İzmir Tulum cheese samples were collected from Aydın province. They stated that identified isolates by using PCR consisted of *lactis* and *cremoris* primers. Because of the production of Tulum cheese by traditional methods in small-scale establishments, starter cultures are generally not used and fermentation depends on indigenous bacterial flora

BERGAMA TULUM CHEESE

Bergama Tulum Cheese (Fig. **3**), being similar to İzmir Tulum Cheese, there are some minor differences in production. Bergama Tulum Cheese is produced using ewe's milk with high yield. When ewe's milk is absent or few, it is mixed with cow's and goat's milk and used in the production of the cheese. In traditional Tulum cheeses, milk is not heated, milk is filtered through the cheesecloth and heated up to the fermentation temperature 28-35°C. Following the rennet addition, renneting procedure lasts between 45 minutes and 1 hour 15 minutes. The curd is broken down to the size of a chickpea using wooden sticks. The curd is rested for 15 minutes and water at 40 - 50 °C is poured onto the curd. The clot is separated from whey using a mesh cloth. Two different methods are adopted to separate the curd from the whey. In the first one, a weight is placed on the curd while in the second whey separation is carried out by hanging the curd in the cheesecloth. In this spontaneous process, the resulting cheese is harder. Bergama Tulum Cheese which was previously ripened in skin-bags is now ripened in tin containers. This ripening period lasts for 18-20 months. Bergama Tulum Cheese is used in a famous local food, the Ayvalık Toast (Karagözlü, 2011). Karagözlü *et al.* (2016), have reported the average dry matter, fat, protein, salinity, milk acid, pH and cholesterol levels in İzmir Tulum Cheese as 606.1 g kg^{-1}, 235.0 g kg^{-1}, 212.5 g kg^{-1}, 31.3 g kg^{-1},1.4, 5.30% and 137.51 mg 100 g^{-1}, respectively.

SEPET CHEESE

Sepet Cheese, produced in the coastal places including Söke, Ayvalık, Dikili, Foça, Urla, Karaburun is produced from the mixture of goat's, ewe's and cow's milk. It is usually pressed and shaped in baskets sown from olive branches. The textural structure obtained from the texture of the basket is preferred in this cheese type. In the production of sepet cheese usually, woolly goat milk is used. In Ödemiş and Söke, a blend of cow's and ewe's milk can be used. In Ayvalık and Dikili, cow milk is usually used in the production of sepet cheese. According to producers, 1 kg sepet cheese can be produced from 9 - 10 l cow's milk or 6 - 7 l goat's milk or 4,5 l ewe's milk (Kamber 2008, Ercan *et al.* 2014).

Fig. (3). Bergama Tulum Cheese (Photographs by Cem Karagözlü).

Non-pasteurized goat's milk was mildly heated and renneted at 28 - 32°C. This temperature is set to 35°C in winter. While breaking the curd, hot water is poured onto the curd and heated to 36-38°C. The precipitated cheese mass is taken in pieces from the bottom of the vat and filled into optionally one, three or five-kilogram-holding baskets, the upper surface is smoothed however no weight is placed on the basket. The baskets in which the cheeses are placed is called "bucket (kova in Turkish)". These baskets are preferably sown using olive branches or straws obtained from a thorny plant growing in marshes. Approximately 15 minutes later the cheese starts to drain from the basket, the basket turned upside down and the cheese removed from the basket. Top and bottom surface of the cheese is salted manually, placed back in the basket and left to drain (Fig. **4**). At the end of 18 hours, the cheese, which can now stay "intact", can be taken out of the basket. It is placed on a veneer in a shady but non-airy place. The cheese is again rubbed with salt. Salting is a very delicate process. The authentic Kelle Sepet Cheese is salted every two days for fifteen days. The salting process is done very carefully since excess salt will for a crust around the cheese. The crust of the cheese is salted every two days stays thin which makes the cheese better. Then, the salting process continues for another month, salting every 10 days. At the end of this period, the cheese leaves its fats, makes a fat-drain, called "malaş" in Turkish. At the end of two months, the cheese can become consumable (Fig. **5**). However, consumption is not recommended before 100 days (Unsal, 1997). The chemical compositions of Sepet Cheese reported by different researchers are given in Table **1**.

Fig. (4). Sepet Cheese Production (Photographs by Zuhal Okuyan).

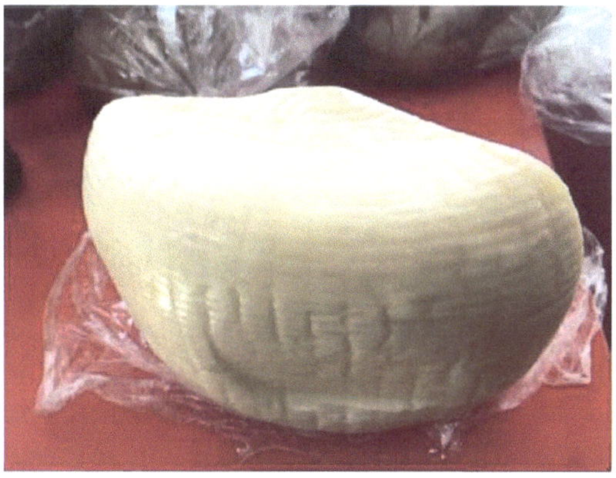

Fig. (5). Sepet Cheese (Photograph by Cem Karagözlü).

Table 1. The chemical composition of Sepet Cheese.

Moisture Content (%)	Fat (%)	Protein (%)	pH	Acidity (Lactic Acid %)	Salinity in Dry Matter (%)	Reference
54.32	24.50	29.34	5.46	0.72	2.05	Karakaş &Korukluoğlu (2015) 4 samples
53.19.	23.86.	18.49.	5.22.	1.42.	2.25.	Kınık *et al.* (1999) 20 samples
54.33.	25.11.	28.99.	5.57.	1.66.	3.85.	Ercan *et al.* (2014) 30 samples

Akpınar *et al.* (2017) have reported the following values for Sepet Cheese; dry matter 51.71 – 62.72%, protein 21.71 – 28.16%, fat in dry matter 32.68 – 44.44%, salt in dry matter 9.94 – 14.46%; pH 4.54 – 4.84. Karagözlü *et al.* (2016) have reported the average dry matter, fat, protein, salinity, milk acid, pH and cholesterol levels in İzmir Tulum Cheese as 588.7 g kg^{-1}, 287.5 g kg^{-1}, 225.2 g kg^{-1}, 32.6 g kg^{-1}, 2.19, 5.24% and 112.24 mg 100 g^{-1}, respectively.

Average total aerobic, lactococci, lactobacilli, enterococci, psychrotrophic bacteria, *Staphylococcus aureus*, yeast, mold, coliform bacteria counts were 7.64±1.18, 7.38±1.10, 7.38±0.99, 6.99±0.99, 5.37±1.15, 1.25±1.72, 3.22±1.25, 0.95±0.961, 2.72±1.82 log cfu g^{-1}, respectively. All bacterial counts increased during cheese production. Then the total aerobic, lactococci, lactobacilli, and enterococci counts decreased slightly during ripening and staphylococci and coliform counts decreased below 10 cfu/g during ripening. Isolates were closely related to *Lactococcus lactis* subsp. *lactis, Lactobacillus casei* subsp. *rhamnosus, L. plantarum,* heterofermentative *Lactobacillus* spp., *Streptococcus thermophilus, Leuconostoc* subsp., *Enterococcus durans* and *E. faceium.* At the beginning of cheese production, the percentages of lactic acid bacteria isolates were 23%, 52.94%, 17.64%, 5.88%, 0.00% for *Lactococcus* spp., *Enterococcus* spp., *Streptococcus* spp., *Leuconostoc* spp., respectively. At the end of ripening, these values reached 0.00%, 60%, 0.00%, 13.33%, 26.67% for *Lactococcus* spp., *Enterococcus* spp., *Streptococcus* spp., *Leuconostoc* spp., respectively (Ercan *et al.* 2014).

KÖY CHEESE

Köy Cheese ("köy" means "village" in Turkish) is not a commonly known cheese type in Turkey. There are some issues that await solution regarding this cheese type since it still preserves a primitive method from production to marketing, it has no standard production method, the characteristics of the cheeses offered to market vary substantially, the production and marketing procedure lacks hygienic conditions and it has no quality control regulations. Depending on the characteristics of the region where it is produced, Köy Cheese can be produced from goat's milk, ewe's milk and cow's milk and their mixtures in certain proportions. Production in the mentioned regions is concentrated in times other than March-July when more milk is given to dairy plants. Köy Cheese is mildly hard, white in color with yellowish outer structure. It is a porous, mildly salty cheese consumed fresh with no ripening in general. On the other hand, it was reported that Köy Cheese salted or kept in brine (Kesenkaş *et al.* 2015).

Toker (2001) in their study on Manisa Köy Cheese, examined 15 Köy Cheese samples and determined the dry matter, protein, salinity, ash, acidity and pH values to be 43.03%, 20.26%, 17.59%, 4.5%, 6.60%, 1.31 and 5.01, respectively. Total microorganism count in the study was determined to be $3.1x10^8$ cfu g^{-1}, coliform bacteria to be 1.5 x 10^6 and yeast to be 2.1 x 10^5 cfu mL^{-1}. Kılıç *et al.* (2003) investigated the properties of Köy Cheeses sold in Istanbul district markets determined the average dry matter, fat protein and salinity values in 15 samples to be 44.08%, 20.50%, 20.50%, 19.30% and 3.42%, respectively. The number of coliform bacteria was found to be 5.20x10^3- 4.30x10^7 cfu g^{-1}. The number of yeast-moulds was found between 5.20x10^3 and 4.30x10^7 cfu g^{-1}. Büyükyörük & Göksoy (2014) did not determine *L. monocytogenes* (0%) in none of the Köy Cheese samples they collected from the local markets, whereas they determined *L. grayi* (14.2%) in one of the samples and *L. ivanovii* (14.2%) in another. Kesenkaş *et al.* (2015) produced Köy Cheeses in three different dairy plants and have reported the dry matter values as 37.82- 42.57% while they determined the fat, protein, salinity values as 16.38–21.25%, 15.9 – 10.81%, 5.80–6.54%, respectively. pH values varied between 5.61 and 6.38 pH on the day the cheeses were produced while they varied between 5.04 and 5.73 at the end of the three-month storage period. *S. aureus* counts of the samples of Köy Cheese were found to be quite high and their number did not fall below 3 log cfu/g during three-month storage. Karagözlü *et al.* (2016) have reported the average dry matter, fat, protein, salinity, milk acid, pH and cholesterol levels in Köy Cheese as 570.8 g kg^{-1}, 160.0 g kg^{-1}, 226.5 g kg^{-1}, 86.4 g kg^{-1}, 3.21, 4.85% and 42.75 mg $100g^{-1}$, respectively.

LOR CHEESE

Lor is a cheese obtained from whey with spreadable consistency and consumed without ripening (Fig. **6**). Lor cheese is produced from whey or by adding about 30% milk to whey. Whey is heated to 50-55 °C and passed through the separator and the remaining fat is taken. Then, lactic acid is added to whey to reduce pH and heated to 80 °C. Salt is added in case of salty Lor Cheese production. Coa - gulated serum proteins begin to clump on the surface. When the lumps on the surface begin to combine, the heating is ended and the curd is taken with perforated buckets or skimmers. Then the curd is drained in a cheesecloth. The drained curd is taken to the cold storage. In traditional dairy plants, whey obtained as a result of cheese production is collected in caldrons and yogurt and lactic acid are added. Then it is heated until it is coagulated, this procedure lasts about 30 minutes. The resulting curd is allowed to stand for 45 minutes to collect curd on the surface of the cauldron. The curd gathered on the surface is collected with ladles and transferred to the filtering cloth and hanged for 24 hours. Some producers produce Lor by adding milk up to one-third of the whey. Unsalted Lor is marketed as "fresh Lor". Its shelf life is short and it is used for breakfast with jam and honey or in "Lor cookies". Lor containing 2-3% salt becomes durable and it has a longer shelf life (Karagözlü, 2014b).

Fig. (6). Lor Cheese (Photograph by Esra Boyacıoğlu).

Ergüllü (1982), in Lor Cheese samples produced with whey obtained from White Cheese production determined the average dry matter as 27.57%, fat as 7.3%, protein as 12.88%, 7.5% lactose and acidity as 3.65 SH. Demirci *et al.* (1991), in 17 Lor samples they collected from the local markets, determined the average dry matter as 25.99%, fat as 5.34%, protein as 13.50%, ash as 1.24%, lactose as 5.91 and acidity as 58.23 SH. The researcher, in the same study, determined $1.3x10^{6}$-$2.9x10^{8}$ cfu g^{-1} total bacterial count and detected coliform group bacteria in 16 of the 17 samples between $1.0x10^{4}$ and $4.0x10^{6}$ cfu g^{-1}. Çiftçioğlu *et al.* (2008) investigated the microbiological and chemical properties of 50 different Lor Cheese samples they collected from different producers. The researchers have reported the moisture as 68.71%, protein as 12.69%, salinity as 2.21%, fat as 23.2%, ash as 4.28%, pH as 4.8 and acidity as 0.59. According to the microbiological results, total mesophilic aerobic bacteria, lactic acid bacteria, coliform, *E. coli*, mold and yeast, coagulase-positive staphylococcus (CPS) and enterococci (ETC) counts were 8.42, 7.65, 3.11, 2.00, 5.52, 4.18 and 5.04 log cfu g^{-1}, respectively. Koçak Kızanklı & Göksoy (2018), in 30 Lor Cheese samples collected in Aydın, determined 9.8 log cfu g^{-1} total mesophilic aerobic bacteria, 5.09 log cfu g^{-1}*Staphylococcus aureus*. The researchers detected coliform group bacteria in 73% of the samples and *E. Coli* in 23% of the samples. None of the lor samples contained Listeria monocytogenes and Salmonella subsp. The researchers attributed the high number of microorganisms in Lor cheese to the re-contamination in the stages after production. Elmas (2014), in 20 Lor Cheese samples collected from local markets in Aydın, determined Listeria *monocytogens* in two of the samples and *L. seeligerdi* in two of the samples. None of the samples contained *Salmonella* spp. Karagözlü *et al.* (2016), have reported the average dry matter, fat, protein, salinity, milk acid, pH and cholesterol levels in Lor Cheese as 340.6 g kg^{-1}, 175.0 g kg^{-1}, 118.1 g kg^{-1}, 11.7 g kg^{-1}, % 4.77, 4.66 and 132.16 mg $100g^{-1}$, respectively. Kavaz *et al.* (2012), in their study on Lor Cheese, determined the average dry matter as 36.60%, fat as 15.94%, ash as 3.30% and salinity in the dry matter as 4.5.

KOPANESTI CHEESE

Local cheese types produced from whey are common. One of them is Kopanesti Cheese. This cheese can be found in Urla, Çeşme, Karaburun and Foça. Its origin is Greek. It is produced in the Greek islands in the Aegean Sea (Akpınar *et al.* 2014). It is usually consumed as a side dish (meze in Turkish and Greek). Kopanesti means "mashed/crushed" in Greek. The whey obtained from Sepet Cheese production is placed in a cauldron and heated. Goat's milk is added to the cauldron in one-fifth of the whey. It is heated for some time and left to cool. The curd is collected on the surface as the whey cools down. This curd is collected using a ladle and drained in cheesecloth for 1-2 days. Then it is taken into earthenware cups, thoroughly kneaded and stored. During storage, it is stirred using a spoon every 1-2 days. Then the curd is salted (Fig. **7**). It is an aromatic and salty cheese.

Fig. (7). Traditional Kopanesti Cheese Production (Photographs by Zuhal Okuyan).

Moisture content, fat-free dry matter, fat, protein, salt, total ash and acidity values of Koponesti Cheese are given in Table **2**. According to Ergülü *et al.* (1998), moisture content of Koponesti Cheese varies between 46.20–65.74% with a mean of 57.78%, and dry matter varies between 34.26–53.80% with a mean of 42.22%.

Table 2. The chemical composition of Kopanesti cheese.

Moisture Content (%)	Fat (%)	Protein (%)	Salinity (%)	Acidity (SH)	Reference
51.60.	16.1.	-	12.	94.	Akgün 1998
57.28.	14.3.	16.18.	6.3.	117.	Ergüllü *et al.* 1998.

The short chain fatty acid contents of Kopanesti Cheese have been identified as: butyric acid, 5.3%; caproic acid, 2.7%; caprylic acid, 3.1%; capric acid, 9.9%; lauric acid, 3.9%; myristic acid, 9.9%; palmitic acid, 23.4%; and stearic acid, 13.1%. The long chain unsaturated fatty acid contents were: oleic acid, 20.3%; linoleic acid, 1.6% and linolenic acid, 1.9% (Ergüllü *et al.* 1998)

Tzanetakis *et al.* (1987), in their study on the microbiological properties of Kopanesti Cheese, reported the main species and counts of these species. The bacterial isolates were identified as *Lactobacillus plantum*, *L. casei* subsp. *casei*, *L. casei* subsp. *rhamnosus*, *L. curuatus*, *L. xylosus*, *L. brevis*, *L. buchneri*, *L. viridescens*, *L. cellobiosus*, *Streptococcus faecalis*, *S. faecium*, *S. faecium* subsp. *casseliflavus*, *S. durans*, *Pediococcus pentosaceus* and *Micrococcus luteus*. Kaminarides *et al.* (1989) stated that the total streptococci, lactobacilli, yeast and molds counts, as well as coliforms, were found in microflora of Kopanesti cheese in their study. They showed that 86 strains of yeast from the cheese samples had been also isolated and identified out of which the following species: *Trichosporon cutaneum (T. beigelil), Kluyveromyces lactis (KI. marxianus* var. *lactis), Saccharomyc*es *exiguus, Saccharomyces cerevisiae, Rhodotorula rubra, (Rh. mudlaginosa), Trichosporon penicilatum (Geotrichum penicillatum), Candida lusitaniae (Clavispora lusitaniae)* and *Debaryomyces hansenii.*Ergüllü *et al.* (1998) showed that the number of aerobic microorganisms, lipolytic bacteria proteolytic bacteria, yeast and mould, coliform bacteria was $3.2x10^5$ cfu g^{-1}, $2.5x10^4$ cfu g^{-1}, $2.7x10^4$cfu g^{-1}, $4.6x10^3$ cfu g^{-1}, and 0 cfu g^{-1}, respectively. The following species of yeast: *Pischia membrane fasciens, Trichon cutaneum, Kluyveromyces lactis, Kluyveromyces marxianus spp. lactis, Saccharomyces cerevisiae, Rhodotorula ruba, Rhodotorula mucilaginosa, Candida lusitanrae and Debaryomyces hanseii* together with various species of penicillin mould, most significantly *Penicilium commune*, play a role in the ripening of Koponesti Cheese. *Pischia membrane* and *Penicilium commune* are the main microorganisms (Kamber, 2008). Karabıyıklı & Karapınar (2007) have reported the lactic acid bacteria they isolated in their study as *Lactobacillus brevis*, *Lactobacillus buchneri*, *Lactobacillus casei*, *Lactobacillus collinoides*, *Lactobacillus johnsonii*, *Lactococcus lactis* subsp. *cremoris*, *Lactobacillus mali*, *Lactobacillus minor*, *Lactobacillus oris*, *Lactobacillus parabuchneri*,

Lactobacillus reuteri, *Lactobacillus sanfrancisco*, *Lactobacillus sharpeae*, *Lactobacillus suebicus*, *Lactobacillus vaginalis* and *Lactobacillus viridescens*.

KIRLIHANIM CHEESE

Kirlihanım cheese produced in Ayvalık is also made from basket lor. Yellow-brown and purple moulds are seen on this cheese type. The whey obtained in Sepet Cheese production is boiled and the curd is collected, salted with fine salt and filled into baskets called "kovalık". This curd dries for 2-3 weeks and ripened for 6-7 months, turning upside down once in a while. The moulds on the cheese are removed and the cheese is then consumed. Sometimes, the cheese is roasted in the oven after moulds are removed. When roasted, a crust forms on the surface and darkens (Fig. **8**). It is called "Kirlihanım (can roughly be translated as the stained lady) due to the moulds on its surface. It is a cheese of irregular shape, with a pungent flavor and aroma. It is low-fat and lightly salted, has dark spots on its surface and resembles chocolate halva in appearance (Kamber, 2008). According to Akgün (1988), the chemical composition of Kirlihanim Cheese was found to be moisture content 46.4%, fat-free dry matter 35.7%, fat 17.9%, protein 10.0% and salt 4.9%.

Fig. (8). Kirlihanım Cheese (Photographs by İsmail Kesebir).

ARMOLA CHEESE

The Seferihisar Armola cheese is said to come from the Albanian culture. It is produced from ewe's or goat's milk. Previously, goat skin-bags used for İzmir Tulum Cheese production was used for the production of this cheese. It is mildly salty and sour with a spreadable consistency, dirty white-yellowish color. The original Armola Cheese was produced by adding yogurt on the cheese crumbs left inside the skin-bag from İzmir Tulum Production and left to rest for a couple of days. Then, goat's milk was added and hanged on a tree for 10-15 days. Today, it is produced by mixing White Cheese crumbs and Lor and filled into skin-bags (Fig. **9**). It is consumed by adding olive, thyme, mint, flaked red pepper on it (Yoldaş, 2017; Durlu – Özkaya & Gün, 2014).

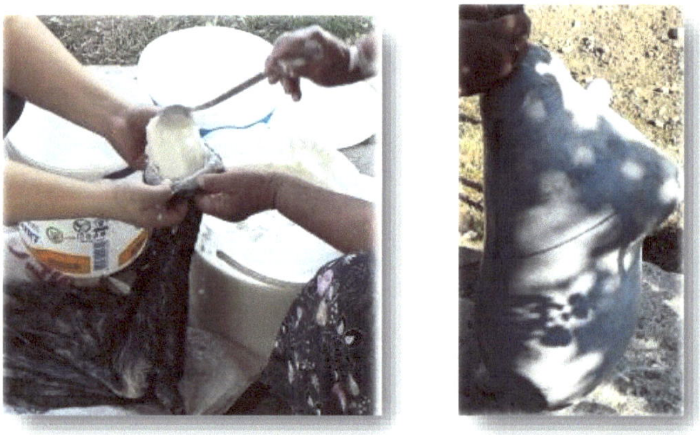

Fig. (9). Armola Cheese production in Seferihisar's Beyler Village (Photographs by Melisa Yoldaş).

For Armola Cheese production, two-year-old goat's skin-bag is shaved and cleaned. The arms of the skin are tied with a rope and inflated leaving the shaved side inside and the hair left on the skin is washed away. The entire surface of the skin is checked for possible holes and salted. To minimize hair contamination, skin-bag is inverted and readied for filling (Yoldaş, 2017).

Regarding the milk used in the production, 10 kg raw cow's milk is pasteurized at 65°C for 30 minutes. In a bucket, 10 kg pasteurized cow's milk, 3 kg whole-fat non-salted Lor Cheese, 5 kg whole-fat yogurt is mixed and salt are added. The prepared mixture is immediately filled into the skin-bag. Following the filling procedure, the skin-bag is airtight-closed. Accordingly, a piece of cloth soaked with olive oil is placed on the opening of the skin-bag. Skin-bag is tightly tied using a rope and left to ripen in a shady, cool and dry place. Armola Cheese should be ripened at least for one month. During this period, with the draining whey, the cheese gets a harder structure. To maintain the ideal consistency of Armola cheese, to improve the structure of the cheese and to prevent the reduction of the amount of cheese in the skin-bag, milk should be added to the skin-bag at certain intervals. Accordingly, the brim of the skin-bag softened by soaking, adding the desired amount of milk and it is tightly closed again. The skin-bag is rubbed to mix the milk with the cheese mass homogeneously. Following this procedure, to prevent worm formation and to maintain durability, skin-bag is rubbed with salt. If this is procedure is repeated frequently, skin-bag can be maintained for a long time with no damages and thus continuity can be maintained in Armola Cheese production (Yoldaş, 2017; Durlu – Özkaya & Gün, 2014).

The average microbial counts were found as follows: total aerobic mesophilic bacteria, 7.82; psychrotrophic bacteria, 6.98; coliform bacteria, 4.56; lactococci, 7.55; lactobacilli, 7.87; enterococci, 6.17; yeast 7.33; mold <1.00; *Staphylococcus* spp., 5.94; and *Listeria* spp., 2.94 cfu/g. The microbial results showed the presence of high counts of microorganisms and the poor hygienic quality of Armola cheese. In order to improve the microbial quality of Armola cheese, it should be produced under good hygienic conditions and stored at refrigeration temperatures until consumption (Orşahin, 2012). Chemical composition of Armola Cheese is given in Table **3**.

Table 3. The chemical composition of Armola Cheese (%).

Moisture Content (%)	Fat (%)	Protein (%)	pH	Acidity (Lactic Acid %)	Ash (%)	Reference
32.26	12.0	13.32	4.22	1.21	3.80	Yoltaş, 2017
37.26	19.52	10.87	4.24	0.95		Orşahin, 2012
27.3	14.0	11.8	4.29	2.09		Karagözlü *et al* (2016)

TIRE ÇAMUR CHEESE

Tire Çamur Cheese produced in Tire can be spread on bread (Fig. **10**). It is obtained by mixing Lor Cheese with the brine of Tulum Cheese. Previously, it was often produced from ewe's and goat's milk whereas now it is produced from cow' milk. It is usually produced by the local people and consumed at breakfast. For Çamur Cheese production, whey obtained during White Cheese production is collected in a container. Milk is added to the container and heated up to 90-95 °C for 30 minutes. The obtained whey (lor) is drained. Then, 2-3% salt is added to the lor and rested for 4-5 hours. At the end of this period, Tulum Cheese brine is added and the mixture is slowly kneaded. It is kept in the refrigerator until the time of consumption (Karagozlu *et al.* 2012).

Ak and Nergiz (1988), in their study, have reported the following values for Çamur Cheese; total dry matter 62.06%, fat 25.82%, protein 6.75%, salinity 3.16%, lactose 1.2%, sodium 8.5 mg / 100 g, calcium 2.4 mg / 100 g, phosphorus 1.9 mg / 100 g and potassium 9.22 mg/100 g. Protein ratio increases as the dry matter content increases. Salinity level varies depending on the brine. Both salting and brining increase the amount of salt. Also, using Tulum Cheese increases the sodium content. In the same study, microbiological contents have been reported as follows; $4.9x10^9$ cfu g^{-1} aerobic microorganisms, $7.4x10^6$ cfu g^{-1} coliform group bacteria and $5.9x10^6$ cfu g^{-1} yeast and mould. The researchers have associated the high microorganism counts with post-production contamination. Karagözlü *et al.* (2016) have reported the average dry matter, fat, protein, salinity, milk acid, pH and cholesterol levels in Tire Çamur Cheese as 282.1 g kg^{-1}, 140.0 g kg^{-1}, 95.1g kg^{-1}, 16.8 g kg^{-1}, 0.48, 5.30% and 85.92 mg 100 g^{-1}, respectively.

Fig. (10). Tire Çamur Cheese production in Tire Boyacıoğlu fabric. (Photograph by Esra Boyacıoğlu).

ÇÖKELEK

Çökelek is a food product with great importance in terms of health and produced from two different raw materials as yogurt and ayran (Fig. **11**). Unlike Lor Cheese, this cheese contains both albumin and casein. Due to its low-fat content, Çökelek Cheese has low-calorie content and it is a good protein and calcium source and therefore recommended by dieticians. Other than the regular Çökelek Cheese marketed in the big cities in Turkey, there are different Çökelek Cheese types with properties unique to the place it is produced, with prolonged shelf-life by preserving in skin-bags or earthenware jars or drying under the sun. Although it has a similar appearance to that of Lor Cheese, they are different cheese products (Karagözlü 2014a). It has different names depending on the place it is produced, such as “Ekşimik”, “Kes”, “Akçakatik”, “Kesmik” or “Urda”, “Minci-Minzi” and “Çökelek” (Kırdar 2003). Three different raw materials can be used in the production including the mixtures of milk, yogurt and whey (Karaca *et al.* 2007).

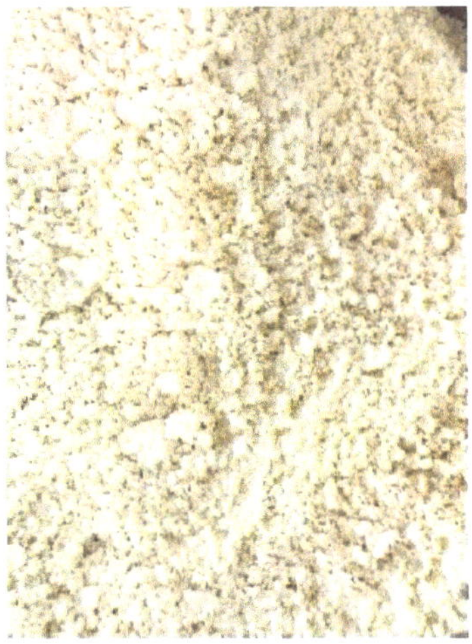

Fig. (11). Çökelek Cheese production in Tire Boyacıoğlu fabric. (Photograph by Esra Boyacıoğlu).

Çökelek Cheese Production from Milk:

Regular raw milk is left to acidify an overnight. The acidified milk is mildly (40-45°C) heated and precipitated on the stove. To accelerate this process lemon juice or lemon salt is added. The water slowly evaporates and the curd precipitates. This curd is collected in cheesecloth and hanged at a high place and left to drain. When the whey is fully drained, the curd is taken from the cheesecloth and salted and usually consumed fresh. Sometimes, for future consumption, it is salted and hermetically filled in earthenware jars or skin-bag and left to ripen in a cold place for 3-4 months.

Çökelek Cheese Production from Yogurt:

The milk is boiled, left to cool to fermentation temperature and spontaneously fermented using yogurt. Water is added and the mixture is churned. The fat obtained churning is removed and the remaining ayran is boiled and salted. It is drained in a cloth bag and the remaining part is consumed as Çökelek.

Çökelek Cheese Production from Yogurt-Whey Mixture:

Yogurt is churned and the fat is removed. The remaining yogurt whey is mixed with cheese whey. The mixture is boiled and salted. It is drained in a cloth bag and the remaining part is consumed fresh (Karagözlü 2014a). Chemical compositions of Çökelek Cheese reported in some studies are given in Table **4**.

Table 4. The chemical composition of Çökelek cheese (%).

Moisture Content (%)	Fat (%)	Protein (%)	Salt	Ash (%)	Reference
20.9 – 39.3	2.5 – 12.2	10.3 – 16.0	0.7 – 4.4	0.8 – 5.7	Adam (1974)
19.8 – 23.0	0.5 – 1.5	-	0.2 – 3.0	-	Eralp & Kaptan (1970)
12.7- 45.0	1.0 – 4.5	-	0.2 – 3.0	-	Kırdar (2003)
20.9 – 55.6	1.5 – 33.0	10.4-18.3	0.7 – 4.7	0.8-5.	Yaygın *et al* (1978)
26.6 – 28.9	4.08 – 5.50	2.02 – 2.2	1.6 – 1.8	-	Şimşek and Sağdıç (2014)
47.85.	19.59.	2.61.		4.78.	Kavaz *et al.* (2012)
17.40	1.2	8.04	1.09	-	Ağaoğlu *et al.* (1997)

In traditional Çökelek production in The Aegean Region, milk is boiled and cooled to incubation temperature, and then inoculated with yogurt culture. Yogurt is left to spontaneous acidification. Water is added to the yogurt, boiled for 5 minutes and left to cool. Then, the curd is drained in a cheesecloth or cloth-bag and optionally, 4-5% salt is added and taken into a cloth bag. It is left to drain for approximately 24 hours. In some regions, hot milk addition provides larger-sized curd particles, resulting in a good curd structure and yield (Durlu – Özkaya & Gün, 2014). Erinç (2006), in their study on Çökelek, has reported the average dry matter as 26.04%, fat as 3.3%, salinity as 0.22%, protein as 15.69%, lactose as 1.95%, water activity as 0.995, total carbonyl content as 553 ppm, titratable acidity as 0.335% and pH as 4.47. The researcher has also reported total mesophilic bacterial count as $3.39x10^8$ cfu g^{-1}, yeast-mould counts as $5.17x10^7$ cfu g^{-1}, coliform bacterial count as $2.53x10^5$ cfu g^{-1} and *Staphylococcus aureus* count as $3.40x10^7$. Öksüztepe *et al.* (2007), on their study on 40 Çökelek Cheese samples, have reported the water activity as 0.95, dry matter as 21.43%, fat as 1.38%, fat in the dry matter as 6.94%, protein as 17.91%, ash as 2.14%, salinity as 0.52%, salinity in dry matter as 2.44%, pH as 3.79 and titratable acidity as 1.25%. Microbiological properties of these samples were as follows; mesophilic aerobic bacterial count $2.87x10^8$ cfu g^{-1}, yeast-mould $3.10x10^7$ cfu g^{-1}, $3.10x10^7$ cfu g^{-1}, *Enterococcus* spp. 15 cfu g^{-1}, *Staphylococcus aureus* $1.04x10^2$ cfu g^{-1}, *Escherichia coli* 1.09x103 cfu g^{-1}, *Staphylococcus-Micrococcus* counts $1.42x10^3$ cfu g^{-1}, *Lactobacillus – Leuconostoc-Pediococcus* counts $2.97x10^7$ cfu g^{-1}, *Lactococcus* $1.03x10^8$ cfu g^{-1} and coliform group bacterial counts $8.53x10^3$ cfu g^{-1}. In the 30 samples of çökelek as an average of total microorganism 8.49±0.79 log cfu g^{-1}, heterofermentative lactic acid bacteria 8.58±0.98 log cfu g^{-1}, psychrophilic bacteria 8.51±0.83 log cfu g^{-1}, proteolytic bacteria 8.49±0.83 log cfu g^{-1}, mould-yeast 6.67±0.86 log cfu g^{-1}, *Staphylococcus aureus* 7.53±1.12 log cfu g^{-1}, *Salmonella* subsp. 7.80±0.64 log cfu g^{-1}, *Bacillus* subsp. 7.70±1.35 log cfu g^{-1}, *Clostridium* subsp. 7.81±1.05 log cfu g^{-1}, *Escherichia coli* % 23.3, coliforms bacteria % 23.3 were determined respectively. In conclusion, the samples of fresh Çökelek Cheeses had very low hygienic quality (Öngenar & Kırbağ, 2009).

İZMIR BIBERLI LOR CHEESE (İZMIR PEPPERED LOR CHEESE)

For the production of İzmir Biberli Lor Cheese bell pepper and chili pepper are sliced or added whole to the Lor cheese and preserved in plastic or earthenware jars (Fig. **12**). Optionally, olive oil is added before consumption (Karagozlu *et al.* 2012).

Fig. (12). Biberli Lor Cheese (Photographs by Cem Karagözlü).

KIRKTOKMAK CHEESE

According to the locals' statements, this cheese type was first produced by the Turkmen tribes migrated to this region from Caucasia. The name "Kırktokmak" comes from the wooden mallet used for continuous churning to separate the fat from whey (Fig. **13**). The raw material for Kırktokmak Cheese is the butter whey separated from butter produced from yogurt. Also, ayran produced from yogurt that could not be marketed is used in the cheese making. Producing butter from yogurt is unique to Anatolia. This is especially common in Western and Southern Anatolia, and the wastes are evaluated in different ways (Kılıç, 2004).

Fig. (13). Kırktokmak Cheese (Photographs by Cem Karagözlü).

The production today is carried out using yogurts produced from cow's, ewe's and goat's milk. The yogurt is placed in the churn. The churning in wooden churns (kovan yayık in Turkish – Fig. **14**) using a special wooden mallet is carried out by moving the mallet up and down and the fat is separated. The fat collected on the top is collected using a ladle. Ayran remaining in the churn is poured in copper containers with wide brims. The churn is heated until the protein is coagulated (following the separation of greenish-yellowish whey, heating is ended). The aqueous curd is poured in cloth-bags and drained for 1-2 hours. Then the curd is salted and nigella is added. The curd is filled into yeanling skin-bags and hermetically closed (Fig. **13**). The curd stays in the skin-bag for 15-20 days. During this time both the whey is drained and the characteristic taste and aroma forms (pre-ripening). In the last step, the curd in the skin-bag is thoroughly crumbled and the lumps are mashed. If redeemed necessary, it is kept under the sun for some further whey removal. Then, it is filled into yeanling skin-bag and the brim is tightly closed. The skin-bag is kept in a cool and dark cave. In summer, these cheeses can be consumed after four months (Kılıç, 2004; Büke, 2017).

Fig. (14). "Kovan" used in Kırktokmak Cheese production (Photographs by Cem Karagözlü).

Özsaatçiler (1983) investigated the characteristics of ten Kırktokmak Cheese samples collected from the region. It was determined that the cheeses were soft and had a color range from white to yellowish. The samples had a yeasty, bitter, sour and acidic taste. In addition to the desired aromas, soily, yeasty and moldy aromas were also reported. In the examined cheeses, the dry matter was between 36.81 and 44.4% with an average of 40.8%. Acidity values were found to be 24.6 SH. The average pH values were determined to be 4.57 in 10 samples. It was determined that protein value ranged from 13.5% to 24.3% in Kırktokmak Cheese samples with an average of 18.3%. Dry matter, with an average of 10.8%, varied between 8.16 and 10.86%. Salinity, with an average of 12.6%, varied between 9.8 and 17.5%. This high salinity was associated with the salting process applied to prevent deterioration during storage. Ripening index varied between 4.2 and 8.5%, with an average of 6.5 (Kılıç, 2004; Büke, 2017).

MILAS KAZIKLI CHEESE

Milas Kazıklı Cheese is produced in the villages of Milas usually from goat's milk in tin containers of skin-bags (Fig. **15**). Raw goat's milk is rennetted using the capricorn abomasus and coagulated in 15 minutes. The curd is collected in a mesh cloth and hanged for one hour to drain the whey. Then, it is cut in pieces and filled either in skin-bags or tin container. Salt is spattered between the pieces. Skin-bag is sealed by sewing, the tin container is sealed shut by welding. Brine is not used. The cheese is kept in a cool place for one month and then offered to market (Büke, 2017).

Fig. (15). Milas Kazıklı Cheese (Photographs by Cem Karagözlü).

BODRUM DESTI (TESTI) CHEESE

Similar to Kırktokmak Cheese, cheeses produced are kept in skin-bags for 15-20 days. Then it is filled in earthenware jars and pressed using a wooden mallet to increase the consistency (Fig. **16**). The brim of the tightly filled jars is covered with vine leaf, covered and tied with a cloth and coated with mud. It is kept upside down in a cool and dark place. Alternatively, the jar is 2/3 buried in the ground (Büke, 2017).

Fig. (16). Bodrum Desti (Testi) Cheese (Photographs by Cem Karagözlü).

AYDIN ÇÖREKOTLU CHEESE

Nigella is used in many local cheese types. It is mostly used in White Cheese, Tulum Cheese, Kırktokmak Cheese and butter production. Aydın Çörekotlu Cheese is one of them. It is produced from the whey of the excess cheese in cheese production. The whey obtained during Kaşar or White Cheese production is mixed with fresh milk and the mixture is boiled. Serum proteins are collected and drained. The obtained curd is mixed with nigella and stirred. For a longer shelf-life, 2-3% salt is added. It is filled in jars or cups, leaving no air inside, and kept in a cool place (Ünsal, 2007).

ÇOBAN CHEESE

It is a cheese type produced from cow's milk specific to the Aegean Region. It is produced in the villages of Manisa, Aydın and Denizli. It is produced by the shepherds (Çoban means shepherd in Turkish). Raw milk is placed in a cooker, rennet is added and stirred. The cooker is closed and covered with a thick cloth and rested for 35 minutes. Then, it is stirred using a wooden spoon and covered again. It is rested for some time and drained using a strainer and squeezed with hand. It is shaped in glob lumps. It is thoroughly salted and drained in a strainer or by hanging. The milk has to be decontaminated and the animal from which the milk is obtained should be healthy. Otherwise, Brusella, Listeria, *E. coli* originated contaminations may occur since it is consumed fresh (Ünsal, 2007).

POSA CHEESE

Posa Cheese is produced in İstanköy (Bodrum) in the west of the region and is only known in this region. According to local people, Atatürk, founder of The Republic of Turkey, liked Posa Cheese very much during a tour and this cheese was sent to him every year upon his request. Red and white wine dregs were used in the production of Posa Cheese. Therefore, it is called "Cheese with red wine dreg" and "Cheese with white wine dreg" according to wine dregs used. In crude cheese, in the beginning, the number of total aerobic microorganisms was 4.0×10^7 cfu g^{-1}, whereas after ripening 7.2×10^6 cfu g^{-1} in the cheese of normal brine, in cheese with red wine dregs 4.5×10^6 cfu g^{-1}, in cheese with white wine dregs 9.0×10^6 cfu g^{-1}. So that the number of yeast and mould is higher in cheeses with wine dregs than in cheeses of brine, because the dregs contain residual yeast from wine-making (İnal, 1998; Kamber, 2008).

CONSENT FOR PUBLICATION

Not applicable.

CONFLICT OF INTEREST

The authors confirm that this chapter contents have no conflict of interest.

ACKNOWLEDGEMENTS

Declared none.

REFERENCES

Adam, RC (1974) *Cheese.* Ege University Agricultural Faculty, Publication number: 176, Ege University Press 109-268. İzmir.

Ağaoğlu, S, Ocak, E & Mengel, Z (1997) Investigations on the microbiological, chemical, physical and organoleptic characteristic of Çökelek Cheeses in Van province. *J of Ataturk Uni Veteri Facu,* 44, 7-13.

Ak, S & Nergiz, C (1998) Investigation of chemical composition and microbiologic quality of Tire Çamur Cheese. In: Demirci, M., (Ed.), *Vth Milk and Milk Products Symposium,* 21-22 May Tekirdağ 79-89. Ankara.

Akgün, C (1988) Traditional Turkish Cheeses and Production Techniques. *Ege University, Agriculture Faculty, Agricultural Products Department, Unpublished Graduation thesis, İzmir.*

Akpınar, A, Açu, M & Uysal, H (2014) Microflora contributed to the ripening of cheeses produced in Aegea on the region. *25th International Scientific - Expert Conference of Agriculture and Food Industry,* 25-7. İzmir, Türkiye

Akpınar, A, Yerlikaya, O, Kınık, Ö, Uysal, H & Korel, F (2017) Physicochemical and sensorial properties of sepet cheeses packaged under different modified atmospheric conditions. *Ege Üniv Ziraat Fak Derg,* 54, 149-55.

Akpınar, A, Yerlikaya, O, Kınık, Ö, Korel, F, Kahraman, C & Uysal, H (2017) Some physicochemical characteristics and aroma compounds of i̇zmir tulum cheese produced with different milk types. *Ege Üniv Ziraat Fak Derg,* 54, 27-35.

Anonymous, (1996) *Salamura Tulum Cheese (İzmir Tulum Cheese); Institute of Turkish Standards: Ankara* TS 11966.

Büyükyörük, S & Göksoy, E (2010) Investigation the presence of listeria spp. from village cheese in aydın province. *Uludag Uni J of the Facu of Veteri Med,* 30, 9-12.

Büke, A (2017) *Personal Communication, 5 September 2017 Bodrum.*

Büyükyörük, S & Soyutemiz, GE (2010) Isolation of *Lactococcus lactis* (*Lactococcus lactis* subsp. *lactis* and subsp. *cremoris*) Strains from Traditionally Manufactured İzmir Tulum Cheese and Identification by Phenotypical and Molecular Technics. *J of Facu of Veteri Med Erciyes Uni,* 7, 81-7.

Çiftçioğlu, G, Erkan, ME, Vural, A & Aksu, H (2008) Assesment of some microbiological and chemical properties of Lor whey cheese. *J Food Agric Environ,* 6, 109-13.

Demirci, M, Şimşek, O & Arıcı, M (1991) An investigation on composition and some microbiological features of Lor marketed in Tekirdağ. *Gida,* 16, 291-4.

Durlu – Özkaya, F & Gün, İ (2014) Aroma Compounds of Some Traditional Turkish Cheeses and Their Importance for Turkish Cuisine. *Food and Nutr Sci,* 5, 425-34. [http://dx.doi.org/10.4236/fns.2014.54050]

Elmas, Ş (2014) An investigation on the presence of Listeria monocytogenes and Salmonella spp. *in white, Tulum and lor cheese sold on local open markets in Aydın province.*Unpublished Postgraduate Thesis, Aydın Adnan Menders University Institute of Science, Aydın.

Eralp, M (1967) *Investagitation of milk and milk products in İzmir* Ankara University Press of Agricultural Fac.. Ankara Üniversitesi Basımevi. 53 sf.

Eralp, M & Kaptan, N (1970) Milk Industry of Antalya City. *AUZFYay 436, Scientific Research and Examine.*A.U.Press 539, Ankara – Turkey.

Ercan, D, Korel, F & Orşahin, H (2014) Microbiological quality of artisanal Sepet cheese. *Int J Dairy Technol,* 67, 384-93.
[http://dx.doi.org/10.1111/1471-0307.12125]

Ergüllü, E, Kinik, O & Akbulut, N (1998) A Study on Koponesti Cheese Production and Features around. In: Demirci, M., (Ed.), *Milk and Milk Products Symposium,* May 21-22, 1998. National Productivity Centre Publications, Tekirdag Ankara 19981-20.

Ergüllü, E (1982) Peynir suyu ve lorun bazı özellikleri üzerine araştırmalar. *Gida,* 7, 63-6. [In Turkish].

Erinç, H (2006) Tokat Yöresinde Üretilen Çökeleklerin Bazı Kimyasal ve Mikrobiyolojik Özelliklerinin Belirlenmesi. *Unpublished Master Thesis.*

Gökovalı, T (1980) Investigations microbiological features of tulum cheeses with starter culture during maturation. *Unpublished PhD Thesis Ege Univ*

Güven, M, Konar, A & Kleeberger, A (1995) Physical, chemical and organoleptic properties of Tulum cheeses ripened in a skin bag and produced from cow, sheep and goat milk. *Cukurova Journal of Agricultural Science,* 19, 203-18.

Hayır Kanaat, M (2016) Türkiye coğrafyası ve Jeopolitiği. *Nobel Akademik Yayıncılık,* 552. İstanbul, Turkey. [In Turkish]

İnal, O (1998) Investigations on production and characteristics of posa cheese. *Unpublished Post Graduate Thesis, Ege University, Institute of Science.*

Kamber, U (2008) The traditional cheeses of Turkey: The Aegean region. *Food Rev Int,* 24, 39-61.
[http://dx.doi.org/10.1080/87559120701762195]

Kaminarides, SE & Anifantakis, EM (1989) Evolution of the microflora of Kopanesti cheese during ripening. Study of the yeast flora. *Lait,* 69, 537-46.
[http://dx.doi.org/10.1051/lait:1989637]

Kara, R (2012) Traditional a cheese: Afyon Tulum Cheese. *Kocatepe Veterinary Journal,* 5, 45-8.

Karabıyıklı, Ş & Karapınar, M (2007) Identification of lactic acid bacteria in the fermentation of kopanesti cheese. *Gida,* 32, 143-9.

Karaca, OB, Ocak, S, Güney, O & Güven, M (2007) Present situation of goat production sector and some typical dairy cheeses in Turkey *3rd Joint Meeting of the Network of Universities and Research Institutions of Animal Science of the South Eastern European Countries.*Thessaloniki10-12 February 2007

Karagözlü, C (2011) Bergama Region Cheeses. *International Bergama Symposium,* 7-9 April 2011Vol I, 424-34.978-605-4597-01-7Lamineks Press, İzmir – Turkey

Karagözlü, C (2014a) *Geleneksel Süt Ürünlerimizden Çökelek; Üretim Yöntemleri ve Yöresel Peynir Üretiminde Kullanımı 4 Geleneksel Gıdalar Sempozyumu 17 – 19 Nisan 2014 Adana* Bildiri Kitabı. Gökçe Matbaacılık Ankara 667-80. ISBN 978-605-01-0603-9.

Karagözlü, C (2014b) *Peyniraltı Suyunun Değerlendirilmesinde Alternatifler: Geleneksel Peynirler ve Peyniraltı Suyu İçecekleri 4 Geleneksel Gıdalar Sempozyumu 17 – 19 Nisan 2014 Adana* Bildiri Kitabı. Gökçe Matbaacılık Ankara. 171-4. ISBN 978-605-01-0603-9.

Karagözlü, C & Atilla, N (2018) Peyniraltı Suyunun Değerlendirilmesinde Alternatifler: Geleneksel Peynirler ve Peyniraltı Suyu İçecekleri. *4 Geleneksel Gıdalar Sempozyumu 17 – 19 Nisan 2014 Adana Bildiri Kitabı Gökçe Matbaacılık Ankara ISBN* 171-4.

Karagözlü, C & Atilla, N (2018) Süt Ürünleri. (Bölüm 8). İzmir'in Lezzet Öyküsü Cilt 1 (Mutfak Kültürü). İzmir İl Kültür Turizm Müdürlüğü. İzmir.

Karagözlü, C, Yerlikaya, O, Akpınar, A, Ünal, G, Ergönül, B, Ender, G & Uysal, H (2016) Cholesterol

Levels and Some Nutritional Parameters of Traditional Cheeses in Turkey. *Ege Üniv Ziraat Fak Derg,* 53, 161-8.

Karakas, R & Korukoglu, M (2006) A Traditional Cheese: Microbiological and Chemical Properties of Sepet Cheese. *Gida,* 31, 169-72.

Kavaz, A, Arslaner, A & Bakırcı, İ (2012) Comparison of Quality Characteristics of Çökelek and Lor Cheeses. *Afr J Biotechnol,* 11, 6871-7.

Kesenkaş, H, Dinkçi, N & Kınık, Ö (2012) Properties of Köy Cheeses produced in different dairies. *Ege Üniv Ziraat Fak Derg,* 49, 167-73.

Kılıç, S & Gönç, S (1990) Investigations on microbiological characteristic of İzmir Tulum Cheese-I. *Ege Üniv Ziraat Fak Derg,* 27, 169-85.

Kılıç, S & Gönç, S (1992) İzmir Tulum peynirinin olgunlaşmasında rol oynayan mikroorganizma gruplarının belirlenmesi üzerine bir araştırma. *Ege Üniv Ziraat Fak Derg,* 29, 71-8.

Kılıç, B, Coşkun, H & Tarakçı, Z (2003) Determination of Changes in İzmir Tulum Cheese During the Maturation Period by Traditional Methods and Culture. *Traditional Milk Products, Milk and Milk Products Symposium* 21-2. Demirci, M., Ed.; Tekirdag, National Productivity Centre Publications, Ankara, 1998; Publication number 621, 43–65.

Kılıç, S (2004) Kırktokmak Cheese. *Geleneksel Gıdalar Sempozyumu Kitabı,* 271-4. 23-24 Eylül 2004 Van.

Kiliç, S, Gönç, S, Uysal, H & Karagözlü, C (1998) Determination of Changes in İzmir Tulum Cheese During the Maturation Period by Traditional Methods and Culture In: Demirci, M., (Ed.), *Milk and Milk Products Symposium,* May 21–22, 1998National Productivity Centre Publications, Tekirdag Ankara199843-65.

Kınık, Ö, Ergüllü, E & Akbulut, N (1999) A study on production and some characteristic of Sepet Cheese. *Gida,* 24, 151-61.

Kırdar, S S (2003) Çökelek Peynirinin Yapılışı ve Özellikleri Üzerine Bir Araştırma. *GAP III Tarım Kongresi, 02-03 Ekim 2003 Şanlıurfa,* 441-3.

Koca, N (1996) Effects of Various Starter Combinations on İzmir Tin Tulum Cheese Quality. *Unpublished Postgraduate Thesis, Ege University Institute of Science.*

Koçak Kızanlık, P & Göksoy, EÖ (2018) Microbiological quality evaluation of various types of cheese. *Journal of Faculty of Veterinary Medicine Erciyes University,* 15, 86-93.

Orşahin, H (2012) Quality characteristics and shelf-life of Armola cheese. *Yüksek Teknoloji Enstitüsü Unpublished Master thesis.*

Öksüztepe, G, Patır, B, Dikici, A & İlhak, Oİ (2009) Microbiological and chemical quality of the vacuum marketed in Elazığ Packed Fresh Cheese. *Fırat University Medical Journal of Health Science,* 23, 89-94.

Önganer, A & Kırbağ, S (2009) Microbiological quality of "cokelek" cheeses as a fresh consuming in Diyarbakir. *Erciyes University J of Inst of Sci and Tech,* 25, 24-33.

Özsaatçılar, E (1983) Kırktokmak Peynirinin Yapılışı ve Özellikleri Üzerinde İncelemeler. *Ege University Agriculture Faculty Unpublished Graduation thesis.*

Swan, S (2009) Türkiye'nin Peynir Hazineleri. *Boyut Yayın Grubu ISBN: 978-975- 23-0067-5 2 baskı pp14 İstanbul, Turkey In Turkish*

Şimşek, B & Sağdıç, O (2012) Effects of starter culture types and different temperatures treatments on physicochemical, microbiological and sensory characteristics, and fatty acid compositions of çökelek cheese made from goat milk. journal of faculty of *Veterinary Medicine. Kafkas University,* 18, 177-83.

Toker, C (2001) *Manisa pazarlarında satılan salamuralı beyaz peynir, salamuralı tulum peyniri ve manisa köy peynirinin bazı kimyasal ve mikrobiyolojik özelliklerinin incelenmesi.*Basılmamış Yüksek Lisans Tezi. Celal Bayar Üniversitesi Fen Bilimleri Enstitüsü, Manisa. (In Turkish)

Tzanetakis, N (1987) Microbiology of Kopanesti, a traditional Greek cheese. *Food Microbiol,* 4, 251-6.

[http://dx.doi.org/10.1016/0740-0020(87)90007-4]

Üçüncü, M (2004) Cheese Technology From A to Z. *Cilt II Meta Basim, Bornova, Turkey* 1236.

Unsal, A (1997) *Turkish Cheeses.*Yapi Kredi Press.. ISBN 975-363-780-2, Istanbul.

Yaygin, H (1971) Investigations on Production and Characteristics of Tulum Cheese. *Ege Üniv Ziraat Fak Derg,* 8, 91-123.

Yaygın, H, Gönç, S, Oktar, E & Ergüllü, E (1978) İzmir İli Sütçülüğü Üzerine Araştırmalar. *EÜ Ziraat Fak Yayınları No: 340 EÜ Ziraat Fak Ofset Yayınevi Bornova.*

Yoldaş, M (2017) Functionally enhanced Armola Cheese Produced. *Unpublished Master of Science Thesis, Ege University Institute of Science,* 92.

CHAPTER 12

Traditional Cheeses of Mediterranean Region in Turkey

Oya Berkay Karaca[1], **Tansu Taspinar**[2,*] and **Mehmet Güven**[2]

[1] *Department of Hotel Management, Karatas School of Tourism and Hotel Management, Çukurova University, 01330 Adana, Turkey*

[2] *Agricultural Faculty, Department of Food Engineering, Cukurova University, 01330 Adana, Turkey*

Abstract: Cheese is an ancient food whose origins predate recorded history. Traditional cheeses are characterized by strong links to their territory of origin and the culture of the community and considered essential food in the Turkish diet, especially in breakfast for centuries. The mediterranean region of Turkey is quite rich with concerning for traditional cheeses. The objectives of the present review were to introduce traditional cheeses produced in the Mediterranean region of Turkey, to characterize the processing stages of Surk cheese (a kind of moulded herb cheese), Tulum cheese (cheese encased in a skin), Sünme cheese (a semi-hard cheese variety, has a fibered structure and consumed for breakfast), Carra cheese (which means "earthenware jug", containing mostly black cumin and/or garden thyme), Künefe cheese (fresh unsalted dessert cheese and has melting properties), Hatay Ezme cheese (Leaf or press cheese), Hatay Dil cheese (a string cheese variety with fibrous texture), Isparta Küp cheese, Isparta Dolaz cheese, Burdur Akçakatık cheese (is produced by drying strained yoghurt with or without adding cloves or black cumin), Burdur Keş cheese (is produced by drying of kneaded Cokelek with several aromatic plants and therefore has a special odor and aroma), Adana Lavaş cheese, Karataş Basma cheese with black cumin and Dilim cheese, Hellim cheese and to describe their compositional, chemical, physical, microbiological and sensory characteristics.

Keywords: Manufacturing technology, Mediterranean region, Traditional cheeses, Turkey.

INTRODUCTION

Classification of many local kinds of cheese produced in Anatolia according to their characteristics, standardization of traditional production technologies, and introduction to the world is important in terms of cheese diversity and continuity

* **Corresponding author Tansu Taspinar:** Agricultural Faculty, Department of Food Engineering Cukurova University 01330 Adana Turkey; Tel: +90 322 338 60 84; Fax: +90 322 338 66 14; E-mail: ttaspinar@cu.edu.tr

Celile Aylin Oluk & Oya Berkay Karaca (Eds.)

of cultural heritage (Durlu-Özkaya & Gün, 2007). To increase the geographic diversity of marked products, traditional cheeses are an important opportunity in countries, and it is a fact that an important element of attraction for cities and regions by using of these cheese as a tourist product (Karaca, 2016). In 2017, 8.978.599 tons of drinking milk was produced by commercial dairies in Turkey. While the cheese produced from cow's milk was 662.151 tons, and cheeses produced from sheep's, goat's, and buffalo's milk and their mixtures were 27.759 tons (TÜİK, 2018) altogether.

Mediterranean region takes its name from the Mediterranean Sea and occupies 15% of the total area of Turkey with its 120.000 square kilometers of land. The main provinces in this region are Adana, Antalya, Burdur, Isparta, Hatay, Kahramanmaraş, Mersin, and Osmaniye (Fig. **1**). Artisanal cheese production was generally limited only within their regions and even unknowns from one city to another within the same area in Turkey. Only the Hatay Künefe (C2018/173, 28.08.2018) and Maraş Finger (C2016/082, 19.09.2016) cheeses in the Mediterranean region were applied to the "Turkish Patent and Trademark Office" for "PGI" (protected geographical indication), but none of them have been registered. Because of this, national and international promotion and industrial production of these cheeses should be increased. "Geographical Indication" and "Designation of Origin" works in Turkey related to traditional cheeses are completed in the fastest way, and "Traditional Specialization Guaranteed Emblem" should be used to increase interest in these products.

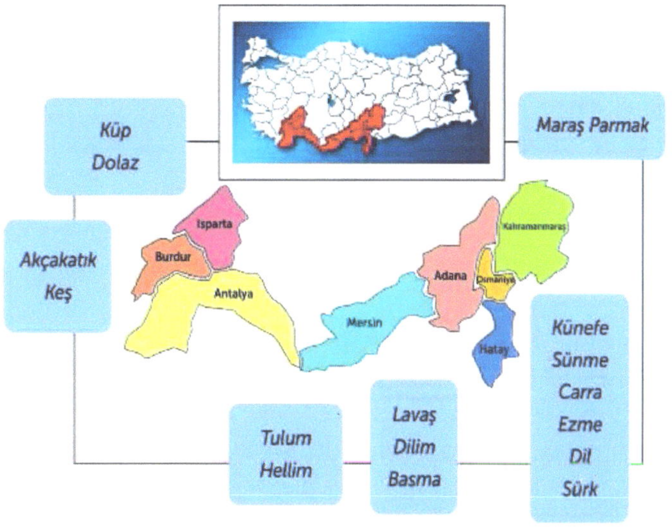

Fig. (1). Some cheeses of Mediterranean region in Turkey.

BURDUR AKÇAKATIK CHEESE

Akçakatık cheese is a traditional yoghurt cheese which is matured in "Karin" (goat or lamb skin sack or stomach) and manufactured in some provinces like Burdur and Isparta in the Mediterranean of Turkey (Kırdar *et al.*, 2017; Şimşek & Tuncer, 2018). Especially this cheese is made in Kozluca and Dinar villages of Burdur. The local people knew this cheese as 'aşçı kaşığı'. Over time the name of cheese was changed to Akçakatık (Kırdar *et al.*, 2012). Akçakatık cheese is a semi-hard cheese. This cheese type has high fat, high dry material content, salty aroma, butter-like aroma, strong acidic taste, homogenous structure. Among these properties, it is especially essential for this kind of cheese with a strong acidity (Şimşek & Gün, 2009).

Akçakatık cheese looks like Labneh and Tulum cheese in appearance (Table **1**, Fig. **2**). However, Akçakatık cheese has a higher dry matter content than the labneh and therefore has a harder structure. The production method of these cheeses is linked to each other, but these are produced locally from concentrated yoghurt with black cumin and clove (Kırdar *et al.*, 2017, Şimşek & Tuncer, 2018). Spring season is preferred for the production of this cheese, but there is no standard production of Akçakatık cheese. Cheese producers at home use some traditional methods, and because of this, the production process can be different. Because of the different manufacturing practices before the absence of control of temperature, likewise absence of control relative humidity during storage cause important variations in the final quality of the product. Cow's milk, goat's milk, or a mixture of these two types of milk is usually used in the production of this cheese (Kırdar *et al.*, 2017). Followed by the addition of carnation or black cumin, drying strained yoghurt Akçakatık cheese is obtained. Akçakatık cheese is different from other cheeses in that it is made from yoghurt, and spice is added during the production process. These spices produce a sharp taste and odor that pleases the cheese (Şimşek & Tuncer, 2018).

There are two ways of consumption of Akçakatık cheese, fresh or ripened. Ripened Akçakatık cheese is produced by drying and maturing in cloth bags the fresh ones. Therefore, after production in order to ensure full maturation, the cheese is buried in goat's or lamb's stomach generally between September and November (Kırdar *et al.*, 2017, Şimşek & Tuncer, 2018). The production method used in traditionally produced Akçakatık cheeses is usually like this (Şimşek & Gün, 2009); the first process used in the Akçakatık cheese production is preparing the stomach. Generally, lamb's stomach is used. The rough dirt of the stomach is to be thoroughly cleaned until the roughness is removed. After rough dirt is removed, stomach is put in the calcium content. After that, the stomach is washed with water and purified from calcium. When the process of calcium removal is finished, the stomach is washed with detergent and soapy water. After it is rinsed with water, it is kept in brine for 2-3 days and inflated. After all the washing operations are finished, it is left to dry for 5-6 days. After 5-6 days, the dry stomach is used for packing the cheese (Kırdar *et al.*, 2017).

Table 1. The Mean Composition Values of Western Mediterranean Cheeses.

	Akçakatık Cheese				Keş			Küp	Çökelek			Dolaz		
Properties*	**A**	**B**	**C**	**D**	**E**	**F**	**G**	**H**	**K**	**L**	**M**	**N**	**O**	**P**
TA (% la)	0.9-2.16	2.40	0.866	2.67	1.85	0.32	0.32	1.9	2.40	0.28	0.49	1.62	2.40	1.14
pH	-	3.96	5.12	3.86	4.31	5.04	-	-	4.30	4.51	3.83	4.58	4.25	-
Dry Matter (%)	24.50-59.63	79.71	73.96	73.38	72.54	55.58	59.50	69.04	27.78	47.85	60.83	52.04	39.71	56.26
Protein (%)	-	-	-	-	4.69	-	19.52	-	-	-	25.95	15.21	-	14.10
P/DM (%)	-	-	-	-	-	19.62	-	-	-	-	-	-	-	-
Fat (%)	10.3-30.2	21.00	25.52	31.86	36.50	22.34	25.00	25	4.33	19.59	15.43	17.70	13.00	20.25
F/DM (%)	-	32.18	34.87	-	50.31	37.95	43.03	36.21	15.78	41.31	24.87	34.52	32.00	-
Salt (%)	5.38-9.12	6.16	0.40	12.43	4.68	2.75	2.96	4.4	1.83	-	2.90	-	4.00	0.00
S/DM (%)	-	4.27	0.56	-	6.45	7.47	5.07	6.37	5.97	6.51	4.85	-	11.00	-
Ash (%)	-	-	-	-	-	-	-	-	-	4.78	4.04	-	-	-
WSN* (%)	-	-	-	0.51	-	-	7.71	-	0.34	-	-	-	-	0.27
TCA/TN (%)	-	-	-	-	-	-	7.67	-	-	1.95	-	-	-	-
PTA/TN (%)	-	-	-	-	-	-	2.92	-	-	5.35	-	-	-	-
RI (%)	-	-	-	18.35	-	-	7.60	-	15.71	2.11	-	10.95	-	12.25

Burdur Akçakatık cheese: A (Kırdar, 2004a), B (Kırdar *et al.*, 2017; n:15), C (Gün and Şimşek, 2006; n:15), D (Şimşek and Tuncer, 2018; n:15)
Burdur Keş cheese: E (Kırdar, 2004b), F (Kırdar *et al.*, 2009; n:20), G (Kırdar, 2012),
Isparta Küp cheese: H (Kırdar, 2003),
Çökelek cheese: K (Şimşek and Gün, 2012), L (Kavaz *et al.*, 2012), M (Çardak, 2012; n:25),
Isparta Dolaz Cheese: N (Şimşek and Sağdıç, 2006), O (Okur and Güzel-Seydim, 2011b), P (Donmez *et al.*, 2005)
*TA, Titratable acidity; P/DM, Protein in dry matter; F/DM, Fat in dry matter; S/DM, Salt in dry matter; WSN, water soluble nitrogen; TCA, 12% trichloroacetic acid-soluble nitrogen; PTA, 5% phosphotungstic acid-soluble nitrogen; CN, Caseine nitrogen; TN, Total nitrogen; RI, Ripening index.

There are two different methods of cheese production. Cow's and/or goat's milk is used as the raw material (Şimşek & Tuncer, 2018). Firstly, the non-skimmed milk pasteurized at approximately 85°C 30 min and then cooled to fermentation temperature 40 °C. At this temperature, strained yoghurt from the previous day is used as a starter culture. The fermentation process takes approximately 2.5-3 hours. At the end of the fermentation process, the yoghurt is kept for one night in a refrigerator temperature (+10°C). The traditional yoghurt production is followed by the production of concentrated yoghurt. The yoghurt which is transferred into cloth bags is drained for 2-3 days until all the serum is removed. Sometimes it can take up to 5 days. After the filtration process, 3-5% salt is added. After salt is added, spices can be added. Salt and spice added yoghurt is filled into the dry stomach. The average weight of the cheese in the sack is about 800 g. Cheeses are buried or kept in a cool place (8-10°C) for 3-4 months. During this process, they are often turned inside out. The reason for using lamb's abdomen is to extend the storage period of cheese. The storage period of cheese in the lambskin sack is around 3-6 months. Besides, during this time, the taste and aroma formation in the cheese is provided. Cheese production is done in spring and summer; the dried product is consumed in the winter months (Kırdar *et al.*, 2017; Kirdar, 2004a).

Fig. (2). Akçakatık Cheese (The packaging by sewing goat's or lamb's stomach) (courtesy of Kırdar, S.S.).

The second method used in cheese production is as follows: Yoghurt production is carried out by using cow or goat milk in a traditional way. After filtration, yogurt is thoroughly kneaded with spice mixtures and hand-formed into pellets. It is left to dry in the sun. The dried product is filled in the stomach and consumed in winter months. Cheese is heated and softened for consumption (Şimşek & Tuncer, 2018).

Kırdar *et al.* (2013) indicated that the mean concentration of Mg, K, Ca, Cu, Mn, Fe, and Zn of the cheeses were 49.62, 417.74, 2842.24, 3.36, 6.15, 12.62 and 13.92 mg/kg, respectively.

Şimşek and Gün (2009) reported that palmitic acid was the highest of the saturated fatty acid and the saturated fatty acid, total unsaturated fatty acid, monounsaturated fatty acid, polyunsaturated fatty acid of Akçakatık cheeses were found as 61.87 ± 4.067, 33.246 ± 2.919, 30.728 ± 2.648 and 2.519 ± 0.516%, respectively.

Şimşek and Tuncer (2018) evaluated that volatile flavor compound analysis showed that acetaldehyde, diacetyl, ethanol, and ethyl acetate are major carbonyl compounds in Akçakatık cheese samples and in ripened cheese were detected on average as 31.62 ± 3.23, 73.42 ± 57.24, 8.04 ± 7.60 and 60.21 ± 68.66 ppb, respectively.

BURDUR KEŞ CHEESE

Keş cheese is known as Keşk, Kurut, Kesuk, Kiş, Kashk, or sometimes even Çökelek (Kırdar, 2012). Keş (Table **1**, Fig. **3**) shelf life is longer because of lower fat and moisture contents (Dervisoglu *et al.*, 2009). The cheese is made from semi-fat yoghurt/ayran or skim milk. It is a cone or irregular cylindrical shape and has a slightly pungent and sour taste. It can be classified as hard cheese (Kamber & Terzi, 2008) and produced by drying of kneaded Çökelek with several aromatic plants (Tarakçı *et al.*, 2010). Thanks to its protein, fat, and dry matter, it has high nutritional valuc (Kırdar, 2012). This cheese can be consumed directly with thin bread or added to soups and pasta (Kırdar, 2012; Kırdar *et al.*, 2018).

Fig. (3). Kes cheese (courtesy of Kırdar, S.S.).

Keş cheese is generally produced in a traditional way in small family businesses for their needs (Dervisoglu *et al.*, 2009). It is produced with self-coagulation of raw milk without any starter culture addition, then cutting coagulum, draining curd, kneading the curd with a mixture of spices such as nigella, red pepper, cumin and black pepper (Yurdakul *et al.*, 2017) are rest of the processes. As the manufacturing procedure is not standardized, the production method depends on the region where the cheese is produced. This kind of cheese making process consists of standard production practices and traditional tools without starter culture addition, so thc microbial population originates from milk and environment (Kırdar *et al.*, 2018).

Firstly, for the removal of unwanted materials, the milk is filtered through a cheesecloth; after that, milk is cooled to 40-45°C and inoculated with yoghurt culture. Inoculated milk is filled into containers and incubated. After incubation is completed, the yoghurt is churned into butter and buttermilk. Until white coagulum floats on the surface, buttermilk is heated to 90-100°C. The curd is transferred into a cheesecloth. To achieve the desired solids level, the weight is adjusted, and the water is removed. The curd is salted (Dervisoglu *et al.*, 2009). After salting, black sesame, capsicum, black, and red pepper are also added, and curd is filled into cotton bags (Kırdar, 2012). The bags are turned over for homogeneous dryness and kept in a room 2-3 days. The curd was shaped into a cone or cylindrical form. The cheese is sometimes dried under the sun or room conditions at the next stage, and after being pressed into a receptacle (Kamber & Terzi, 2008) it is stored in a refrigerator for 4-5 months (Dervisoglu *et al.*, 2009). When the cheese is ready to be consumed moldy, it is placed in pots after shaping, buried underground, and molded (Karaca & Güven, 2004a). Çakır *et al.* (2009) determined that mean composition of Keş: 61.59% dry matter, 13.66% ash, 32.42% protein, 6.30% fat, 13.26% salt, 0.22% titratable acidity, 3.81 pH and 5.76% ripening index. The predominant free fatty acids in the Keş samples were palmitic, oleic, stearic, and myristic acids, respectively. Kilicel *et al.* (2008) found that Zn, Fe, Cu, and Mn in Keş Cheese were 14.10, 15.70, 20.0, and 0.81 μg/g.

ISPARTA (POTTERY) KÜP CHEESE

Küp cheese (Table **1**) is widely produced in provinces like Ankara, Nevşehir, and Yozgat. This cheese type is matured underground in pots, and its name comes from this procedure and packaging method (Koluacik *et al.*, 2015; Kaptan, 2016). Küp cheese is a semi-hard cheese. It has a quite favorable taste, crumbly texture, and white-cream color (Kaptan, 2016).

Like other traditional cheeses, Küp cheese doesn't have a standard production method. Küp cheese is usually produced in small farms. Sheep, goat, or cow milk can be used. Lactic starters are not used. For cheese production, firstly, rennin and salt (3%) are added to the raw milk. After coagulation, the curd is cut into small pieces (1 cm^3) and put into small cotton bags. Whey is drained by putting weight over the bags. This is done at room temperature and lasts about 24 hours. After the whey has been removed, the curd is kneaded with rock-salt and once again put into larger cotton bags and left under pressure 12-16°C. Afterward, the curd is compressed into the pot whose bottom is salted, and also some salt is sprinkled on the surface. The pot (earthenware materials) is then turned upside down and buried underground. Küp cheese is ripened at least five months at 4-6°C in the caves and subsoil (Kaptan, 2016; Kamber & Terzi, 2007). Küp cheese can also be consumed while ripening but generally consumed after-ripening (Kamber & Terzi, 2007).

ISPARTA DOLAZ (TORT) CHEESE

Dolaz cheese is produced from the whey of Beyaz cheese manufactured from cow, ewe and goat milk by especially the Yuruks (Turkish nomads in Anatolia) living in the rural areas of Isparta, Antalya and Afyonkarahisar cities (Şimşek & Sağdıç, 2006; Yerli *et al.*, 2018). Dolaz cheese (Table **1**) has a soft granular structure similar to Çökelek cheese, characteristic odor, and unique flavor and dark yellowish-light brown color (Okur & Güzel-Seydim, 2011a). Dolaz cheeses are traditionally produced from raw milk without using starter cultures (Donmez *et al.*, 2005). Materials used traditionally in the production of Dolaz cheese; whey (53%), buttermilk (9.5%), milk (10%), yoghurt (25%), curd cheese (2 kg) (Okur & Güzel-Seydim, 2011b). Dolaz cheese production is shown in Fig. (**3**). It is stored in bags made of skin or plastic containers. The production method of Horç cheese produced by nomads in Erdemli district of Mersin is similar to this cheese. It is a dark or light brown product obtained by boiling the whey for a long time and draining it from cloth bags. Yuruks consume it in winter after frying in cooking oil, in a patty, in breakfast, and soups (Uçgun & Işık, 2018).

Şimşek and Sağdıç (2006) indicated that the average counts of total aerobic bacteria, yeast and mold, psychrophilic bacteria and *Enterobacteriaceae* counts were determined to be 5.41±1.14, 4.13±1.26, 3.24±1.19 and 1.50±0.68 log cfu/g and *enterococci, lactobacilli* and *lactococci* microorganisms in the Dolaz cheese were found to be 3.25±1.80, 5.06±0.79 and 5.12±0.93 log cfu/g, respectively.

The saturated fatty acid level of Dolaz cheeses (Figs. **4** & **5**) was 60.80-68.66%. (Donmez *et al.*, 2005). Acetaldehyde, acetone, ethanol, acetic acid, diacetyl and 1-butanol contents of Dolaz cheese were 0.0526 mg/kg, 0.0018 mg/kg, 2.61 mg/kg, 1.15 mg/kg, 0.0079 mg/kg and 0.0046 mg/kg, respectively (Okur & Güzel-Seydim, 2011a). The total bacterial, the yeast-mold, the coliform, *Lactobacillus*, *Lactococcus* counts of the Dolaz cheeses determined varied between 7.68-8.23 log cfu/g, 6.90-7.37 log cfu/g, 2.83-3.84 log cfu/g, 7.87-8.08 logcfu/g, and 7.63-8.17 log cfu/g by Okur and Güzel-Seydim (2011a). Okur and Güzel-Seydim (2011b) found that in Dolaz cheeses 2.16 mg/g Ca, 11.73 mg/g Na, 0.27 mg/g Mg, 0.01 mg/g Zn and P; 66.43 L*, 11.82 a* and 24.45 b* values are found. According to Donmez *et al.* (2005), among 20 different cheese samples, Dolaz cheese has the highest conjugated linoleic acid level based on fat content, and sample weights and the cholesterol level as 147.69 mg/100 g.

Fig. (4). Dolaz Cheese (courtesy of İ.Gün).

Whey (adding of yogurt, whole milk, *etc*)
Boiling with stirring (4-6 h)
Increase in consistency until its colour turns yellow-light brown (10-15 h)
Cooling (20°C, left room temperature during 12 h)
Drainage in cloth bags (3-4 day)
Drying and grinding of the filtrate and adding salt
Holding in cloth bags (5 day)
Pressing the leather bags ("Tulum" or "Tuluğ"; leather used are derived from goat skin and "White leather" is called hairless skin) or plastic containers
Storage (+4 °C, 3-4 m)

Fig. (5). Dolaz cheese production steps (Şimşek and Sağdıç, 2006; Okur and Güzel-Seydim, 2011b).

ÇÖKELEK CHEESE

Çökelek is a traditional Turkish acidic cheese (Simsek & Sagdıc, 2010) (Table **1**). Firstly, White cheese and Tulum cheese varieties are the most consumed cheeses in Turkey. After that, Çökelek cheese and Lor cheese are the most consumed ones. Çökelek cheese is generally manufactured in family corporations according to traditional methods (Kavaz *et al.,* 2012). Çökelek cheese has a similar appearance to Lor cheese, so these two kinds of cheese can be confused with each other. Even in some regions, Lor cheese is referred to as Çökelek cheese, but these cheeses differ significantly from each other (Çardak, 2012). This cheese is produced in various ways and given different names in Turkey. It is known as

'Eksimik', 'Tortu' and 'Cabaaltı' in Western Anatolia, 'Süt Koptu', 'Akkatık', 'Kesik', 'Torak', 'Urda', 'Süt Kırması' in Mediterranean (Kavaz *et al.,* 2012) and 'Jaji', 'Jajik', 'Cacik', 'Torak', 'Kurç', 'Kurçi' and 'Çotran' in Eastern Anatolia, 'Minci', 'Minzi', 'Yağlı', 'Beğerim' in Northern Anatolia (Karaca & Güven, 2004a). The dairy products produced from yoghurt such as Çökelek, tarhana, yayik butter are very consumed, popular products in the Middle East (Şimşek & Sağdıç, 2012). Because Çökelek cheese is tasty, nutritious, and economical, it is consumed a lot. Yoghurt that is too sour to consume can be utilised by making Çökelek cheese (Simsek & Sağdıç, 2010).

Çökelek cheese is also produced with milk beside the yoghurt. Çökelek has been made from cow, goat, ewe milk for a long time (Şimşek & Sağdıç, 2012). Çökelek cheese production methods are not standardized, but it is easy, and this kind of cheese has a low or nonsalt proportion. Çökelek cheese is obtained by boiling milk, yoghurt or ayran. Çökelek cheese curd obtained by heating of yoghurt aroma can be different from the one obtained by heating of milk (Durlu-Özkaya & Gün, 2014). Çökelek cheese applies heat treatment to milk, contains casein and serum proteins. Çökelek cheese has regional characteristics, can be consumed fresh or can be ripened in earthenware or animal skin called 'tulum' and then kept and protected under the sun (Kavaz *et al.,* 2012). There are some types in which wild herbs are added in order to give them a unique taste and odor (Durlu-Özkaya & Gün, 2014). In Antalya province (Tulum Çökelek with black cumin), after the production of butter, the remaining ayran is boiled, and the clot formed as a result of heating is transferred to the cloth bags. After filtration, salt, black cumin, and starch (10%) are added to curd, placed into the tulum and nylon bags, and left to mature in cold rooms (Kırdar, 2005). There are dimethyl sulfide, ethyl butyrate, ethyl pentanoate, ethyl hexanoate, methional, butyric acid, phenylacetaldehyde, and 2-phenyl-ethanol and hexanoic acid as aroma active compounds in Çökelek cheese made from milk, yoghurt and tuluk yoghurt (Gün *et al.,* 2012). The results of some studies on Çökelek cheeses showed that myristic, palmitic, stearic, and oleic acids are the main fatty acids present in this type of cheese (Şimşek & Sağdıç, 2010; 2012).

KÜNEFE CHEESE

This cheese type is similar to commonly known white cheese (Table **2**, Fig. **6**). Künefe cheese is mostly made from goat milk. Cow milk is also used. Sometimes the mixture of both milk can be used. The cheese production special care is taken to use milk containing a high level of fat. It is made without the addition of salt and without being given a particular shape. Künefe cheese is sold fresh as soft cheese for a special dessert Künefe, and it is used for production. This cheese has good melting properties (Karaca & Kırdar, 2010).

Yuvalama, Dil, and Sünme cheeses are produced by making some modifications like stretching, kneading, shaping on Künefe cheese, so this cheese is the main component for other cheese varieties (Fig. **4**). These cheese types are also produced in the same region. A starter culture isn't used in the Künefe cheese production method.

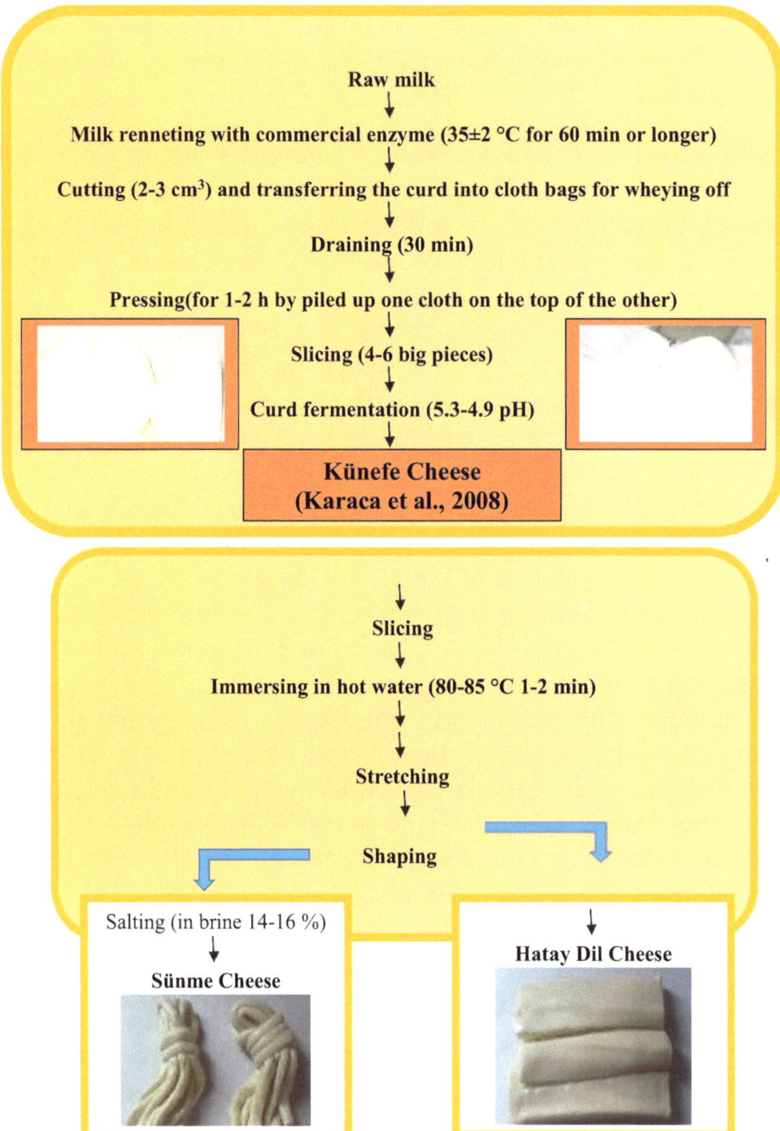

Fig. (6). Traditional Hatay Künefe, Sünme and Dil cheese production methods.

Table 2. The Mean Composition Values of Eastern Mediterranean Cheeses.

	Dil		Ezme	Künefe		Sünme				Carra				Sürk						Lavaş	Dilim
Properties*	**A**	**B**	**C**	**D**	**E**	**F**	**G**	**H**	**I**	**K**	**L**	**M**	**N**	**O**	**P**	**R**	**T**	**U**	**V**	**Y**	**Z**
TA (% la)	1.16	-	0.65	0.63	1.18	2.0	0.27	1.41	0.76	0.85	-	-	0.45	-	1.14	1.67	1.4	-	1.61	1.01	-
pH	5.62	5.64	6.25	5.36	5.32	-	5.05	5.60	5.57	5.63	-	5.24	5.77	-	4.94	4.60	5.8	3.87	4.59	5.49	5.33
Dry Matter (%)	48.55	52.75	49.23	46.43	36.52	63.1	54.09	53.74	51.73	53.43	52.43	58.74	57.97	39.57	44.32	45.45	49.8	27.4	75.83	51.25	52.65
Protein (%)	20.12	24.62	19.48	19.47	15.71	21.4	24.38	26.12	22.84	18.86	23.09	-	-	23.15	19.02	15.64	26.4	-	14.99	20.33	23.48
P/DM (%)	41.45	-	39.60	42.15	-	-	-	48.51	44.13	-	-	-	-	-	-	-	-	-	-	-	44.60
Fat (%)	19.80	19.25	23.00	24.19	18.14	31.7	21.45	20.07	23.75	24.86	25.67	26.77	-	6.50	8.99	6.60	14.7	-	13.75	22.13	24.25
F/DM (%)	40.73	-	46.67	52.07	-	-	39.66	37.20	45.91	46.65	-	-	-	-	19.62	-	29.0	-	18.13	-	46.06
Salt (%)	5.38	0.92	3.96	0.24	0.51	9.3	6.59	2.65	3.23	8.83	6.35	7.82	7.24	4.87	8.35	4.95	5.4	0.2	4.80	5.85	2.96
S/DM (%)	11.14	-	8.04	0.53	-	-	12.30	4.98	6.26	16.73	-	-	-	-	18.72	-	-	-	6.31	-	5.62
Ash (%)	6.66	-	5.58	1.88	1.58	1.9	-	8.54	-	-	-	-	-	-	-	6.33	-	1.3	-	8.10	-
WSN* (%)	0.353	-	0.269	-	-	-	-	0.657	0.416	0.70	1.75	-	-	1.72	1.58	-	-	0.23	0.15	0.27	-
TCA/TN (%)	-	-	-	-	-	-	-	-	1.49	0.46	1.21	-	-	1.08	2.48	-	-	-	-	-	-
PPN (%)	-	-	-	-	-	-	-	-	-	0.28	0.63	-	-	0.63	1.69	-	-	-	-	-	-
PTA/TN (%)	-	-	-	-	-	-	-	-	0.75	-	0.37	-	-	0.75	0.28	-	-	-	-	-	-
CN (%)	2.80	-	2.78	-	-	-	-	3.44	3.17	-		-	-	-	-	-	31.7	-	-	2.91	-
RI (%)	11.64	-	8.85	-	-	-	-	16.60	11.86	21.84	48.33	-	-	47.62	52.91	-	-	-	6.45	8.71	-

Dil Cheese: A (Karaca *et al.,* 2009a; n=26), B (Kılıc and Işın, 2004)
Ezme Cheese: C (Karaca *et al*., 2009b; n=30)
Künefe Cheese: D (Karaca *et al*., 2008), E (Akın *et al*., 2014)
Sünme Cheese: F (Biçer *et al*., 1995), G (Mutlu, 2000; n:20), H (Karaca and Güven, 2004b; n:30), I (Mutluer *et al.,* 2014)
Carra Cheese: K (Konar and Güler, 1998; n:30), L (Guler, 1999), M (Aygun *et al*., 2005; n:50), N (Güner and Telli, 2011; n:30)
Surk Cheese: O (Guler, 1999), P (Güler, 1999; n: 36), R (Güler, 2014; n:12), T (Durmaz *et al.,* 2004, n: 25), U (Masatcıoğlu and Avşar, 2005),
V (Sağdıç *et al*., 2005)
Lavaş Cheese: Y (Güzeler *et al.,* 2011)
Dilim Cheese: Z (Karaca and Oluk, 2018)
*TA, Titratable acidity; P/DM, Protein in dry matter; F/DM, Fat in dry matter; S/DM, Salt in dry matter; WSN, Water soluble nitrogen; TCA, 12% trichloroacetic acid-soluble nitrogen; PTA, 5% phosphotungstic acid-soluble nitrogen; CN, Caseine nitrogen; PPN, Proteose peptone nitrogen; TN, Total nitrogen; RI, Ripening index.

For the cheese production, firstly, the raw milk is coagulated with commercial rennet at 35±2°C at least 60 min. After coagulation, the curd is cut into small pieces approximately 2-3 cm^3. The curd is put into the cheesecloth, and whey is separated within 30 minutes. After separation, the cheesecloth is tied very tightly for 30 min again. After that, raw cheese is cut into 4-6 big pieces. Cheeses are then left to ferment. In summer, the fermentation process can be conducted at room temperature for 6 h. If it is winter, the fermentation process can be conducted in 12 h or 24 h +4°C in the refrigerator. The fermentation process is terminated by determining that the pH value of the curd has reached 4.9-5.3. After the fermentation is completed, the curd is stretched in hot water for 1-2 min. The hot curd is kneaded and then stretched by hand until it becomes shiny, smooth, and elastic. Curd shaped and shaped curd is then chilled in cold water to harden, and salted for several days in cool brine. Lastly, the cheese is packaged with vacuum plastic. Its yield is about 10% (Karaca & Ocak, 2016; Karaca & Kırdar, 2016).

SÜNME CHEESE

One of the cheeses produced and consumed especially in Hatay province is Sünme cheese (Table **2**, Fig. **6** and **7**). Sünme cheese is generally consumed in breakfast. It has a semi-hard texture and fibered structure. Originally, it is made from goat's milk, but cow's milk can also be used (Karaca & Kırdar, 2010).

Fig. (7). Sünme cheese.

Sünme cheese has the same production method as Künefe cheese. After Künefe cheese is obtained with some processing like stretching, shaping, salting in brine, Sünme cheese can be obtained. Thanks to these stages, the cheese has gained fibrous structure, stretching, and melting properties similar to Pasta filata cheeses (Karaca & Kırdar, 2016). The Künefe curd is cut into slabs that are kneaded and stretched in 80-85°C water at 1-2 min. The hot curd is kneaded and then stretched until it becomes elastic, smooth, shiny, and this stretching process is repeated until the curd has a fibrous character. The stretched curd is given string shapes by folding up and then tying a knot to the top of the cheese. The shaped curd is then chilled in cold water to harden and salted for several days in cool brine, which is 14-16% concentrated. After brining, the cheese is packaged with a vacuum plastic or can be stored in a freezer (Karaca & Ocak, 2016).

For the Sünme cheese production, firstly, the raw milk is pasteurized at 68°C 10 min. After that, milk is cooled to 32±1°C. 1% starter culture and 0.02% $CaCI_2$ added. Milk is pre-ripened 30 min. Coagulation complete in 50 min. After coagulation, the curd cut into small pieces approximately 2-3 cm^3. The curd put into cheesecloth, and whey is separated within 30 minutes. After separation pressing process applied for 20 min. After that, fermentation stopped at 5.2 pH, and raw cheese is cut into 3-3.5 cm cubic shapes. After the fermentation is complete, the curd is immersed and kneaded in hot water, which is 80±1°C for 5-6 min. The hot curd is stretched and shaped. Shaped pieces of curd salted in 14-16g 100 g^{-1} NaCl. Lastly, the cheese is packaged and ripened 6±2°C for 90 days (Karaca & Güven, 2004b).

HATAY DİL (STRING) CHEESE

Dil cheese is a pasta filata type cheese produced in Turkey (Kılıc & Isın, 2004). Dil cheese produced in many regions of Turkey is a type of cheese produced in industrial as well as being a traditional product (Karaca *et al.,* 2009a). Dil cheese is a fresh variety and unripened cheese (Table **2**, Figs. **6** and **8**). Dil cheese can be consumed as both table cheese or used for pizza production like Mozzarella cheese (Irkin, 2010). The shape of this cheese resembles a tongue because its name is 'dil' which means tongue in Turkish. Dil is a semi-hard cheese with a characteristic fibrous texture (Hayaloglu *et al.*, 2008). The fibrous texture is a characteristic of Dil cheese (Kılıc & Isın, 2004).

Fig. (8). Dil cheese.

It is usually made from pasteurized cow's milk and inoculated with thermophilic lactic cultures. During the production of Dil cheese, firstly, milk is coagulated at 33–35°C for approximately 60 min. The coagulum is cut into small cubes, rested for 5-10 min, and a portion of whey is removed, and the curds are held for a further 30 min. Afterward, the remaining whey is removed, and the curds are pressed until the desired acidity is reached. Desired acidity value is pH 5.0-5.2. The acidified curd is cut into pieces and cooked (hot water or 5% NaCl brine at 80-90°C). The hot curd is kneaded and stretched to form longitudinal fibers. The elastic curd is extended manually, cooled in cold water, and cut into cylindrical pieces, 2-3 cm in diameter, and 8-15 cm long. The pieces are immersed in 10% NaCl brine at 25°C. Salted Dil cheese is packaged under vacuum and sold (Hayaloglu *et al.*, 2008).

In Hatay region, after the fermentation (a pH in the range of 5.1-5.3) of Künefe cheese, boiling process is conducted. According to the given shape, cheeses are named as Dil, Sünme and Örgü cheese in the region. Blocks of curd obtained by cheddaring are placed into a twin-screw mixer containing water or brine at about 80°C and kneaded and stretched to provide longitudinal fibers. Elastic curd is extended manually into a long mold, cooled with cold water, and cut into strings. Strings of cheese could be salted in brine according to the manufacturing procedure. Several cheese strings are packaged in plastic packages under vacuum and stored at 4°C for 3-4 months (Karaca *et al.*, 2009a; Karaca & Kırdar, 2010; Karaca & Kırdar, 2016). Usually, Dil cheese is sold as soon as possible after the manufacture and consumed without ripening (Irkin, 2010; Karaca *et al.,* 2009a). Shelf-life of Dil cheese is reported as 3-4 months by manufacturers (Kılıc & Isın, 2004).

HATAY EZME CHEESE

Ezme Cheese, also known as Yaprak (Leaf) Cheese, or Baskı (Press) Cheese is a traditional cheese in Turkey (Table **2**). It is popular in Hatay province and neighboring provinces (Karaca *et al*., 2009b). It is made from cow's milk. Ezme cheese is in the form of thin slices (0.5-1 centimeters thick), either square, rectangular (4-6 cm), or round (with a diameter of approximately 10-15 centimeters) (Karaca *et al*., 2009b; Kamber, 2008a). It has white color. It also has a soft consistency. Hatay Ezme Cheese is a non-porous, non-crumbing, elastic, and homogenous cheese variety. Hatay Ezme Cheese is boiled after pressure and shaped by applying the crushing process with weight, and the characteristic feature of this cheese is that the curd being subjected to fermentation is kneaded in hot water and then pressed underweight and shaped. Ezme cheese can be consumed fresh as well as salted in jars and can be consumed after a short period of ripening (Karaca *et al*., 2009b).

In the production of Ezme Cheese (Fig. **9**), the fresh cheese is made like the Künefe Cheese production method (Kamber, 2008a). Following milking, filtered milk is fermented with commercial rennin to complete clotting at 1.5-2 hours. This process is done at summer milking temperature, and in winter, milk is heated to 35°C. The resulting clot is disintegrated with the help of a scoop and taken 10 minutes later into the tulbents in the cases, which in this case can be wood or chromium. Here, the four corners of the tulbent are brought loose, knotted, and the case is removed. Once it has been filtered for half an hour of pestle water in its own condition, it is compressed by the tulbent when it comes to sediment. After half an hour, the cheeses are cut into 4 to 6 large pieces and left for fermentation. Fermentation conditions are about 6 hours at room temperature in summer, 12 hours in winter, or 24 hours in the refrigerator at +4°C. This form of raw cheese obtained at this stage is also considered as "Künefe Cheese". The fermented cheeses are boiled for 5-6 minutes with boiling water heated in copper boilers.

After the kneading process, the cheese curd is placed under the proper large marble weights and is left in order to drain the water until it reaches a certain thickness. After the printing process, the cheeses are kept in brine, thus increasing its strength time. From time to time, the black cumin is handled, and the cheese is crushed in this way, and the cheese paste is placed in glass jars in brine (Karaca *et al.*, 2009b; Karaca & Kırdar, 2010).

Fig. (9). Ezme Cheese.

It has been determined that Hatay Ezme cheeses produced from untreated milk have different physical and chemical properties by using different raw materials and different production techniques and standard quality cannot be obtained in the region (Karaca *et al.*, 2009b). This cheese is widely consumed in the region in breakfast and as a fried dish (Kamber, 2008a).

CARRA (TESTİ) CHEESE

Carra is a traditional cheese (Fig. **10**) which contains black cumin (*Nigella sativa*) and/or garden thyme (*Thymus vulgaris*) and is one of the most commonly consumed cheeses in the Antakya (Antioch) region, located in southern Turkey (Karaca *et al.*, 2007; Karaca & Kırdar, 2016). Carra means earthenware jug, and this cheese is also known by the name Testi Cheese (Table **2**). It is a mixture of white cheese and skim milk cheese called 'Çökelek' (Aygun *et al.*, 2005; Güner & Telli, 2011; Kamber, 2008a). Mostly goat's milk is used in the production rarely raw cow's milk is used. It is produced in private houses, small dairy plants, or in farms and sold in blocks (Aygun *et al.*, 2005; Güner & Telli, 2011; Hayaloglu & Fox, 2008). Its texture is semi-hard or hard, its color is generally white, and has a sharp flavor and taste due to spices added to the curd (Kamber, 2008a).

Traditionally, commercial culture is added to warm milk. After approximately 1 h, the formed curd is cut into small pieces, and the whey is separated, and the pressing process is conducted for 30 min. Curd is sliced into 1 cm thick pieces, and salt is put between the slices, and the curd is left for hardening for 2-3 days. Meanwhile, Çökelek cheese is used as a secondary ingredient. Salt is added into Çökelek cheese, and cheese is placed in a cotton bag and put under pressure to release its water. Dried black cumin and/or garden thyme is added and mixed with the çökelek cheese. The salt ratio is approximately 4%, and the spice ratio is 5%. The two kinds of cheese, one obtained directly from milk and the other spiced çökelek, are pressed into a glazed earthware jug called 'Carra' in altering layers. When the jug is completely filled, it is turned upside down and kept for 3-4 days. The top of the jug is sealed with a mixture of salt, olive oil, wood ash, and water. The jug is then buried underground for ripening for at least 3 months. The cheese is usually buried in August and taken out in January (Durmaz 2009; Karaca *et al.,* 2004a; Kamber, 2008a; Ocak *et al.,* 2009).

The composition of Carra cheeses (Fig. **10**) is similar to Tulum cheese. Carra cheese has a higher amount of fat, protein, and salt and has a lower amount of moisture content than Çökelek and Lor cheeses (Kamber, 2008a). There is no standardized processing method for Carra cheese production. Due to the lack of standard manufacturing procedures, the production method may vary, and there may be significant variations in cheeses chemical composition (Hayaloglu & Fox, 2008). Aygun *et al.* (2005) examined the microbiological quality of 50 Carra cheese produced in the Antakya region and found total mesophilic bacteria, 1.87x108 cfu/g; *Staphylococcus aureus*, 2.51x103 cfu/g; Enterococci, 4.40×10^5 cfu/g; *Enterobacteriaceae,* 5.60×10^5 cfu/g; yeast and mould, 4.80×10^7 cfu/g; coliform, 1.02×10^4 cfu/g and *Escherichia coli*, 4.27×10^3 cfu/g. Hayaloglu and Kirbag (2007) specified that the intestinal origin of these bacteria and their presence in cheese should be considered as evidence of poor manufacturing conditions and lack of hygiene. *Listeria* (*L*) *monocytogenes* in Carra cheese made from raw milk could survive for 90 days of ripening, although low pH and high number of lactic acid bacteria at the early stages of ripening inhibited the development of *L. monocytogenes* in Carra cheese (Durmaz, 2009).

Fig. (10). Carra Cheese.

SÜRK CHEESE

Sürk, which means "Çökelek" in Arabic, a special spicy and a curd cheese variety, produced in all seasons of the year and made of all sorts of milk (Aygun *et al.*, 2009; Masatcıoğlu & Avşar, 2005; Sağdıç *et al.*, 2005). With its pear-like form and brick red color, Sürk Cheese has a distinctive appearance (Table **2**). Fresh cheeses are tile red colored and very stiff whereas aged cheeses are brown inside and moldy (Kamber, 2008a). There isn't any standardized manufacturing method of Sürk cheese. The Sürk cheese (Fig. **11**) is generally produced on family farms or small dairy plants (Aygun *et al.*, 2007; Karaca & Kırdar, 2016). Fresh cheese, skim milk cheese, buttermilk, and whey can be used to produce Sürk cheese (Karaca & Kırdar, 2016). Firstly one of these to be used is boiled for approximately 30 min. After that, excess whey is drained by putting under pressure for 5-6 h. Various spices like cumin, black pepper, peppermint, thyme, cinnamon, chili pepper, ginger, garlic, and salt are added. The salt percentage is 5%, chili pepper percentage is 2%, garlic percentage is 1%, and others 0.1-0.3% (Aygun *et al.,* 2007). These spices give the distinctive flavor of Sürk cheese and may have a preservative effect and can play an important role in promoting human health due to their anticancer, antioxidative, and anti-inflammatory properties (Güler, 2014). The mixture is kneaded and given a pear-like shape. Each ball weighs 150-200 g, and the size is 5-7 cm. These balls can be consumed fresh or ripened. Sürk cheese is usually consumed fresh. If it is going to consume fresh, to prevent mold growth, cheese balls can be kept in olive oil, or their surface can be covered with olive oil. If it is going to be consumed ripened, cheese balls are stored for 30 days at room temperature to promote mold growth. Mold growth is desired by some consumers due to its enhanced levels of flavor and the aroma (Sakin *et al.,* 2018; Güler, 2014; Karaca & Kırdar, 2016). Lactic, acetic, propionic, oxalic, formic, citric, pyruvic, orotic, hippuric, and uric acids were identified in the Sürk cheeses. The main volatile compounds in a total of 134 of Sürk cheese were found to be carvacrol, γ -terpinene, p-cymene, hexanoic acid, octanoic acid, decanoic acid, butanoic acid, and eugenol (Güler, 2014).

Fig. (11). Sürk Cheese

Aygun *et al.* (2009) indicated that the occurrence amounts exceeding the maximum tolerance limit (250 ng/kg) established in Turkey of AFM1 in 13.3% of 120 samples Sürk cheese samples might be considered as a possible risk for consumer health. Aflatoxins and ochratoxin concentrations of Sürk have been determined below the official Turkish standard, and it can be consumed without caution (Sakin *et al.*, 2018). The high acidity of Sürk cheese inhibited the growth of *Streptococcus (S) aureus* under anaerobic conditions, and the survival of *S. aureus* decreased during storage (Masatcıoğlu & Avşar, 2005). *Tyrophagus putrescentiae* was found in 8.44% of 450 samples (Aygun *et al.*, 2007).

HELLİM CHEESE

Hellim cheese is a traditional cheese in Cyprus (Ergönül *et al.,* 2011) and it is produced mainly in Cyprus, after that in Turkey, Greece, and Middle Eastern countries like Syria, Lebanon (Hayaloglu *et al.*, 2008). In Southern Cyprus, it is known as Halloumi cheese, in Northern Cyprus, it is known as Hellim cheese (Garanti, 2016). The cheese has international merit (Hayaloglu *et al.*, 2008). Because Hellim cheese has a complex historical background, it is hard to prove that it belongs to some country or nation. This cheese's complex history dates back to Roman and Egyptian civilizations. At present it is considered a local delicacy in Cyprus (Garanti, 2016).

Hellim cheese is a firm pickled, popular, widely consumed dairy product (Ergönül *et al.,* 2011). Hellim has a characteristic taste, semi-hard and elastic texture, easily sliceable form, a slight rind, and no holes or openings. Hellim cheese is made from sheep or goat milk or a mixture of these. This cheese can also be made from cow milk (Hayaloglu *et al.*, 2008; Ergönül *et al.,* 2011; Garanti, 2016). The starter culture is not used. Its color, depending on the milk used, varies from white through yellowish (Hayaloglu *et al.*, 2008). There are two types of Hellim cheese; fresh and mature. It can be preserved and mature in brine, can be eaten raw, fried, grilled, and oven cooked (Garanti, 2016). Hellim cheese can be flavored by the addition of fresh mint leaves (Kamber, 2008a).

The production process of Hellim cheese is coagulation, processing of the curd, taking the nor cheese, cooking, salting, folding, packaging. Firstly fresh raw milk is fermented. The fermented milk is left for approximately 30-45 minutes for complete coagulation. After that, the coagulum is cut into 1 cm cubes, and the curd is allowed to rest for 10-15 min. Afterward, the curd is heated at 40°C for 10-15 minutes. To drain its liquid off, the curd is put under pressure. This process takes about half an hour. The raw cheese is cut in the form of moulds measuring 10×10×2 centimeters. The moulds are thrown into the whey and kept boiling at 90-95°C for 15 minutes.

The cheese molds, which rise to the surface during boiling are removed, salt and mint are added, and Hellim is formed. If the cheese is to be preserved, it is placed in brine. If it is to be consumed fresh, the cheese is offered for consumption after being left to stand for 7-8 days. The convenience packaged Hellim Cheese has a life span of around 2 years at a temperature of 4°C (Kamber, 2008a). About one kilogram of cheese will be obtained from nine liters of milk (Öztürk *et al.,* 2014). The whey obtained during its production is used in the production of Nor Cheese (Kamber, 2008a).

Hellim cheese is consumed by locals and also exported to numerous countries (Garanti, 2016). Since it is very popular in the region, it is also produced in Turkey. Hellim cheese is consumed in Turkey under 7 different trademarks, and some of them are being produced in Turkey. Others are imported from Southern Cyprus (Ergönül *et al.,* 2011). If this cheese is stored below 12°C, it will be kept for several months. If the cheese is stored after salting in plastic bags without brining at about 10°C, the shelf life will be two or three months (Öztürk *et al.,* 2014).

Keles *et al.* (2001) reported that the mean values of the composition of Hellim cheeses made from goat's milk were 39.40% moisture, 11.98% fat-free dry matter, 27.42% fat, 5.46% ash, 4.83% salt, and 4.85 pH; Hellim cheeses made from cow's milk were 40.67% moisture, 15.47% fat-free dry matter, 25.20% fat, 6.36% total ash, 4.99% salt, and 4.86 pH. Moisture content, fat-free dry matter, fat, protein, salt and pH of Hellim cheese were 46.22%, 20.65%, 25.57%, 19.25%, 3.91%, and 5.97, respectively (Güley & Akbulut, 2001). Milci *et al.* (2005) reported that the mean values of the composition of Hellim cheeses made from sheep's milk were 49.90% moisture, 30.46% fat-free dry matter, 19.44% fat, 22.69% protein, 5.81% ash, 4.24% salt, and 5.47 pH; Hellim cheeses made from cow's milk were 50.82% moisture, 27.07% fat-free dry matter, 23.75% fat, 16.73% protein, 6.75% ash, 6.27% salt, and 5.30 pH; Hellim cheeses made from goat's milk were 50.20% moisture, 31.20% fat-free dry matter, 19% fat, 22.37% protein, 6.46% ash, 5.51% salt, and 5.32 pH.

TULUM CHEESE

Tulum cheese is one of the most popular and important Turkish cheese variety. This cheese undergoes a certain ripening process, after which it is consumed immediately like some other cheeses. Its name comes from the material which is used for packaging and ripening and means animal skin-bag (Fig. **12**). Nowadays, wooden earthenware or plastic containers and cloth bags are also used for the ripening of Tulum cheese. Tulum cheese is matured at least 90-100 days in these skin bags in caves (Adanacioglu & Albayram, 2012; Kinik *et al.,* 2005; Şengül *et al*., 2014; Kırdar *et al*., 2015).

Tulum cheese characterized by its high nutrition level, high dry matter, sharp and butter-like aroma, high-fat content, semi-hard texture, white to cream-like color. Goat and sheep skin contributes to forming the characteristic flavor of Tulum cheese; also fat plays an important role in determining the characteristic taste, aroma, and texture of Tulum cheese. Another factor is the amount of salt added. Salt amount affects the color. Its shape and size varies (Şengül *et al.*, 2014; Tarakçı & Durmuş, 2016; Oluk *et al.,* 2014a; Sert *et al.,* 2014; Kamber, 2008b).

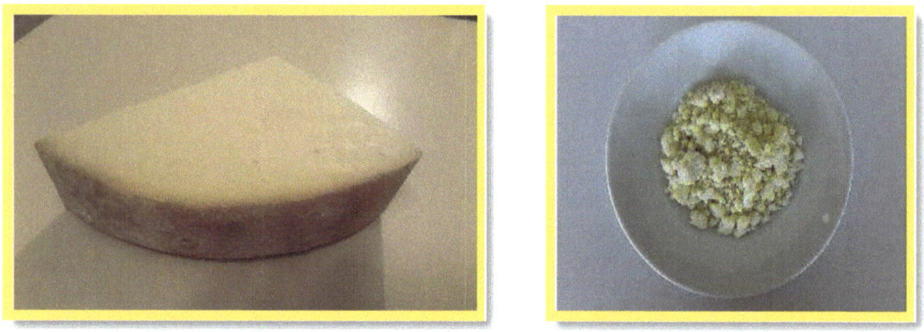

Fig. (12). Tulum Cheese.

Tulum cheese is produced all over the country, but production techniques and ripening times vary significantly. Varieties of Tulum cheese are referred to by the names of the regions where they are produced. Generally, in villages, it is produced in animal skins and called Tulum Cheese (Adanacioglu & Albayram, 2012; Oluk *et al.,* 2014b, Kamber, 2008b).

Physical, chemical, and microbiological properties of Tulum cheese depend on some factors, and these are; quality of raw milk, production procedure, production conditions, storage material, and storage conditions. It is hard to produce a standard quality of the product in villages or homes (Sert *et al.,* 2014). Tulum Cheese is made of raw milk, and goat's, sheep's or cow's milk can be used. Sometimes a combination of these kinds of milk can also be used (Yogurtcu & Tuncer, 2012; Sert *et al.,* 2014). Firstly, production is usually carried out between March and July (Kamber, 2008b). Raw milk is heated at 63°C for 30 min and cooled to approximately 35°C. 125 ml rennet per 1000 L is added, and the coagulation step is started. It generally takes 45-50 min. After coagulation, the obtained curd is pressed, and the whey is separated. The teleme is cut into small slices and salted. Cheese samples are firstly pre-ripened at 15°C for one week after pre-ripening are stored and ripened at 4-6°C for 3 months. Because of the absence of alternative packaging materials, animal skin was probably used, but nowadays, for preserving and ripening the cheese beside the Tulum, hardened plastic barrels, tins, bins, and cloth materials are used. (Durlu-Özkaya & Gün, 2014; Karabey *et al.*, 2018; Celik & Tarakcı, 2017).

Tulum cheese can be consumed at breakfast, in salads, in patty, or like these foods. Tulum cheese is not consumed fresh; it is consumed after ripening for at least 3 months. Proteolysis is an important and principal step. It plays an important role in flavor and texture development. During the ripening period, biochemical processes like proteolysis, glycolysis, and lipolysis occur and allow the taste, and aroma development in cheese. At the end of this process, the cheese has the desirable taste, aroma, and texture (Durlu, Özkaya & Gün, 2014; Şengül *et al.*, 2014).

ADANA LAVAŞ CHEESE

This cheese may have been taught by the French in World War I, and "La vache" means cow in French. Lavaş cheese (Fig. **13**), which is produced in local dairy and small household businesses, is being produced in modern facilities from pasteurized milk instead of raw milk, and consumption in the region has been widespread in recent years. The raw milk is filtered and then pasteurized at 65°C for 10 minutes. After pasteurization, it is cooled to 33-35°C. Milk is fermented with a starter culture to clot in 40-60 minutes. The formed clot is disintegrated and heated to 38-40°C, and the whey is filtered and left for 2-3 hours. When the pH of the curd is 5.15-5.25, the cheese is sliced and boiled in water at 70-80°C. The hot cheese dough is portioned into pieces of 200-250 grams, flattened by shape, and left in cold water for 2-3 hours. Then it is taken from cold water, salted in 16-20% brine for 12-14 hours, and then stored at 4-6°C after packing (Güzeler *et al.*, 2011).

Fig. (13). Adana Lavaş cheese.

KARATAŞ DİLİM CHEESE AND BASMA CHEESE WITH BLACK CUMIN

Traditional production methods and compositional properties of Dilim (sliced) cheese and Basma cheese with black cumin are given in Fig. (**14**) and Table **2**.

Both of them are produced and consumed extensively in the Karataş district of Adana province, Gölkaya, and Bahçe villages (Karaca & Oluk, 2018).

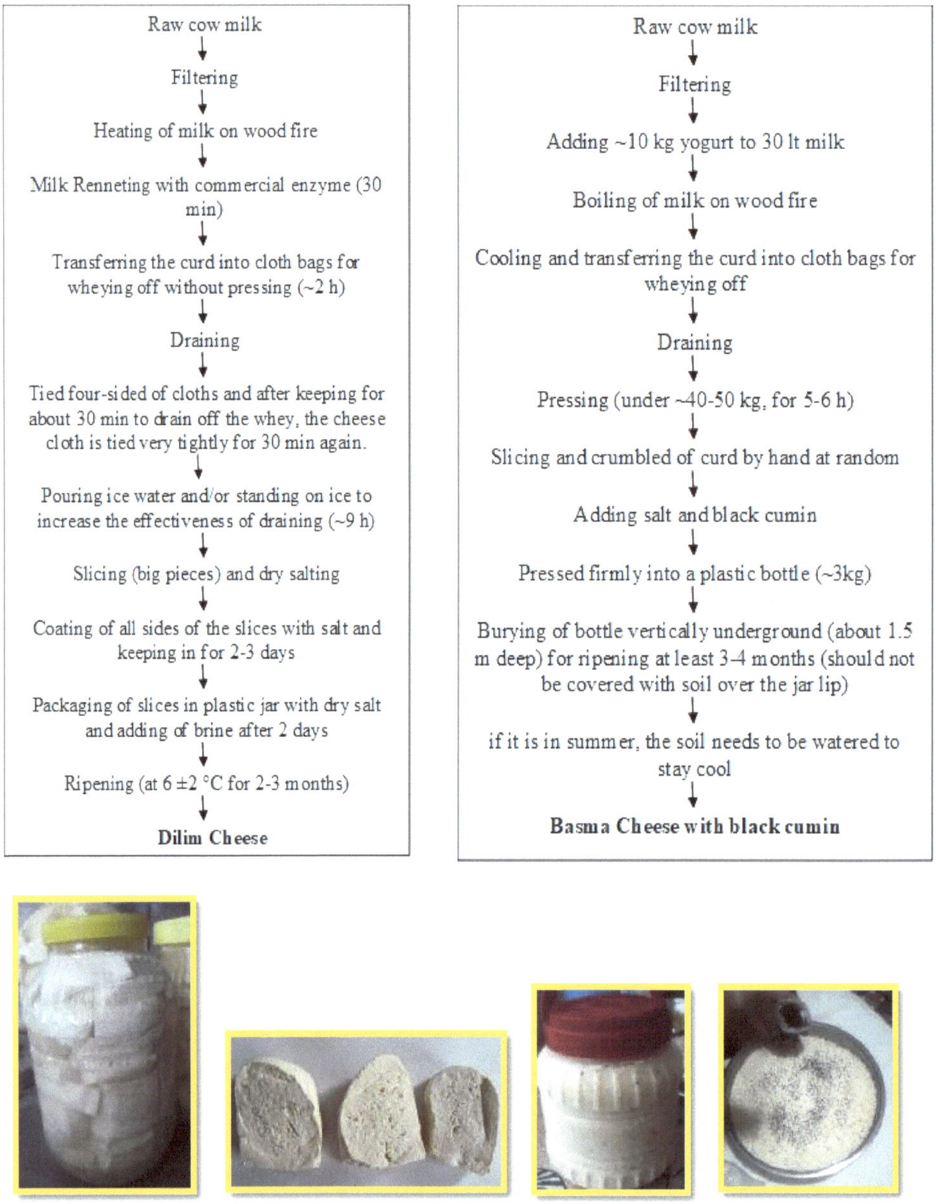

Fig. (14). Manufacturing steps of Karataş Dilim and Basma Cheese with black cumin.

MARAŞ PARMAK (FINGER) CHEESE

Parmak (Finger or Maraş cheese) cheese is especially produced in Kahramanmaraş city of Mediterranean region. The production scheme of Maraş Finger cheese is given in Fig. (**15**). Parmak Cheese has a non-porous elastic structure, white color, slightly-salty taste. It is a semi-hard cheese (Tekinşen, 2001). It is made by boiling and kneading like Pasta filata or Kashkaval cheeses. Pasteurized sheep's, goat's and sometimes cow's milk are used in this cheese production (Kamber, 2008a). The curd is boiled in the whey, and cooked curds are shaped by hand and dry-salted. Parmak Cheese is put in brine (12-13%) and ripened at 6-10°C for 2 months. Moisture, fat in dry matter, salt in dry matter, and pH of Parmak cheese range from 42-45%, 40-41%, 14-15%, and 5.0-5.1, respectively (Hayaloglu *et al.*, 2008). Tekinşen *et al.* (1999) reported that the mean values of the composition of Parmak cheeses made from sheep's milk were 41.11% moisture, 19.83% fat, 6.62% salt, 7.09% ash, 0.37% la and 5.13 pH at 90th days of ripening.

Fresh raw milk
Heating (65°C for 30 min)
Cooling (42-45°C)
Addition of $CaCl_2$ (0.02%)
Addition of starter cultures (0.2%)
Cooling (30°C)
Addition of rennet (10 g/100 kg milk) (strenght 1:10.000)
Coagulation (45 min)
Curd cutting (3 cm×3 cm×3 cm)
Curd ladling in metal hoops and curd draining under pressure (8 kg for 1 h)
Teleme cutting (5×3 cm, 5×2 cm)
Teleme boiling and forming (75°C for 10 minute)
Dry salting (1% w/w salt/cheese) for 24 h
Fill up containers with 13% brine
Stroge (at 7°C)

Fig. (15). Maraş Parmak cheese production steps (Kamber, 2008a).

CONSENT FOR PUBLICATION

Not applicable.

CONFLICT OF INTEREST

The authors confirm that this chapter contents have no conflict of interest.

ACKNOWLEDGEMENTS

Declared none.

REFERENCES

Adanacioglu, H & Albayram, Z (2012) A Conjoint analysis of consumer preferences for traditional cheeses in turkey: a case study on Tulum cheese. *Korean Journal Food Scinece An.* 32, (4), 458-466. [http://dx.doi.org/10.5851/kosfa.2012.32.4.458]

Akın, MB, Akın, MS, Daşnik, F, Alibekiroğlu, R & İzol, G 'Şanlıurfa'da satılan Künefe peynirlerinin bazı özellikleri ve Künefelerin güvenirliliği', 4. Geleneksel Gıdalar Sempozyumu, 17-19 Nisan 2014, Adana, pp. 362-365.

Aygun, O, Aslantas, O & Oner, S (2005) A survey on the microbiological quality of Carra, a traditional Turkish cheese. *J Food Eng,* 66, 401-4. [http://dx.doi.org/10.1016/j.jfoodeng.2004.04.013]

Aygun, O, Yaman, M & Durmaz, H (2007) A survey on occurrence of *Tyrophagus putrescentiae* (Acari: *Acaridae*) in Surk, a traditional Turkish dairy product. *J Food Eng,* 78, 878-81. [http://dx.doi.org/10.1016/j.jfoodeng.2005.11.029]

Aygun, O, Essiz, D, Durmaz, H, Yarsan, E & Altintas, L (2009) Aflatoxin M1 levels in Surk samples, a traditional Turkish cheese from southern Turkey. *Bull Environ Contam Toxicol,* 83, 164-7. [http://dx.doi.org/10.1007/s00128-009-9765-x] [PMID: 19448962]

Biçer, O, Güler, MB, Keskin, M & Kaya, Ş (1995) *Goat production, some traditional goat milk products with special referance to Hatay region of Turkey.* Seminar on Production &Utilization of Ewes & Goats Milk, Creete, Greece.

Celik, O & Tarakcı, Z (2017) The effects of starter cultures on chemical, biochemical and sensory properties of low-fat Tulum cheeses during ripening. *Int J Dairy Technol,* 70, 583-91. [http://dx.doi.org/10.1111/1471-0307.12377]

Çakır, İ, Coşkun, H, Akoğlu, İT, İşleyen, MF, Kıralan, M & Bayrak, A (2009) Introducing a traditional dairy product Keş: chemical, microbiological, and sensorial properties and fatty acid composition. *J Food Agric Environ,* 7, 116-9.

Çardak, AD (2012) Microbiological and chemical quality of Çökelek cheese, Lor cheese and Torba (strained) yoghurt. *Afr J Microbiol Res,* 6, 7278-84.

Dervisoglu, M, Tarakci, Z, Aydemir, O, Temiz, H & Yazici, F (2009) A survey on selected chemical, biochemical and sensory properties of Kes cheese, a traditional Turkish cheese. *Int J Food Prop,* 12, 358-67. [http://dx.doi.org/10.1080/10942910701784688]

Donmez, M, Kemal Seckin, A, Sagdic, O & Simsek, B (2005) Chemical characteristics, fatty acid compositions, conjugated linoleic acid contents and cholesterol levels of some traditional Turkish cheeses. *Int J Food Sci Nutr,* 56, 157-63. [http://dx.doi.org/10.1080/09637480500131137] [PMID: 16009630]

Durlu-Özkaya, F & Gün, İ (2007) 'Anadolu'da Peynir Kültürü', ICANAS 38, Uluslararası Asya ve Kuzey Afrika Çalışmaları Kongresi, Eylül 2007, Ankara, pp.485.

Durlu, Özkaya (2014) Aroma aroma compounds of some traditional Turkish cheeses and their importance for Turkish cuisine. *Food Nutr Sci,* 5, 425-34. [http://dx.doi.org/10.4236/fns.2014.54050]

Durmaz, H, Tarakci, Z, Sagun, E & Aygun, O (2004) Surkun kimyasal ve duyusal nitelikleri. *Firat Universitesi Saglik Bilimleri Dergisi,* 18, 85-90.

Durmaz, H (2009) Fate of *Listeria monocytogenes* in Carra cheese during manufacture and ripening. *J Food Saf,* 29, 253-60.
[http://dx.doi.org/10.1111/j.1745-4565.2009.00154.x]

Ergönül, B, Ergönül, PG & Seçkin, AK (2011) Chemical and textural attributes of Hellim (Halloumi) cheese marketed in Turkey. *Mljekarstvo,* 61, 168-74.

Garanti, Z (2016) Marketing Hellim/ Halloumi cheese: a comparative study of northern and southern Cyprus *Proceedings of the International Conference, Economic Science for Rural Develpoment,* 21-22 April 134-134.

Güler, MB (1999) *Production of Hatay region Surk (Küflü Çökelek) and Carra (testi) cheeses, characteristics and possibilities of standardization* Cukurova University, Institute of Science, Department of Food Engineering, Adana 116.

Güler, Z (2014) Profiles of organic acid and volatile compounds in acid-type cheeses containing herbs and spices (Surk cheese). *Int J Food Prop,* 17, 1379-92.
[http://dx.doi.org/10.1080/10942912.2012.697957]

Gün, İ & Şimşek, B (2006) 'Burdur ilinde üretilen Akçakatık peynirlerinin yağ asidi düzeyinin belirlenmesi', Türkiye 9. Gıda Kongresi, 24-26 Mayıs 2006, Bolu, Turkey, pp: 511-512.

Gün, I, Güneşer, O, Karagül-Yüceer, Y, Güzel-Seydim, Z, Torun, F & Çakıcı, S (2012) 'Determination of aroma active compounds of Çökelek cheese produced from different methods by gas chromatography olfactometry (GCO)', III. Traditional Foods Symposium, Konya, 10-12 Mayıs 2012, p. 531.

Guner, T & Telli, N (2011) A survey on the presence of *Listeria monocytogenes* in various semi-hard cheeses from different regions of Turkey. *J Anim Vet Adv,* 10, 1890-4.
[http://dx.doi.org/10.3923/javaa.2011.1890.1894]

Güzeler, N, Saydam, İB & Karaca, OB (2011) Çukurova bölgesi peynirlerinin üretim yöntemleri ve bilesim özellikleri. *Akademik Gıda,* 9, 25-9.

Hayaloglu, AA & Kirbag, S (2007) Microbial quality and presence of moulds in Kuflu cheese. *Int J Food Microbiol,* 115, 376-80.
[http://dx.doi.org/10.1016/j.ijfoodmicro.2006.12.002] [PMID: 17258341]

Hayaloglu, AA, Ozer, BH & Fox, PF (2008) 'Cheeses of Turkey: 2. varieties ripened under brine', Dairy Science and Technology, EDP Sciences/Springer, 88, (2), 225-244.
[http://dx.doi.org/10.1051/dst:2007014]

Hayaloglu, AA & Fox, PF (2008) Cheeses of Turkey: 3. Varieties containing herbs or spices. *Dairy Science and Technology.* EDP Sciences/Springer.
[http://dx.doi.org/10.1051/dst:2007015]

Irkin, R (2010) Determination of microbial contamination sources for use in quality management of cheese industry: "Dil" cheese as an example. *J of Cons Protec and Food Safety,* 5, 91-6.
[http://dx.doi.org/10.1007/s00003-009-0525-y]

Kamber, U & Terzi, G (2007) The traditional cheeses of Turkey: Central Anatolian region. *Food Rev Int,* 24, 74-94.
[http://dx.doi.org/10.1080/87559120701764548]

Kamber, U (2008) The traditional cheeses of Turkey: Mediterranean region. *Food Rev Int,* 24, 119-47. a
[http://dx.doi.org/10.1080/87559120701764571]

Kamber, U (2008) The traditional cheeses of Turkey: The Aegean region. *Food Rev Int,* 24, 39-61. b
[http://dx.doi.org/10.1080/87559120701762195]

Kamber, U & Terzi, G (2008) The traditional cheeses of Turkey: middle and eastern Black Sea region. *Food*

Rev Int, 24, 95-118.
[http://dx.doi.org/10.1080/87559120701764555]

Kaptan, B (2016) Prevalence of *Listeria* spp and *L. monocytogenes* in home made pottery cheese. *Tekirdag Ziraat Fak Derg,* 13, 80-7.

Karabey, B, Eroglu, D, Vural, C, Ozdemir, G, Yerlikaya, O & Kinik, O (2018) Determination of the microbial flora in traditional İzmir Tulum cheeses by Denaturing Gradient Gel Electrophoresis. *J Food Sci Technol,* 55, 956-63.
[http://dx.doi.org/10.1007/s13197-017-3003-z] [PMID: 29487437]

Karaca, OB & Güven, M (2004) Çökelek peyniri ve çökelek kullanılarak yapılan bazı yöresel peynir çeşitleri *Traditional Food Symposium,* 23-24 September 2004 Van, Turkey242-6. a

Karaca, OB & Güven, M (2004) Chemical and organoleptical properties and made of Hatay Sünme cheese *Traditional Food Symposium.*23-24 September 2004 Van, Turkey b

Karaca, OB, Ocak, S, Guney, O & Guven, M (2007) Present situation of goat production sector and some typical dairy cheeses in Turkey. *3rd Joint Meeting of the Network of Universities and Research Institutions of Animal Science of the South Eastern European Countries.* Thessaloniki 10–12 February.

Karaca, OB, Güven, M, Mutluer, U & Saydam, İB (2008) Hatay Künefe Peynirinin yapılışı ve özellikleri Turkey*Food Congress,* 21-23 May 2008 Erzurum, Turkey, 745-748.

Karaca, OB, Saydam, IB, Güven, M & Güzeler, N (2009a) Hatay Dil peynirinin bazı kalite özellikleri. *II International Traditional Foods Semposium,* Van, Turkey May 27-29734-8.

Karaca, OB, Saydam, IB, Güven, M & Güzeler, N (2009b) Hatay Ezme peynirinin bazı kalite özellikleri. *II International Traditional Foods Semposium,* Van, Turkey May 27-29153-7.

Karaca, OB & Kırdar, SS (2010) Traditional cheese types in the Mediterranean region of Turkey *11th Egyptian Conference for Dairy Science and Technology,* 1-3 November 133-143.

Karaca, OB (2016) The Importance of Our Traditional Cheeses in Gastronomy Tourism. *J of Tour and Gast Stud,* 4, 17-39.
[http://dx.doi.org/10.21325/jotags.2016.12]

Karaca, OB & Kırdar, SS (2016) Traditional dairy products in Hatay province. *Agriculture & Food,* 4, 339-47.

Karaca, OB & Ocak, S (2016) 'Dairy goat products of east Mediterranean region of Turkey: Künefe and Sünme cheeses', *Scientific Papers Series D. Animal Science*, Vol. LIX. *University of Agronomic Sciences and Veterinary Medicine of Bucharest Faculty of Animal Science Bucharest,* LIX, 288-95.

Karaca, OB & Oluk, CA (2018) Some artisanal cheeses in Turkey: Karataş district. *The International Conference on Raw Materials to Processed Foods,* 229.

Kavaz, A, Arslaner, A & Bakirci, İ (2012) Comparison of quality characteristics of Çökelek and Lor cheeses. *Afr J Biotechnol,* 11, 6871-7.

Kılıc, M & Isın, TG (2004) Effects of salt level and storage on texture of Dil cheese. *J Texture Stud,* 35, 251-62.
[http://dx.doi.org/10.1111/j.1745-4603.2004.tb00836.x]

Kırdar, SS (2003) *'Isparta Küp peyniri üzerine bir araştırma', GAP III Tarım Kongresi, 02-03 Ekim* Şanlıurfa 445-8.

Kırdar, SS (2005) Burdur yöresi geleneksel peynirleri: Çökelek peyniri *1st Burdur Symposium,* Burdur, Turkey, 16-19 November 1176-1179.

Kırdar, SS, Kurşun, Ö & Özrenk, E (2009) Keş peynirinin kimyasal ve mikrobiyolojik özellikleri. *II Geleneksel Gıdalar Sempozyumu* 27-29 Mayıs 2009, Van, Turkey 130.

Kırdar, S (2012) A Survey on Chemical, biochemical and microbiologic characteristics of a traditional dairy product in Mediterrean region: Kes. *J Anim Vet Adv,* 11, 330-4.

[http://dx.doi.org/10.3923/javaa.2012.330.334]

Kırdar, SS, Ocak, E & Köse, S (2013) Mineral and trace metal levels of Akçakatık cheese collected from Mediterranean region-Turkey. *Asian J Chem,* 25, 1643-6.

Kırdar, SS, Kose, Ş, Gun, İ, Ocak, E & Kursun, Ö (2015) Do consumption of Kargi Tulum cheese meet Daily requirements for minerals and trace elements? *Mljekarstvo,* 65, 203-9. [http://dx.doi.org/10.15567/mljekarstvo.2015.0307]

Kırdar, S, Köse, Ş, Yurdakul, Ö & Ocak, E (2017) A Survey on the microbiological and chemical characteristics of Akçakatık cheese in the west Mediterranean region. *European International Journal of Science and Technology,* 6, 25-35.

Kırdar, SS, Kursun-Yurdakul, O, Kalıt, S & Kalıt, TM (2018) Microbiological changes throughout ripening of Keş cheese. *J Cent Eur Agric,* 19, 61-7. [http://dx.doi.org/10.5513/JCEA01/19.1.2024]

Kilicel, F, Tarakci, Z, Sancak, H & Durmaz, H (2008) Survey of trace metals in Turkish Kes cheese. *Rev Anal Chem,* 27, 101-10. [http://dx.doi.org/10.1515/REVAC.2008.27.2.101]

Kinik, O, Gursoy, O & Seckin, AK (2005) Cholesterol content and fatty acid composition of most consumed Turkish hard and soft cheeses. *Czech J Food Sci,* 23, 166-72. [http://dx.doi.org/10.17221/3387-CJFS]

Kirdar, SS (2004a) A research on Akçakatık cheese. *Geleneksel Gıdalar Sempozyumu* 23-24 Eylül, Van, Turkey 354-6.

Kirdar, SS (2004) A reasearch on Kes cheese. In recent developments in dairy science and technology *International Dairy Symposium,* Süleyman Demirel University, Isparta, Turkey 239-41. b

Koluacik, A (2015) Aflatoxin M1 determination in traditional Küp cheese samples of Turkey using immunoaffinity column and high-performance liquid chromatography. *Turkish Journal of Agriculture-Food Science and Technology,* 3, 916-9. [http://dx.doi.org/10.24925/turjaf.v3i12.916-919.504]

Konar, A & Güler, MB (1998) Hatay Cara (Testi) Cheese, Chemical composition and level of proteolysis *Traditional Dairy Products and Milk and Dairy Products Symposium,* 21-22 MayTekirdag, 145-153.

Masatcioğlu, TM & Avşar, YK (2005) Effects of flavorings, storage conditions, and storage time on survival of *Staphylococcus aureus* in Sürk cheese. *J Food Prot,* 68, 1487-91. [http://dx.doi.org/10.4315/0362-028X-68.7.1487] [PMID: 16013393]

Mutlu, G (2000) The chemical composition of sünme cheese in Hatay region of Turkey , *Blacksea and Central Asaian Symposium on Food Technology.* October 12-16 Ankara, Turkey

Mutluer, M, Güven, M, Karaca, OB & Saydam, İB (2014) Farklı oranlarda tuz içeren salamuralarda depolanan çiğ ve ısıl işlem uygulanmış sütten üretilen sünme peynirlerinin fiziksel, kimyasal ve duyusal özellikleri. *4 Geleneksel Gıdalar Sempozyumu* Çukurova Üniversitesi, 17-19 Nisan, Adana, Turkey 750-4.

Ocak, S, Karaca, OB & Güney, O (2009) Traditional dairy goat products produced in the Mediterranean region of Turkey. XVII. International Congress of Mediterranean Federation of Health and Production of Ruminants Perugia, Italy27-30 May 2009.

Okur, ÖD & Güzel-Seydim, Z (2011) Geleneksel Dolaz peynirinin üretim yönteminin, mikrobiyal ve uçucu aroma bileşen içerikleriyle duyusal özelliklerinin belirlenmesi. *Gida,* 36, 83-8. (a).

Okur, ÖD & Güzel-Seydim, Z (2011) Geleneksel Dolaz peynirinde bazı karakteristik özelliklerin belirlenmesi. *Ege Univ Ziraat Fak Derg,* 48, 113-7. (b).

Oluk, AC, Güven, M & Hayaloglu, AA (2014) Proteolysis texture and microstructure of low-fat Tulum cheese affected by exopolysaccharide-producing cultures during ripening. *Int J Food Sci Technol,* 49, 435-43. (a).

[http://dx.doi.org/10.1111/ijfs.12320]

Oluk, AC, Guven, M & Hayaloglu, AA (2014) Influence of exopolysaccharide-producing cultures on the volatile profile and sensory quality of low-fat Tulum cheese during ripening. *Int J Dairy Technol,* 67, 265-76. (b).
[http://dx.doi.org/10.1111/1471-0307.12118]

Öztürk, B, Çelik, F, Çelik, Y, Kabaran, S & Zaver, T (2014) To determine the occurence of aflatoxin m1 (afm1) in samples of Cyprus traditional cheese (Halloumi): a cross sectional study. *Kafkas Univ Vet Fak Derg,* 20, 773-8.

Sağdıç, O, Şimşek, B & Padem, H (2005) Some characteristics of surk, a traditional Turkish cheese. *Arch Lebensmittelhyg,* 56, 13-6.

Sakin, F, Tekeli, İO, Yipel, M & Kürekci, C (2018) Occurrence and health risk assessment of aflatoxins and ochratoxin a in Surk, a Turkish dairy food, as studied by HPLC. *Food Control,* 90, 317-23.
[http://dx.doi.org/10.1016/j.foodcont.2018.03.012]

Sert, D, Akin, N & Aktumsek, A (2014) 'Lipolysis in Tulum cheese produced from raw and pasteurized goats' milk during ripening'. *Small Rumin Res,* 121, 351-60.
[http://dx.doi.org/10.1016/j.smallrumres.2014.06.006]

Şengül, M, Erkaya, T, Dervişoğlu, M, Aydemir, O & Gül, O (2014) Compositional, biochemical and textural changes during ripening of Tulum cheese made with different coagulants. *Int J Dairy Technol,* 67, 373-83.
[http://dx.doi.org/10.1111/1471-0307.12120]

Şimşek, B & Sağdıç, O (2006) Isparta ve yöresinde üretilen dolaz (tort) peynirinin bazı kimyasal ve mikrobiyolojik özellikleri. *Süleyman Demirel Univ Fen Bilim Enst Derg,* 10, 346-51.

Şimşek, B & Gün, I (2009) Free fatty acid composition of Akçakatik cheese, a traditional Turkish dairy product. *Asian J Chem,* 21, 5923-8.

Simsek, B & Sagdıc, O (2010) Determination of fatty acids and chemical characteristics of Cokelek cheese from cows milk using of *L. helveticus* and/or yoghurt bacteria. *Food Sci Technol Res,* 16, 179-84.
[http://dx.doi.org/10.3136/fstr.16.179]

Şimşek, B & Sağdıç, O (2012) Effects of starter culture types and different temperatures treatments on physicochemical, microbiological and sensory characteristics, and fattty acid compositions of Çökelek cheese made from goat milk. *Kafkas Univ Vet Fak Derg,* 18, 177-83.

Şimşek, B & Tuncer, Y (2018) Some properties of fresh and ripened traditional Akcakatik cheese. *Korean Journal for Food Science of Animal Resources,* 38, 110-22.
[PMID: 29725229]

Tarakçı, Z, Dervişoğlu, M, Temiz, H, Aydemir, O & Yazıcı, F (2010) Review on Kes cheese. *Gida,* 35, 283-288..

Tarakçı, Z & Durmuş, Y (2016) Effects of packaging materials on some ripening characteristics of Tulum cheese. *Mljekarstvo,* 4, 293-303.
[http://dx.doi.org/10.15567/mljekarstvo.2016.0405]

Tekinşen, OC, Atasever, M, Keleş, A & Uçar, G (1999) İnek ve Koyun Sütü Kullanımının ve Farklı Tuzlama Tekniklerinin Maraş Peynirinin Bazı Kalite Niteliklerine Etkisi. *J Vet Anim Sci (Lahore),* 23, 213-26.

Tekinşen, KK (2001) *Maraş peyniri üretiminde baskılama ağırlığı ve haşlama suyu sıcaklığının standardizasyonu üzerine araştırmalar.* Selçuk Üniversitesi, Sağlık Bilimleri Enstitüsü, Besin Hijyeni ve Teknolojisi Anabilim Dalı, Doktora Tezi, Konya.

TÜİK (2018) *Turkish Statistical Institute.* http://www.tuik.gov.tr/PreTabloArama.do. Acessed on October 22th, 2018.

Uçgun, D & Işık, N (2018) Geleneksel bir lezzet: Horç peyniri a traditional flavor: Horc cheese. *Güncel Turizm Araştırmaları Dergisi,* 2, 177-88.

Yerli, Y, Şen, A & Özbay, M (2018) Dolaz peyniri üzerine nitel bir araştırma: Yalvaç örneği. *Güncel Turizm Araştırmaları Dergisi,* 2, 630-6.

Yogurtcu, N & Tuncer, Y (2012) Antibiotic susceptibility patterns of *Enterococcus* strains isolated from Turkish Tulum cheese. *Int J Dairy Technol,* 66, 236-42. [http://dx.doi.org/10.1111/1471-0307.12014]

Yurdakul, ÖK, Kırdar, SS, Doğru, AK & Keyvan, E (2017) Survival of *Escherichia coli* O157:H7 during processing and ripening period of Keş Cheese. *Benha Veteri Med J,* 32, 83-8. [http://dx.doi.org/10.21608/bvmj.2017.31117]

CHAPTER 13

Microbiological Quality of Traditional Turkish Cheeses

Zerrin Erginkaya[1,*], **Gözde Konuray**[1] and **Mihriban Korukluoğlu**[2]

[1] *Department of Food Engineering, Faculty of Agriculture, Cukurova University, Turkey*

[2] *Department of Food Engineering, Faculty of Agriculture, Uludag University, Turkey*

Abstract: Traditional foods are foodstuffs that have been produced for hundreds of years. Besides, being recognised by geography, climate and agricultural production possibilities in all cultures, it is mostly influenced by traditional lifestyle Each of these foodstuffs are original products, created without the use of modern technology, simply by artificially ingraining the key factors of food preservation with available means. Ensuring the preservation and continuity of our traditional food products is one of our important social and cultural heritage from the past; it is important that traditional food products at national and regional scales are identified, inventories are removed, hygienic conditions and modern industrial methods are used to produce safe food, and it is important to move to the future.

The benefits and risks of traditional cheese, especially produced from raw milk, are a complicated issue and have been argued through the years.

The microbiota of traditional cheese varies depending on the raw milk, inoculation applications, environmental factors, cheese production methods and hygienic condition of the environment. Cheese microbiota gives information about the sensory, hygienic, health benefits and possible risks of the product.

In traditional cheeses, depending on the production method, as well as Gram-positive and Gram-negative bacteria, several yeasts and molds, especially lactic acid bacteria have been isolated and identified in many research studies. A variety among cheeses, produced from raw milk, occurs due to microorganism diversity, numerical differences, depending on the microbiota involved in the process of maturation, thereby showing changes in some of its characteristics (such as aroma formation, antimicrobial activity). Therefore, the taste of cheese produced from raw milk is more dense and richer than the processed ones.

In addition, microbiota found in traditional cheese is known to positively affect the aroma development and production of antimicrobial metabolites against pathogens and spoilage microorganisms in the same environment. On the other hand, other allergic

[*] **Corresponding author Zerrin Erginkaya:** Department of Food Engineering, Faculty of Agriculture, Cukurova University, Turkey; Tel: +90 322 3386173-2174; Fax: +90 322 3386614; E-mail: zerriner@cu.edu.tr

Celile Aylin Oluk & Oya Berkay Karaca (Eds.)

reactions, toxic and chemical compounds, which are important with regard to food safety, are substantial for health, besides the microbial risks based on raw milk.

For these reasons, the scientific data on the production and consumption of traditional cheeses should be accurately investigated in terms of benefits, harm and possible new risks, using the omic system as well as cheese ecosystem now and in the future.

Keywords: Cheese microbiota, Lactic acid bacteria, Turkish cheese, Traditional cheese.

INTRODUCTION

The cheese was first produced 8000 years ago in the Mediterranean coasts, in the south of Turkey. Today, cheese production is one of the most developed biotechnology application (Gobbetti *et al.*, 2015). Cheese contains main milk components such as protein (casein), lactose and oil and is well known as an important calcium and phosphorus source (Temiz & Kılıç, 2016). The important stages of the cheese production are the use of rennet, coagulation, curd processing, draining of whey, salting and ripening (Asensio *et al.*, 2015; Karabey *et al.*, 2018). Coagulator, microorganisms (starter or non-starter microbiota and seconder microorganisms) and enzymes are inherently found in milk affecting all of the above mentioned stages (Hayaloglu & Kirbag, 2007). Important microbiological and biochemical changes take place in the ripening period (Hayaloglu & Kirbag, 2007; Levante *et al.*, 2017). Even though cheese is produced with the same material, it gains various characteristics due to the different manufacturing stages (Karabey *et al.*, 2018). The formation of cheese flavor changes depending on the ripening conditions, milk type (raw or pasteurized), milk composition (lipid, protein and lactose contents) and the use of starter or non-starter bacteria (Durlu Özkaya & Gün, 2014). The aroma development of cheese is affected by several biochemical alterations such as lipolysis, proteolysis, lactose fermentation and volatile component formation. The type of milk, rennet properties, the difference between the production processes and ripening conditions, the addition of spice (mint, black sesame, red pepper) and different herbs give its own characteristics to the cheese (Durlu Özkaya & Gün, 2014).

Cheese plays a rather important role in Turkish cuisine. It takes place in the traditional Turkish breakfast, cooked foods (pastry, pita, spicy Turkish omelet) and used to prepare salads and some traditional desserts. There are many different kinds of cheeses that have various taste, shape and texture in Turkey (Durlu Özkaya & Gün, 2014). In addition to the production of more than 50 types of cheese, Turkey is an important country with its considerable amount of cheese production (Ozmen Togay *et al.*, 2016; Temiz & Kılıç, 2016; Karabey *et al.*, 2018). In 2017, 688,061 tons of cheese was produced in Turkey and it has been reported that 96% of these cheese is made from cow milk (Salum *et al.*, 2018). Manufactured cheese meets the local demand but the recognition of local cheese is quite low in other regions (Temiz & Kılıç, 2016). Various cheese types can be found in small-scale enterprises in Turkey, which is produced through traditional techniques and spontaneous fermentation (Sönmez *et al.*, 2018). White cheese, kashar, tulum cheese, curd cheese and cottage cheese are the principal cheeses and sheep, cow and goat milk are used in the production. In addition, local cheeses are also produced such as abaza, herby cheese, mihalic, string cheese, sirvatka, hellim, circassian cheese, basket cheese, ezine cheese, cayir cheese and civil cheese (Ozmen Togay *et al.*, 2016; Temiz & Kılıç, 2016).

White cheese is the most important cheese produced in Turkey (Öner *et al.,* 2006; İspirli *et al.*, 2017). The annual production of white cheese represents 60% of the total cheese production (Salum *et al.*, 2018). White cheese is a soft or semi-soft cheese (Öner *et al.,* 2006). In addition to industrial manufacture, it can be produced in small scale (İspirli *et al.*, 2017). Traditionally, goat and sheep milk is used in production. In the case of an inadequate supply of milk, cow milk or a mixture of these milk is used in the production (Hayaloglu *et al.*, 2002; Salum *et al.*, 2018). Although it has a soft texture, it gains a semi-soft or semi-hard texture after 3 months of ripening in brine (Öner *et al.,* 2006).

White cheeses are quite diverse and known in different names such as feta (Greece), domiati (Egypt), queso blanco (USA) and bjalo salamureno sirene (Bulgaria) (Sofu & Yeşim Ekinci, 2016). Turkish white cheese is a brined cheese and generally white, salty, acidic and has a non-porous texture (Salum *et al.*, 2018). Starter cultures are not used in traditional production or in small-scale production (Öner *et al.,* 2006; Salum *et al.*, 2018). It is generally produced with raw milk in small-scale enterprises (Hayaloglu *et al.*, 2002; Öner *et al.,* 2006). The microbial composition of the cheese changes in ripening period. As the lactic acid bacteria (LAB) is more dominant than others in ripening, it is considered to have an effect on the ripening characteristic (Öner *et al.,* 2006). While *Enterococcus faecalis* and *E. faecium* are important strains at the beginning of the ripening period, *Lactococcus lactis* ssp. *lactis* is the dominant strain. In addition, *Lactobacillus casei*, *L. plantarum*, *L. fermentum*, *L. brevis*, *L. lactis* and *Leuconostoc mesenteroides* ssp. *dextranicum* also could be found in the microbiota. In the later stages of ripening, lactococci decrease and lactobacilli increase. *L. casei* and *L. plantarum* become dominant strains (Çelik & Uysal, 2009).

Kashar cheese is the other most important cheese among traditional Turkish cheese (Yuvaşen *et al.*, 2018). It is produced with raw sheep milk, raw cow milk or a mixture of these (Aydemir *et al.*, 2015). It is classified as a "pasta-filata" cheese (Yuvaşen *et al.*, 2018). It has similarities with Mozzarella, Caciocavallo, Kashkaval, Regusono and Provolone. Kashar cheese is produced traditionally in Kars city located in the North East of Turkey (Aydemir *et al.*, 2015; Yuvaşen *et al.*, 2018). It has a cylindrical shape with 30 cm diameter, 20 cm height and 12 kg weight. Generally, kashar cheese is produced with traditional techniques but the ripening period is an expensive and labor-intensive process (Aydemir *et al.*, 2015). In the small scale production, kashar cheese is produced without the addition of starter. For the traditional production of kashar cheese, raw milk is filtered through cheesecloth and transferred into the cans. Subsequently, milk is heated up to 30-33°C and rennet is added. After the coagulation of the milk, the curd is cut into little pieces (brass size) and transferred into the cheesecloth. Weights are placed on the curd in order to filter the remaining whey. After the pressing stage, the curd is cut and grated. The grated curd is cooked in water at 70-80°C for 2-3 min in a metal basket. The cooked curd is kneaded like dough then transferred into a round metal container and stored for 12-24 h. After the addition of salt to the surface, cheese is placed on wooden shelves in the ripening room. Cheese is periodically inverted and stacked. After the first ripening period, which takes 3-4 weeks at the room temperature, cheese is stacked into a bag and stored from 1 month up to 2 years for the second ripening in a cold room (3-4°C) (Aydemir *et al.*, 2015). No starter culture is added during these two maturation periods. Kashar cheese gains its traditional characteristics due to the non-starter LAB and some fungi species contribute to the characteristics of cheese, which are resistant to cooking heat and can be found in the raw milk or contaminated from the environment during processing and storage (Yuvaşen *et al.*, 2018).

Tulum cheese is a popular Turkish cheese. It is traditionally produced from raw milk (Bayar & Özrenk, 2011). It is produced in every region of Turkey, except for Thrace region (Bayar & Özrenk, 2011; Karabey *et al.*, 2018). The quality of tulum cheese is not up-to-the-markas the production and maturation techniques are different in each region (Bayar & Özrenk, 2011). Tulum cheese has more than 10 kinds, which include Erzincan tulum cheese, Karin kaymagi cheese, Izmir tulum cheese, Simi tulum cheese, Divle tulum cheese and Cimi cheese (Bayar & Özrenk, 2011; Karabey *et al.*, 2018). Tulum cheese compounds can show the difference as there is no standard production method (Bayar & Özrenk, 2011). The ripening time is long. Sheep, cow or goat milk is used in the production. Its color is white or cream. Tulum cheese has a crumbly and semi-hard or hard texture. It has a piquant and greasy taste. This cheese is called tulum because its ripening period occurs in a goat or sheepskin (Karabey *et al.*, 2018). The goatskin is more durable than sheepskin, thats why goatskin is more preferred. Generally, cheese is placed inside the skin. Although in the Aegean region, the skin is shaved and turned inside out, then cheese is placed into it (Bayar & Özrenk, 2011). Skins are air and water permeable. It gives hard texture and characteristic taste and flavor (Karabey *et al.*, 2018). In the past, goatskin is commonly used in the ripening period, but due to the difficulty in obtaining the skin, nowadays alternative containers are being used, which are made from wood, plastic or soil (Bayar & Özrenk, 2011; Karabey *et al.*, 2018).

Izmir tulum cheese is produced with sheep milk or goat milk and a mixture of sheep/cow milk. It is produced in West Anatolia (Izmir, Aydin, Mugla, Manisa, Denizli) and other regions of Turkey. One of the differences between Izmir tulum cheese and other tulum cheese is that it has high salt content and it is matured in brine (Karabey *et al.*, 2018).

Divle tulum cheese is a semi-hard cheese and made from unpasteurized semi-skimmed sheep milk. Traditionally, it is produced in Karaman in spring. Divle cave is located in the south of Karaman and has 70 m depth. After the cheese is produced without a starter culture of fungal spore addition, it is transferred in goatskin bags and ripened in Divle cave for 4 months (Öztürkoğlu Budak, 2016). Due to the ripening conditions of the cave (85-90% moisture content at 8-10°C), Divle tulum cheese has its own characteristics. The goatskin bags are blue-green colored at the beginning of the ripening period but it turns red after 60 days of ripening. Mold, yeast and coryneform bacteria comprise the cheese microbiota, which could be found inherently in the cave and cheese production environment. Furthermore, non-starter LAB, which can be found in raw milk, plays an important role in the ripening period and contributes to the characteristic aroma of cheese. This semi-hard cheese has a crumbly texture, distinctive taste and smell (Öztürkoğlu Budak, 2016).

Curd cheese has various names according to its production regions such as basket curd, herby curd, sirvatka, şor and Tire mud cheese (Kavas & Kavas, 2016). It is a soft cheese and produced with whey (Temiz *et al.*, 2009). At first, whey is heated up to 50-55°C. After the oil is separated, it is heated up to 80°C and 2% salt is added. The temperature is increased to 90-95°C. The coagulants of curd cheese are drained with cheesecloth at 25°C for 12 h (Irkın, 2011). Generally, it is consumed as unripened (Temiz *et al.*, 2009). The natural microbiota is not resistant to the production temperature. Thus, if contamination occurs after production, it is a safety risk for cheese (Kavas & Kavas, 2016). Curd cheese is generally sold and transferred as unpacked. As a result, it is exposed to non-sterilized conditions in the market. Therefore, the shelf life of curd cheese is less than 7 days (Temiz *et al.*, 2009).

Cottage cheese is a white colored cheese. It has a crumbly texture and characteristic flavor. It is one of the five main cheese produced in Turkey, although it is traditionally produced in small-scale enterprises (Celem *et al.*, 2018). The composition of cottage cheese changes according to raw material, production technique and consuming type (fresh or ripened) (Kavas, 2017; Celem *et al.*, 2018). Cottage cheese is generally produced by boiling the yogurt and filtering it with cheesecloth (Özbek & Güzeler, 2017; Celem *et al.*, 2018). Milk, buttermilk or whey mixed with yogurt is used as raw material. Butter, milk, salt and various herbs could be added in the different stages of production, for the purpose of enrichment of the aroma compounds (Celem *et al.*, 2018). Calcium and water content are high and oil content is low in cottage cheese (Kavas, 2017). It contains significant amounts of casein and whey proteins. It is an important animal protein source due to its low calorie and being cheap. It can be consumed as plain or sweetened (with jam or grape molasses) and with pies or salads (Celem *et al.*, 2018).

Civil cheese is a traditional cheese and produced in the eastern regions of Turkey. Whey and skimmed milk are used in production (Cakmakci *et al.,* 2015). After the coagulation of milk, heat treatment is applied. The product gains a filamentous texture as a result of continuous mixing during the coagulation step of milk. It has low oil content and flexible texture due to the constant kneading in the production stage. Lately, it is drawing attention with its nutrient value with low oil and high protein contents (Cakmakci *et al.,* 2015).

Sürk cheese is a traditional cheese and produced in the south of Turkey. It is made with the addition of various herbs and spices. The curd, which is used in the production, is obtained by mixing water and yogurt and then boiled for 30 min. After the draining of whey, diverse spices (mint, thyme, cumin, black pepper, cinnamon, red pepper and ginger) and herbs are added and the cheese is formed in a conical shape by hand (Kürekci *et al.*, 2016).

THE MICROBIOTA OF TRADITIONAL CHEESES

The cheese microbiology is an important issue. It affects the complex and dynamic process and several food systems, such as milk type, chemical composition, enzyme addition, ripening time and technology (Gobbetti *et al.*, 2015). The microbiota of cheese produced with raw milk is quite complex. The compounds of the milk, which is used in the production of traditional cheese, contribute to the characteristics of cheese (Öner *et al.,* 2006; Lazzi *et al.*, 2016). As the ripening period is extended, the microbial balance changes depending on the environmental conditions and microbial interaction. Thus, the characteristics of cheese are affected (Lazzi *et al.*, 2016).

In order to achieve standardization in the production of safe and high-quality food products, high amounts of microorganisms are added for the activation of fermentation. For example, the production is made with milk or whey, which is taken from the previous production. With this method, indigenous and the most adapted microorganisms are chosen and used in the fermentation (Bassi *et al.*, 2015).

The selection of starter culture is one of the most important stages in the production of cheese from pasteurized milk. The use of starter culture in cheese production affects many factors such as microorganism activity, yeast retention in the clot, yield, dry matter, acidity and biochemical changes that occur in the ripening period. Generally, cultures with high proteolytic activity are chosen for the production of ripened and hard textured cheese type, while the cultures with low proteolytic activity are used in the production of fresh cheese (Çelik & Uysal, 2009).

The cheese made from raw milk has a quite diverse and rich microbial ecosystem (Aydemir *et al.*, 2015). This diversity affects the quality of cheese significantly (Aydemir *et al.*, 2015; Czaran *et al.*, 2018). The fermentation in cheese production is maintained with the bacteria found in the raw milk, which could be found indigenously or inoculated as a starter culture. Cheese microbiota also plays a significant role for acidification parameters, aroma development and the inhibition of undesirable microorganisms (Bassi *et al.*, 2015). The coagulation temperature of milk allows the growth of many microorganisms. However, the temperature inhibits most of these microorganisms while the curd is occurring. The cooking temperature and curd formation have an important effect on cheese characteristics (Aydemir *et al.*, 2015). The microbiota of cheese during ripening plays a more important and critic role in cheese characteristics. The water content, salt content, organic acid and nitrate presence, redox potential and ripening temperature are the main environmental factors that affect the microorganism growth (Aydemir *et al.*, 2015).

Two types of starter cultures are used in the cheese production. The first of these is the identified starter culture. The other one is the undefined starter culture. Identified starter culture generally consists of one or more culture which has known characteristics. These strains are usually obtained from identified starter cultures. Many LAB strain identified as starter culture such as *L. delbrueckii* subsp. *lactis*, *L. lactis* subsp. *lactis, L. lactis* subsp. *cremoris*, *Leuconostoc* spp., *Streptococcus thermophilus* and *L. helveticus* (Öztürkoğlu Budak, 2016).

The ripened cheese microbiota consists of LAB, starter LAB, non-starter LAB and LAB that are contaminated from the environment. Each of these microorganisms plays a different role in cheese production (Lazzi *et al.*, 2016). In most of the ripened cheese microbiota, non-starter LAB count is quite high. These microorganisms, unlike starter LAB, do not contribute to acid production during cheese production. These microorganisms have important effects on the formation of aroma in matured cheeses (Aydemir *et al.*, 2015).

The physicochemical characteristics of traditional cheeses are usually generated by non-starter LAB. They are responsible for the maturation of cheese. In this group, heterofermentative lactobacilli are the dominant microorganisms. Besides, enterococci and pediococci can be found in the microbiota (Yuvaşen *et al.*, 2018).

Apart from the non-starter LAB, yeast species are also found in traditional cheeses during maturation, especially on the cheese surface. These yeasts affect the physicochemical characteristics of cheese and they are generally associated with the desired cheese characteristics. In some cheeses, yeast occurrence could lead to cheese spoilage. Desirable effects of yeast are associated with biochemical properties that affect the ripening of cheese. Some of the yeast strains that are found in cheese microbiota, depend on the physicochemical properties of cheese such as pH value and salt content, ripening and storage conditions (Yuvaşen *et al.*, 2018).

Molds can be found in traditional cheese microbiota as well as non-starter LAB and yeasts (Yuvaşen *et al.*, 2018). Generally, *Penicillium* strains are found in various cheese microbiota. Mold associated mycotoxin is one of the biggest problems in the food industry. However, different types of cheeses are produced by using certain mold strains that are non-toxic or low toxic as starters. It is stated that these molds contribute to the physicochemical and sensory properties of cheese at the ripening stage. For this reason, it is important to determine the type of mold in traditional cheeses and to know their biochemical functions and toxin production characteristics (Yuvaşen *et al.*, 2018).

FOOD SAFETY AND MICROBIAL QUALITY CRITERIA

Cheese is an important dairy product with high nutritional value and many kinds. The microbiological quality of cheese consumed in daily life is very important (Kaynar, 2011). Food safety concerns also include dairy products and cheese (Le *et al.*, 2014).

Today, food-borne diseases are one of the most common problems in the world. It could be originated from microorganisms (bacteria, virus, fungi and protozoa) and non-microbial originated such as toxic plants, heavy metals, pesticides and herbicides. However, it is stated that the most dangerous ones are caused by pathogenic microorganisms. Consumption of food contaminated with viruses, bacteria, parasites, toxins, metals and prions leads to food-borne diseases. Some symptoms may cause mild gastrointestinal disorders, while others may cause life-threatening neurological, liver and kidney problems (Akkaya & Alişarlı, 2006).

In the past, *Mycobacterium bovis* and *Brucella abortus* were known as pathogens that cause problems in dairy products. Nowadays, dairy animals in countries with advanced dairy producers are completely protected from these pathogens. The most concerned pathogens are *Listeria monocytogenes,* enterotoxic strains of *Escherichia coli* (such as *E. coli* O157:H7), *Staphylococcus aureus, Salmonella* spp., and *M. paratuberculosis*. These pathogens cannot survive in low pH (5.3), high salt content (5-10% salt) and in the presence of bacteriocin. Alternatively, cheese can be pasteurized or pasteurized milk can be used in production. If cheese pH increases during maturation or stays below the desired value, it may be identified as the riskiest cheese in terms of pathogenic bacteria (Fox *et al.*, 2017). *L. monocytogenes*, *Salmonella* spp. and *S. aureus* cause worldwide concern in terms of food safety (Ozmen Togay *et al.*, 2016). The pathogens in raw milk such as *E. coli* O157:H7, *L. monocytogenes, Salmonella* spp. and *S. aureus* can survive during cheese production (Le *et al.*, 2014).

Salmonella spp. growth in dairy cattle, poultry and naturally infected in bird droppings among other animals, water, wastewater, food processing environment, in food and feed for long periods is an issue which should be discussed in detail from the perspective of public health (Akkaya & Alişarlı, 2006).

The most important reasons for the emergence of cheese-borne diseases are contamination after the pasteurization, improper pasteurization tools or procedures, and the use of raw milk in production. Although pasteurization is an effective method for eliminating food-borne pathogens, some dairy products cannot be pasteurized (Oliver *et al.*, 2005). Raw milk production without pathogenic bacteria is not possible. Nevertheless, milk with relatively low pathogen content can be obtained from healthy animals. Any pathogen in milk can be killed by pasteurization or novel alternative methods can be removed by bactofugation or microfiltration; pathogen growth may be prevented by the use of bacteriocin or acid-producing starter cultures (Fox *et al.*, 2017).

Microorganisms found in cheese can cause changes in the structure and can lead to food poisoning. As a result of various studies conducted on the microbiological quality of cheese in Turkey, it has been determined that compliance with hygienic conditions has an effect on the microbiological quality of cheese at all the stages from production to consumption (Kaynar, 2011).

In the research studies, it has been observed that insufficient microbiological quality of the cheese has resulted from the lack of importance given to the water, staff, quality control and the tools and equipment used up to the point where it meets with the consumer. For this reason, it is necessary to raise awareness of cheese producers, distributors and consumers about hygiene (Kaynar, 2011).

SOURCE OF MICROORGANISM CONTAMINATION AND PATHOGENS

Cheese produced by traditional methods could be contaminated with pathogenic microorganisms because of the raw milk, milk-producing animal, environment of the production stages and after the production (Çelik & Uysal, 2009; Fox *et al.*, 2017). Sanitation processes such as milk pasteurization play an important role in the control of pathogenic bacteria (Morea *et al.*, 2009). The microorganism limits set for cheese by Turkey is shown in Table **1**.

The milk used in cheese production is an important ingredient of dairy products and could be contaminated from different sources during the process and transport (Dığrak & Özçelik, 1996). Although pasteurized milk or heat-treated milk is used in many cheese production, most of the cheeses are produced from raw milk (Fox *et al.*, 2016). Various microorganisms can be found in milk obtained from dairy cows and also in milk products produced with this milk (Oliver *et al.*, 2005). Some of these microorganisms are saprophyte and can produce metabolites, that cause bad taste and aroma, by utilizing nutrients such as protein, fat and carbohydrate found in cheese. As a result, deterioration signs can be seen on cheese such as bitterness, putrefaction and it leads to economic losses. Some of these microorganisms can cause various diseases, such as typhoid, paratyphoid, tuberculosis and brucellosis, and food poisoning which can result in death. Among these microorganisms, the most harmful group is coliform bacteria (Dığrak & Özçelik, 1996).

The presence of coliform bacteria is considered to be an indicator of the insufficient sanitation and pasteurization applications or contamination after pasteurization (Akel & Alemdar, 2016). The coliform bacteria convert milk sugar to acid and gas. It changes the taste and aroma of cheese because the gas is collected inside and it causes the formation of pores. The pores are able to cover the entire surface of the cheese and may also occur in certain places (Dığrak & Özçelik, 1996).

The formation of vinegar acid in cheese, H_2S production of *Citrobacter* and indole production by *E. coli* from proteins causes undesirable taste and aroma in cheese. Coliform bacteria are able to cause swelling and explosions in the cheese can. Some species are the most harmful group of bacteria in cheese technology because of their pathogenic effect (Dığrak & Özçelik, 1996). On the other hand, typhoid, paratyphoid and enteritis bacteria can be found considering the presence of coliform bacteria and regarding that, there may be a risk of infection, the harmful effects of coliform bacteria might emerge. It shows pathogen outbreaks especially in children, and some serological types of *E. coli* can cause diarrhea in newborn infants (Dığrak & Özçelik, 1996). *E. coli* is a bacterium that shows the local pollution in terms of food microbiology. In addition to its significant importance as a fecal indicator, the role of different biotypes in many outbreaks in recent years has led to the importance of pathogenicity. The presence of *E. coli* in food indicates direct or indirect fecal contamination and the risk of enteric pathogens. *E. coli* can be found in soft, semi-hard and fresh cheeses and can survive for a long time even though there is a decrease in living cell number during the ripening period (Akel & Alemdar, 2016). *E. coli* O157:H7 and *L. monocytogenes* are considered as two of the most important food pathogens in milk and dairy products (Langa *et al.*, 2018).

E. coli and *Enterobacter aerogenes* were found to be responsible for early blowing. They are associated with taste and aroma disorders. It was also determined that *E. coli* and *E. aerogenes* form gas in cheeses and could develop even in the environments containing more than 10% salt. Cheese can be contaminated with these bacteria from raw milk which are not found in the pasteurized milk. Fresh cheese can be contaminated at the production stage. For this reason, the presence of coliform bacteria in foods indicates poor hygiene conditions, inadequate and inaccurate heat treatment applications and post-heat treatment contamination (Kaynar, 2011).

Among the *E. coli* serotypes, which causes diarrhea, the most important one is *E. coli* 0157:H7 serotype. *E. coli* 0157:H7 is a pathogenic bacterium that can be found in the intestines of animals such as cow, sheep and it can contaminate milk and dairy. It is found especially in raw milk, yogurt and cheese made from raw milk (Kaynar, 2011).

Another microorganism that can be found in cheese is *L. monocytogenes.* It is determined that milk and milk products are contaminated from environmental materials and processing surfaces. *L. monocytogenes* was isolated from milk processing machines, cheese brine materials, presses and cheese washing systems in the plant (Kaynar, 2011). *L. monocytogenes* is a pathogenic bacteria that can be commonly found in the environment such as soil, water, plants, wastewater, animal feed and silage. From these sources, it can be transferred to animals and environment *via* feces, blood and milk of these animals. Foodstuffs can be contaminated during production, transportation and consumption due to inadequate sanitation (Akkaya & Alişarlı, 2006). Dairy products could be contaminated by some pathogens such as *L. monocytogenes* developed in the post-pasteurization environment (Oliver *et al.*, 2005).

Table 1. Microorganism limits set for cheese by Turkey (Regulation on Turkish Food Codex Microbiological Criteria, Official Gazette of Publication 29.12.2011–28157 https://www.tarimorman.gov.tr/Belgeler/eng/Legislation/regulation_microbiological_criteria.pdf

Type of Cheese	Microorganisms/Toxins	Sampling Plan		Limits	
		n	c	m	M
Cheese (all cheeses except for melted cheese)	Coagulase positive staphylococci	5	2	10^2	10^3
	Salmonella	5	0	0/25 g-mL	
	L. monocytogenes	5	0	0/25 g-mL	
Melted cheeses and melted cheese products	*Staphylococcal enterotoxins*	5	0	absence in 25 g	
	E. coli ([1])	5	0	$<10^1$	
	L. monocytogenes	5	0	0/25 g-mL	

([1]) Most Possible Number (MPN) Method

Salmonella spp. can be transferred to the food in various ways. In particular, milk, egg and meat can be contaminated due to use of the *Salmonella* carrier farm animals in production. For this reason, animal food is one of the most common *Salmonella* spp. containing food. Especial studies carried out about cow milk show that raw milk is an important source of contamination for *Salmonella* spp. (Akkaya & Alişarlı, 2006; Acaröz *et al.*, 2018). Many factors such as inappropriate hygienic conditions in the farm, contamination risks from food production staff and consumption of raw milk and dairy products can be the source of *Salmonella* infections (Acaröz *et al.*, 2018). *Salmonella* is a mesophilic bacteria which can live in a wide temperature range and easily adapt to extreme environmental conditions. The growth temperature range is between 2 and 54°C; it can also develop in stored food at 2-4°C due to its psychrotrophic character. As a result of the acid resistance of *Salmonella*, the products obtained by using starter culture form a suitable environment for the development of *Salmonella*. The presence of acid-resistant *Salmonella* in foods leads to an increase in public health risk (Akkaya & Alişarlı, 2006).

Micrococci and staphylococci are very important because of their presence in the natural flora of skin, mouth and respiratory system and role of some strains in food poisoning. They can be found especially in low-acid cheese and they can maintain their vitality during the maturation process. Although these microorganisms are eliminated by pasteurization, the produced toxin at the beginning is resistant to temperature and can exist in cheese for months or even years (Akel & Alemdar, 2016). Staphylococci is common in nature and it is the causative agent of many diseases observed in humans and animals. In particular, *S. aureus* synthesizes enterotoxin which is highly toxic and effective in the intestinal flora and cause food-borne intoxications. It is considered as an indicator microorganism in case of infection from the food processor or prolonged waiting at room temperature. Staphylococci found in cheese was also reported as able to survive during ripening (Akel & Alemdar, 2016). The most common poisoning outbreaks in dairy products associated with *S. aureus* is cheese. *S. aureus* is found naturally in raw milk, but it can also found in pasteurized milk and other dairy products due to insufficient pasteurization or contamination from equipment and personnel (Kaynar, 2011).

Clostridium perfringens can contaminate foods via direct contact with soil, water and fecal waste. It is naturally found in the intestines of healthy humans and animals. The enterotoxin production by *C. perfringens* may cause mastitis in animals and gastroenteritis in humans. *C. perfringens* is derived in 95% of the food poisoning cases as a result of keeping the food warm or cooling it slowly after cooking (Akel & Alemdar, 2016).

Milk used in the cheese production should be pasteurized under the appropriate conditions for the purpose of elimination of these food pathogens. After this process, hygienic conditions should be fulfilled and contamination should be prevented. The awareness of cheese producers should be raised (Dığrak & Özçelik, 1996; Acaröz *et al.*, 2018).

The microbiological quality of cheese is affected by water, tool and equipment used up to the point where consumer meets the product, underestimation of hygiene, insufficient personnel and quality control in the sale points (Kaynar, 2011). For some kinds of cheese, yeasts and molds can cause economic and sensory problems. In dairy products, especially in cheese, molds and yeasts can contaminate the environment (walls and shelves of ripening room), air, equipment and tools, water, milk and brine (Godic Torkar & Vengust, 2008). As a result, attention should be paid to milking hygiene and follow-up of mastitis. Milk should be kept in the cold at all stages from milk transport to processing. GMP (Good manufacture practice) and HACCP (Hazard Analysis and Critical Contol Points) system should be established and put into practice in all stages from milking to cheese production (Akkaya & Alişarlı, 2006).

IMPORTANCE OF LACTIC ACID BACTERIA

LAB are Gram-positive, catalase-negative, non-spore forming and facultative anaerobic microorganisms (Yerlikaya, 2014; Domingos-Lopes *et al.*, 2017; Erginkaya *et al.*, 2017). They ferment glucose and produce lactic acid as the main product. They are found indigenously in raw milk microbiota (Domingos-Lopes *et al.*, 2017). They reduce carbohydrates to lactic acid heterofermentatively or homofermentatively (Şimşek *et al.*, 2016). Their homofermentative strains produce acetic acid, formic acid, ethanol and carbon dioxide as a by-product beside lactic acid (Giraffa *et al.*, 2011; Şimşek *et al.*, 2016).

LAB are generally considered as safe (GRAS) bacteria that are not harmful to health (Kıvanç & Kovancı, 2017). It has been reported that LAB have probiotic properties and are used in probiotic mixtures for human consumption (Sönmez *et al.*, 2018). For centuries, LAB have been used in fermentation for the production of dairy products such as cheese, yogurt and kefir (Milani *et al.*, 2016; Erginkaya *et al.*, 2017; Kıvanç & Kovancı, 2017). LAB contribute significantly to sensory, biochemical and aroma characteristics of the product. It is stated that microorganisms indigenously present in the product can improve the safety, technological and sensory properties of fermented products and even extend their shelf life (Domingos-Lopes *et al.*, 2017). The LAB found in cheese are shown in Table **2**

LAB is an industrially important group in terms of fermented food production. They can produce lactic acid and some volatile compounds by lactose fermentation (Gezginc *et al.*, 2015). They reduce the pH due to the production of lactic acid and other acids (Akel & Alemdar, 2016). The features such as acid production, proteinase and peptidase activities, autolysis and inhibiting agent production of LAB contribute to their use a as starter culture (Bozoudi *et al.*, 2015). LAB are of critical importance for maintaining biochemical reactions such as proteinase, lipolysis and glycolysis (Şimşek *et al.*, 2016). LAB are autolyzed by losing their cellular integrity under various conditions with their peptidoglycan hydrolase enzymes. As a result of autolysis, the cellular enzymes of LAB are released into cheese matrix. In the early stages of maturation, the early autolysis of microorganisms (starter and non-starter microbiota) combines the intracellular microbial enzymes with substrates in a shorter time and in this case, the formation of flavor compounds in cheese is accelerated. Therefore, autolysis can shorten the ripening time of cheeses. In addition to accelerating maturation, autolysis has positive effects on cheese flavor (Şimşek *et al.*, 2016).

The antifungal activity of LAB depends on their production of organic acid, competitiveness for nutrients and synthesis of antifungal compounds. Metabolites have various affects, such as preventing or delaying the development of the filament and the development of spore against molds. Therefore, LAB can be used as an alternative to chemical preservatives (Kıvanç & Kovancı, 2017).

LAB found in raw milk and cheese microbiota play an important role in the control of pathogen growth, the formation of nutritional and sensory characteristics, during the ripening of traditional cheese (Milani *et al.,* 2016). LAB found in dairy are separated into two groups as starter LAB and non-starter LAB (Vandera *et al*., 2019).

Starter LAB are responsible for the conversion of lactose to lactate during the fermentation of milk and as a result of this, pH decreases. The starter LAB species vary according to the desired type of cheese production (Czaran *et al*., 2018). *S. thermophilus*, *L. lactis, L. helveticus, L. delbrueckii* and a few *Leuconostoc* spp. are found in the starter LAB group (Vandera *et al*., 2019).

Raw milk and whey are accepted as the main source of non-starter LAB (Durlu Ozkaya *et al*., 2001; Gobbetti *et al*., 2015). The non-starter LAB in cheese produced with raw milk can be found inherently or could be contaminated from the environment (Bozoudi *et al*., 2015; Vandera *et al*., 2019). The non-starter LAB are very diverse (Vandera *et al*., 2019). The most important non-starter LAB group found in cheese is mesophilic lactobacilli (Gobbetti *et al*., 2015). The most common non-starter LAB found in cheese are facultative heterofermentative LAB such as *L. casei* , *L. paracasei* , *L. rhamnosus*, *L. plantarum,* *L. paraplantarum, L. pentosus* and obligatory heterofermentative LAB such as *Leuconostoc, Weisella* and *Lactobacillus* spp., *Enterococcus* spp., *E. faecium*, *E. faecalis* and *E. durans* (Vandera *et al*., 2019).

Generally, non-starter LAB contaminate after pasteurization of the milk. The source of these non-starter LAB can be the tools and equipment used in production and other components (Nieto-Arribas *et al*., 2009). It plays an important role in the development of desired physicochemical and technological properties of cheese (İspirli *et al*., 2017). As the maturation takes longer time, the contribution of these bacteria to cheese increases. Most of the non-starter LAB group usually consist of *L. casei, L. paracasei, L. plantarum* and *L. curvatus* (Nieto-Arribas *et al*., 2009). Enterococci are used in cheese production and as a member of the non-starter LAB. It has important roles during ripening (İspirli *et al*., 2017).

Among the LAB, the most involved ones in the cheese production are *Enterococcus* spp. (*E. faecium* and *E. faecalis*), *Lactococcus* spp. (*L. lactis*), *Lactobacillus* spp. (*L. plantarum, L. fermentum, L. paraplantarum*, *L. casei* and *L. brevis*), *Leuconostoc* spp. (*L. mesenteroides* and *L. pseudomesenteroides*) and *Streptococcus* spp. (*S. thermophilus*) (Milani *et al.,* 2016).

In cheese technology, LAB play a role both as starter and non-starter cultures and provide important contributions to maintaining of biochemical reactions which are necessary for cheese ripening (Şimşek *et al*., 2016). In order to obtain cheese in original taste and aroma, it is stated that the LAB should be added as pure culture and milk should be pasteurized (Akel & Alemdar, 2016).

Enterococci

Enterococci are Gram-positive commensals that can be colonized in the digestive tract of humans and most of the animals (Psoni *et al.*, 2009; Yüksel *et al.*, 2015; İspirli *et al.*, 2017). They are found in many foods such as milk, cheese, meat and vegetables (Yüksel *et al.*, 2015). Especially, they are found in high amounts in foods of animal origin such as fermented sausages and cheeses (Psoni *et al.*, 2009). In many traditional cheeses, enterococci create natural flora, taste, structure and textural properties of the product, although they are not identified as GRAS (generally recognized as safe) (Andrighetto *et al.*, 2001; Yerlikaya, 2014).

They have an important role in the fermentation process because they are able to survive and have high fermentation ability in many foods. In addition, enterococci can be found in cheeses obtained from raw and pasteurized milk due to their high-temperature resistance (İspirli *et al.*, 2017). *Enterococcus* are a dominant type of bacteria in the microbiota of traditional cheeses produced from raw milk (Yüksel *et al.*, 2015). They play an important role in the microbiota during the ripening period (Ozmen Togay *et al.*, 2016). They contribute to the sensory properties of the final product with diacetyl production *via* their proteolytic, esterolytic activities and citrate metabolism (Psoni *et al.*, 2009; Yüksel *et al.*, 2015). Enterococci play an important role in white cheese production and affect product quality (İspirli *et al.*, 2017). They help to improve the quality of cheese as a starter culture used in cheese production (Psoni *et al.*, 2009). The metabolic and technological properties of many strains of *E. faecalis*, *E. faecium* and *E. durans*, which are frequently found in dairy products, have an impact on the typical taste and structure of the cheese. They also produce esterases such as diacetyl and acetoin, which play an important role in aroma formation (Andrighetto *et al.*, 2001).

They are suitable for being used in fermentation applications and can be used as probiotics with their properties (Yüksel *et al.*, 2015). In addition, technological properties of enterococci such as enterocin and bacteriocin production are an important characteristic of this species. It inhibits the development of some pathogens in the fermentation and maturation stages of cheese and other foods (İspirli *et al.*, 2017).

Enterococci produce bacteriocin with antibacterial properties against food-borne Gram-positive pathogens such as *L. monocytogenes, S. aureus, Bacillus cereus* and spores of *C. botulinum*. On the other hand, hospital-derived enterococci such as *E. faecium* and *E. faecalis,* have a virulence gene and antibiotic resistance (Yüksel *et al*., 2015).

Lactococci

Lactococci are an important group of the LAB. For centuries, they are used for the production of cheese and other fermented dairy products (Delgado & Mayo, 2004; Psoni *et al*., 2007). They are preferred due to their ability to produce acids, their resistance to phages and their adaptability (Fox *et al*., 2016).

L. lactis is a LAB used as a starter culture for the production and maturation of cheese in the dairy industry throughout the world (Florez *et al*., 2008). It constitutes the majority of commercial starter culture (Alegria *et al*., 2010). It is stated that *L. lactis* has an important role in the formation of the aroma and texture of cheese (Florez *et al*., 2008; Pedersen *et al*., 2016). The development of *L. lactis* in milk contributes some properties such as preventing the spoilage bacteria and pathogens, clot formation and biochemical conditions for ripening. However, *L. lactis* can be found on the membrane surfaces of cows as saprophytes and it can gain resistance against antibiotics (Florez *et al*., 2008).

Lactococci in cheese produced in small-scale enterprises are usually characterized as *L. lactis* and *L. lactis* subsp. *lactis*, predominantly (Gaya *et al*., 1999). Among lactococci, *L. lactis* subsp. *lactis* and *L. lactis* subsp. *cremoris* are technologically important (Delgado & Mayo, 2004; Leroy & De Vuyst, 2004).

Lactobacilli

Lactobacilli are an important, heterogeneous group of the LAB and used in the food fermentation process. They have been used as a starter for dairy products, fermented vegetables, fish and sausages for many years (Giraffa *et al*., 2011). Lactobacilli can be found in plants, animals and raw milk. They are mostly involved in the ripening period in cheese production (Bulut *et al*., 2005). Some lactobacilli are defined as probiotic due to its beneficial effect on human health. Lactobacilli are one of the most important starters used in dairy products such as cheese and fermented milk. They are used in cheese production as a starter and complementary strain. Lactic acid production provides a rapid decline in the pH of the raw material. The most common lactobacilli are *L. paracasei* and *L. rhamnosus.* It is stated that they are mostly isolated from cheese (Giraffa *et al*., 2011). *L. reuteri* is a LAB found in the digestive systems of humans and animals. It is frequently encountered in probiotic and fermented products. It produces reuterine, which is known as an antimicrobial compound (Langa *et al*., 2018). *L. plantarum* is facultative heterofermentative lactobacillus. It can be found in fermented milk, meat and vegetable products. In addition, it can be found in several kinds of cheese. It is reported that it can survive in the digestive tract conditions and colonize in the intestines of humans and other mammals (Zago *et al*., 2011). *L. plantarum* can produce conjugated linoleic acid (Ribeiro *et al*., 2018). *L. acidophilus* has positive effects on health and has shown antagonistic activity against food-borne disease agents (Kasımoğlu *et al*., 2004). Lactobacilli are reported to have antibacterial activity on *Yersinia enterocolitica*, *E. coli* and *B. cereus* (Durlu Ozkaya *et al*., 2001).

Others

Leuconostoc spp. can not grow in milk like *Lactococcus* spp., but they can continue to grow at low pH value. For this reason, they develop ones the cheese production process is completed. The role of *Leuconostoc* spp. primarily inhibits the lactococci strains which are citrate-positive. Thus, gas production is maintained for a long time to ensure the formation of pores. Other functions of *Leuconostoc* spp. are, to prevent the development of harmful bacteria, ethanol production to be converted to ethyl ester during heterofermentative fermentation and to influence the redox potential of the cheese matrix (Pedersen *et al*., 2016).

S. thermophilus is widely used in the production of dairy products such as fermented milk, yogurt and cheese (Mora *et al*., 2002; Proust et al., 2018). It has been used in commercial starter cultures commonly due to its characteristics such as rapid growth, reducing the pH, changing the microbial population with the decrease of pH (Tarrah *et al*., 2018).

IMPORTANCE OF YEASTS

The yeast found in cheese was first discovered in the beginning of 1900s (Welthagen & Viljoen, 1998). The presence of yeast in food is commercially important because they can cause spoilage and undesirable fermentations (Fleet, 1990). The yeast number found in cheese changes according to the cheese type (Welthagen & Viljoen, 1998). Generally, yeasts are found in high concentrations in dairy products. They adapt well to the environment rich in protein, lipid, sugar and organic acid (Corbaci et al., 2012).

The importance of yeast depends on the types of cheeses (Welthagen & Viljoen, 1998). The most isolated yeasts are *Debaryomyces hansenii* and *Candida lipolytica* from cheese (Fleet, 1990). Some yeasts lead to spoilage in cheese, while certain yeast strains contribute positively to the development of aroma (Fleet, 1990; Romano *et al*., 1996; Welthagen & Viljoen, 1998; Pereira-Dias *et al*., 2000). In the first days of maturation, the number of yeast on the cheese surface increases very quickly and reaches the maximum value after 10 days. This number can be as high as 10^6-10^9 CFU/g (Wyder & Puhan, 1999).

Table 2. Lactic acid bacteria found in cheese.

Lactic Acid Bacteria		Advantages	Antimicrobial Activity	Reference
Enterococci	*E. faecalis* *E. faecium* *E. durans*	Natural flora, taste, structure and textural properties Enterocin and bacteriocin production.	*L. monocytogenes, S. aureus, Bacillus cereus* and spores of *C. botulinum*	Andrighetto *et al.*, 2001; Yerlikaya, 2014; Yüksel *et al.*, 2015; İspirli *et al.*, 2017
Lactococci	*L. lactis* *L. lactis subsp. lactis* *L. lactis subsp. cremoris*	Acid production Resistant to phages Adaptability Formation of the aroma and texture	Preventing the spoilage bacteria and pathogens	Gaya *et al.*, 1999; Delgado & Mayo, 2004; Leroy & De Vuyst, 2004; Florez *et al.*, 2008; Fox et al., 2016; Pedersen *et al.*, 2016
Lactobacilli	*L. paracasei* *L. rhamnosus* *L. reuteri* *L. plantarum* *L. acidophilus*	Reuterine production Linoleic acid production	Antibacterial activity on *Yersinia enterocolitica, E. coli* and *B. cereus*	Durlu Ozkaya *et al.*, 2001; Giraffa *et al.*, 2011; Zago *et al.*, 2011; Langa *et al.*, 2018; Ribeiro *et al.*, 2018
Others	*Leuconostoc spp.*	Ethanol production	Preventing the development of harmful bacteria	Pedersen *et al.*, 2016
	S. thermophilus	Rapid growth Reducing the pH	Changing the microbial population with the decrease of pH	Tarrah *et al.*, 2018

The sensory properties of the cheese are affected by yeasts due to the fact that the yeast found in cheese is able to hydrolyze milk fat (Pereira-Dias *et al.*, 2000). At first, they metabolize the lactic acid and increase pH (Romano *et al.*, 1996; Godic Torkar & Vengust, 2008). Whereby, acid-sensitive bacterial flora that does not develop below pH 5.5, starts to develop. Then, the first aroma compounds are formed by yeasts, which are going to affect the final product with their proteolytic and lipolytic activities (Romano *et al.*, 1996). On the other hand, special strains used in soft cheese production or starter cultures used in maturation and aroma formation contribute significantly with their metabolic characteristics (Godic Torkar & Vengust, 2008).

Yeasts can develop in conditions that are not suitable for many bacteria, and play an important role in the spoilage of dairy products. It may cause cheese degradation due to their ability to grow at low temperature, low pH, low water activity and high concentration of salt (Welthagen & Viljoen, 1998). Yeasts do not cause spoilage in dairy products unless they ferment lactose (Godic Torkar & Vengust, 2008). Excessive gas production, fruit aromas, increase in acidity, change in texture, the bitter and unsavory taste can be given as examples to yeast spoilage in cheese (Welthagen & Viljoen, 1998; Pereira-Dias *et al.*, 2000). It is stated that the color loss observed on cheese surface may be due to the yeast activity (Pereira-Dias *et al.*, 2000). However, in some cases, yeasts have advantageous properties such as inhibiting the growth of undesired microorganisms, supporting the development of starter culture by proteolytic and lipolytic activity, increasing pH by metabolizing lactic acid, contributing to fermentation and maturation by secreting growth factors (Welthagen & Viljoen, 1998).

D. hansenii is a typical yeast used for cheese production (Deiana et al., 1984; Corbaci et al., 2012). It has significant biotechnological potential in the food and agricultural industry. It can be commonly found in cheese and sausages (Corbaci et al., 2012). It acts synergistically with the lactic microbiota. It has proteolytic and lipolytic activity (Deiana et al., 1984). It inhibits undesirable microorganism growth found in brine by metabolizing protein and oil (Corbaci et al., 2012). It contributes to the maturation of some types of cheeses. It has the ability to use acetate, lactate, citrate, lactose and galactose being one of the starter culture components used in cheese production (Deiana et al., 1984; Corbaci et al., 2012). It was stated that the most effective strain for deacidification of cheese surface is *D. hansenii* (Wyder & Puhan, 1999).

IMPORTANCE OF MOLDS

In some types of cheese, molds can cause economic and sensory problems. Molds can be found in raw milk and not be able to survive after the pasteurization procedure. It can contaminate pasteurized milk during production. Mold contamination to dairy products, especially to cheeses, can be derived from the environment (such as walls and shelves of the maturation chamber), air, material, water, milk and brine during production. In addition to the production of mycotoxins and its harmful effects on health, the presence of wild-type molds in cheese is not desirable since they may affect its sensory properties (Godic Torkar & Vengust, 2008).

Molds have a more complex enzyme system compared to bacteria. They contribute to proteolysis and lipolysis by their enzymes in the mold-ripened cheeses. *Penicillium* spp. is found to be the dominant species in cheese ripened with mold (Hayaloglu & Kirbag, 2007). In kashar cheese, *P. verrucosum* is found to be the dominant species (Kıvanç, 1997). However, the growth of mold in some cheese can cause economic losses and quality problems. Some molds can produce toxic secondary metabolites such as mycotoxins and cause undesirable aroma formation. As a result, these organisms can be of great risk to public health (Hayaloglu & Kirbag, 2007).

CONSENT FOR PUBLICATION

Not applicable.

CONFLICT OF INTEREST

The authors confirm that this chapter contents have no conflict of interest.

ACKNOWLEDGEMENTS

Declared none.

REFERENCES

Acaröz, U, Kara, R, Gürler, Z, Arslan-Acaröz, D & Zemheri, F (2018) Afyonkarahisar'dan Toplanan Çiğ Manda Sütlerinde Salmonella spp. Varlığının Araştırılması. *Kocatepe Vet J,* 11, 180-5. [http://dx.doi.org/10.30607/kvj.417136]

Akel, E & Alemdar, S (2016) 'Van'da Tüketime Sunulan Feta Peynirlerinin Mikrobiyolojik ve Kimyasal Kalitesi' *Türk Tarım – Gıda Bilim ve Teknoloji Dergisi,,* 999-1005. [http://dx.doi.org/10.24925/turjaf.v4i11.999-1005.950]

Akkaya, L & Alişarlı, M (2006) Afyonkarahisar'da Tüketime Sunulan Peynirlerde Listeria monocytogenes ve Salmonella spp. Varlığının Belirlenmesi. *YYÜ Vet Fak Derg,* 17, 87-91.

Oliver, SP, Jayarao, BM & Almeida, RA (2005) Foodborne pathogens in milk and the dairy farm environment: food safety and public health implications. *Foodborne Pathog Dis,* 2, 115-29. [http://dx.doi.org/10.1089/fpd.2005.2.115] [PMID: 15992306]

Alegría, A, Delgado, S, Roces, C, López, B & Mayo, B (2010) Bacteriocins produced by wild Lactococcus lactis strains isolated from traditional, starter-free cheeses made of raw milk. *Int J Food Microbiol,* 143, 61-6. [http://dx.doi.org/10.1016/j.ijfoodmicro.2010.07.029] [PMID: 20708289]

Andrighetto, C, Knijff, E, Lombardi, A, Torriani, S, Vancanneyt, M, Kersters, K, Swings, J & Dellaglio, F (2001) Phenotypic and genetic diversity of enterococci isolated from Italian cheeses. *J Dairy Res,* 68, 303-16. [http://dx.doi.org/10.1017/S0022029901004800] [PMID: 11504393]

Asensio, CM, Grosso, NR & Rodolfo Juliani, H (2015) Quality preservation of organic cottage cheese using oregano essential oils. *Lebensm Wiss Technol,* 60, 664-71. [http://dx.doi.org/10.1016/j.lwt.2014.10.054]

Aydemir, O, Harth, H, Weckx, S, Dervişoğlu, M & De Vuyst, L (2015) Microbial communities involved in Kaşar cheese ripening. *Food Microbiol,* 46, 587-95. [http://dx.doi.org/10.1016/j.fm.2014.10.002] [PMID: 25475332]

Bassi, D, Puglisi, E & Cocconcelli, PS (2015) Comparing natural and selected starter cultures in meat and cheese fermentations. *Curr Opin Food Sci,* 2, 118-22. [http://dx.doi.org/10.1016/j.cofs.2015.03.002]

Bayar, N & Özrenk, E (2011) The effect of quality properties on Tulum cheese using different package materials. *Afr J Biotechnol,* 10, 1393-9.

Kavas, G (2017) Çörek Otu ve Tarçın Uçucu Yağ İlaveli Yumurta Beyazı Protein Tozu Esaslı Filmlerin

Çökelek Peyniri Muhafazasında Kullanımı. *Ege Univ Ziraat Fak Derg,* 54, 439-46.
[http://dx.doi.org/10.20289/zfdergi.386549]

Bozoudi, D, Kotzamanidis, C, Hatzikamari, M, Tzanetakis, N, Menexes, G & Litopoulou-Tzanetaki, E (2015) A comparison for acid production, proteolysis, autolysis and inhibitory properties of lactic acid bacteria from fresh and mature Feta PDO Greek cheese, made at three different mountainous areas. *Int J Food Microbiol,* 200, 87-96.
[http://dx.doi.org/10.1016/j.ijfoodmicro.2015.02.008] [PMID: 25702882]

Bulut, C, Gunes, H, Okuklu, B, Harsa, S, Kilic, S, Coban, HS & Yenidunya, AF (2005) Homofermentative lactic acid bacteria of a traditional cheese, Comlek peyniri from Cappadocia region. *J Dairy Res,* 72, 19-24.
[http://dx.doi.org/10.1017/S0022029904000536] [PMID: 15747727]

Cakmakci, S, Gurses, M, Hayaloglu, AA, Cetin, B, Sekerci, P & Dagdemir, E (2015) Mycotoxin production capability of Penicillium roqueforti in strains isolated from mould-ripened traditional Turkish civil cheese. *Food Addit Contam Part A Chem Anal Control Expo Risk Assess,* 32, 245-9.
[http://dx.doi.org/10.1080/19440049.2014.997808] [PMID: 25580944]

Celem, E, Celik, OF & Tarakci, Z (2018) The effects of ripening on some physicochemical and microbiological characteristics of Çökelek cheeses: A market survey of fresh and skin-ripened Çökelek. *J Cent Eur Agric,* 19, 335-48.
[http://dx.doi.org/10.5513/JCEA01/19.2.2155]

Cogan, TM & Guinee, TP (2017) *Factors That Affect the Quality of Cheese.*Cheese. Elsevier Ltd.

Corbaci, C, Ucar, F & Yalcin, H (2012) Isolation and characterization of yeasts associated with Turkish-style homemade dairy products and their potential as starter cultures. *Afr J Microbiol Res,* 6, 534-42.

Czárán, T, Rattray, FP, Møller, CODA & Christensen, BB (2018) Modelling the influence of metabolite diffusion on non-starter lactic acid bacteria growth in ripening Cheddar cheese. *Int Dairy J,* 80, 35-45.
[http://dx.doi.org/10.1016/j.idairyj.2017.12.010]

Deiana, P, Fatichentf, F, Farris, GA, Mocquop, G, Lodi, R, Todesco, R & Cecchi, L (1984) Metabolization of lactic and acetic acids in Pecorino Romano cheese made with a combined starter of lactic acid bacteria and yeast *Lait,* 64, 380-94.
[http://dx.doi.org/10.1051/lait:1984640-64231]

Delgado, S & Mayo, B (2004) Phenotypic and genetic diversity of Lactococcus lactis and Enterococcus spp. strains isolated from Northern Spain starter-free farmhouse cheeses. *Int J Food Microbiol,* 90, 309-19.
[http://dx.doi.org/10.1016/S0168-1605(03)00323-4] [PMID: 14751686]

Dığrak, M & Özçelik, S (1996) Elazığ'da satışa sunulan peynirlerden izole edilen koliform grubu bakterilerin tanımlanması. *Gida,* 21, 1-, 3-7.

Domingos-Lopes, MFP, Stanton, C, Ross, PR, Dapkevicius, MLE & Silva, CCG (2017) Genetic diversity, safety and technological characterization of lactic acid bacteria isolated from artisanal Pico cheese. *Food Microbiol,* 63, 178-90.
[http://dx.doi.org/10.1016/j.fm.2016.11.014] [PMID: 28040167]

Durlu-Özkaya, F, Xanthopoulos, V, Tunail, N & Litopoulou-Tzanetaki, E (2001) Technologically important properties of lactic acid bacteria isolates from Beyaz cheese made from raw ewes' milk. *J Appl Microbiol,* 91, 861-70.
[http://dx.doi.org/10.1046/j.1365-2672.2001.01448.x] [PMID: 11722664]

Durlu-Özkaya, F & Gün, İ (2014) Aroma compounds of some traditional Turkish cheeses and their importance for Turkish cuisine. *Food Nutr Sci,* 5, 425-34.
[http://dx.doi.org/10.4236/fns.2014.54050]

Erginkaya, Z, Turhan, EU & Tatlı, D (2018) Determination of antibiotic resistance of lactic acid bacteria isolated from traditional Turkish fermented dairy products. *Majallah-i Tahqiqat-i Dampizishki-i Iran,* 19, 53-6.
[PMID: 29805464]

Fleet, GH (1990) Yeasts in dairy products. *J Appl Bacteriol,* 68, 199-211.
[http://dx.doi.org/10.1111/j.1365-2672.1990.tb02566.x] [PMID: 2187843]

Flórez, AB, Ammor, MS & Mayo, B (2008) Identification of tet(M) in two Lactococcus lactis strains isolated from a Spanish traditional starter-free cheese made of raw milk and conjugative transfer of tetracycline resistance to lactococci and enterococci. *Int J Food Microbiol,* 121, 189-94.
[http://dx.doi.org/10.1016/j.ijfoodmicro.2007.11.029] [PMID: 18068255]

Fox, PF, Guinee, TP, Cogan, TM & McSweeney, PLH (2016) Factors that affect cheese quality *Fundamentals of Cheese Science,* 533-542.

Fox, PF, Cogan, TM & Guinee, TP (2017) 'Factors that affect the quality of cheese', *Cheese: Chemistry. Physics and Microbiology,* 25, 617-41.

Gaya, P, Babín, M, Medina, M & Nuñez, M (1999) Diversity among lactococci isolated from ewes' raw milk and cheese. *J Appl Microbiol,* 87, 849-55.
[http://dx.doi.org/10.1046/j.1365-2672.1999.00932.x] [PMID: 10692072]

Gezginc, Y, Topcal, F, Comertpay, S & Akyol, I (2016) Erratum to "Quantitative analysis of the lactic acid and acetaldehyde produced by Streptococcus thermophilus and Lactobacillus bulgaricus strains isolated from traditional Turkish yogurts using HPLC". *J Dairy Sci,* 99, 1694.
[http://dx.doi.org/10.3168/jds.2016-99-2-1694] [PMID: 29079328]

Giraffa, G, Chanishvili, N & Widyastuti, Y (2010) Importance of lactobacilli in food and feed biotechnology. *Res Microbiol,* 161, 480-7.
[http://dx.doi.org/10.1016/j.resmic.2010.03.001] [PMID: 20302928]

Gobbetti, M, De Angelis, M, Di Cagno, R, Mancini, L & Fox, PF (2015) Pros and cons for using non-starter lactic acid bacteria (NSLAB) as secondary/adjunct starters for cheese ripening. *Trends Food Sci Technol,* 45, 167-78.
[http://dx.doi.org/10.1016/j.tifs.2015.07.016]

Hayaloglu, AA, Guven, M & Fox, PF (2002) Microbiological, biochemical and technological properties of Turkish White cheese "Beyaz Peynir.". *Int Dairy J,* 12, 635-48.
[http://dx.doi.org/10.1016/S0958-6946(02)00055-9]

Hayaloglu, AA & Kirbag, S (2007) Microbial quality and presence of moulds in Kuflu cheese. *Int J Food Microbiol,* 115, 376-80.
[http://dx.doi.org/10.1016/j.ijfoodmicro.2006.12.002] [PMID: 17258341]

Irkin, R (2011) Shelf-life of unsalted and light "lor" whey cheese stored under various packaging conditions: Microbiological and sensory attributes. *J Food Process Preserv,* 35, 163-78.
[http://dx.doi.org/10.1111/j.1745-4549.2009.00469.x]

İspirli, H, Demirbaş, F & Dertli, E (2017) Characterization of functional properties of Enterococcus spp. isolated from Turkish white cheese. *Lebensm Wiss Technol,* 75, 358-65.
[http://dx.doi.org/10.1016/j.lwt.2016.09.010]

Karabey, B, Eroglu, D, Vural, C, Ozdemir, G, Yerlikaya, O & Kinik, O (2018) Determination of the microbial flora in traditional İzmir Tulum cheeses by Denaturing Gradient Gel Electrophoresis. *J Food Sci Technol,* 55, 956-63.
[http://dx.doi.org/10.1007/s13197-017-3003-z] [PMID: 29487437]

Kasimoğlu, A, Göncüoğlu, M & Akgün, S (2004) Probiotic white cheese with *Lactobacillus acidophilus. Int Dairy J,* 14, 1067-73.
[http://dx.doi.org/10.1016/j.idairyj.2004.04.006]

Kavas, G & Kavas, N (2016) Use of egg white protein powder based films fortified with sage and lemon balm essential oils in the storage of lor cheese. *Mljekarstvo,* 66, 99-111.

Kaynar, P (2011) Ülkemiz Peynirleri Üzerine Mikrobiyolojik Araştırmalar. *Turk Mikrobiol Cem Derg,* 41, 1-8.

Kivanç, M (1992) Fungal contamination of Kashar cheese in Turkey. *Nahrung,* 36, 578-83. [http://dx.doi.org/10.1002/food.19920360609] [PMID: 1491710]

Kıvanç, M & Kovancı, P (2017) 'Fermente Gıdalardan İzole Edilen Laktik Asit Bakterilerinin Antifungal Aktivitesinin Belirlenmesi', *Gıda / J. Food,* 42, 477-84.

Kürekci, C, Arkadaş, M & Avşar, YK (2016) Occurrence, genetic characterization and antimicrobial resistance of extended spectrum β-lactamase producing Escherichia coli isolated from Sürk samples, a traditional turkish cheese. *J Food Meas Charact,* 10, 709-14. [http://dx.doi.org/10.1007/s11694-016-9355-7]

Langa, S, Martín-Cabrejas, I, Montiel, R, Peirotén, Á, Arqués, JL & Medina, M (2018) Protective effect of reuterin-producing *Lactobacillus reuteri* against *Listeria monocytogenes* and *Escherichia coli* O157:H7 in semi-hard cheese. *Food Control,* 84, 284-9. [http://dx.doi.org/10.1016/j.foodcont.2017.08.004]

Lazzi, C, Povolo, M, Locci, F, Bernini, V, Neviani, E & Gatti, M (2016) Can the development and autolysis of lactic acid bacteria influence the cheese volatile fraction? The case of Grana Padano. *Int J Food Microbiol,* 233, 20-8. [http://dx.doi.org/10.1016/j.ijfoodmicro.2016.06.009] [PMID: 27299670]

Le, S, Bazger, W, Hill, AR & Wilcock, A (2014) Awareness and perceptions of food safety of artisan cheese makers in Southwestern Ontario: A qualitative study. *Food Control,* 41, 158-67. [http://dx.doi.org/10.1016/j.foodcont.2014.01.007]

Leroy, F & De Vuyst, L (2004) Lactic acid bacteria as functional starter cultures for the food fermentation industry. *Trends Food Sci Technol,* 15, 67-78. [http://dx.doi.org/10.1016/j.tifs.2003.09.004]

Levante, A, De Filippis, F, La Storia, A, Gatti, M, Neviani, E, Ercolini, D & Lazzi, C (2017) Metabolic gene-targeted monitoring of non-starter lactic acid bacteria during cheese ripening. *Int J Food Microbiol,* 257, 276-84. [http://dx.doi.org/10.1016/j.ijfoodmicro.2017.07.002] [PMID: 28735145]

Milani, E, Shahidi, F, Mortazavi, SA & Saeedi, M (2017) Isolation And Identification Of Lactic Acid Bacteria In Kurdish Cheese During Ripening Using 16s rRNA Gene Sequence Analysis. *J Food Process Preserv,* 41e13009 [http://dx.doi.org/10.1111/jfpp.13009]

Mora, D, Fortina, MG, Parini, C, Ricci, G, Gatti, M, Giraffa, G & Manachini, PL (2002) Genetic diversity and technological properties of *Streptococcus thermophilus* strains isolated from dairy products. *J Appl Microbiol,* 93, 278-87. [http://dx.doi.org/10.1046/j.1365-2672.2002.01696.x] [PMID: 12147076]

Morea, M, Baruzzi, F & Cocconcelli, PS (1999) Molecular and physiological characterization of dominant bacterial populations in traditional mozzarella cheese processing. *J Appl Microbiol,* 87, 574-82. [http://dx.doi.org/10.1046/j.1365-2672.1999.00855.x] [PMID: 10583686]

Nieto-Arribas, P, Poveda, JM, Seseña, S, Palop, L & Cabezas, L (2009) Technological characterization of Lactobacillus isolates from traditional Manchego cheese for potential use as adjunct starter cultures. *Food Control,* 20, 1092-8. [http://dx.doi.org/10.1016/j.foodcont.2009.03.001]

Öner, Z, Gül Karahan, A & Aloğlu, H (2006) Changes in the microbiological and chemical characteristics of an artisanal Turkish white cheese during ripening. *Lebensm Wiss Technol,* 39, 449-54. [http://dx.doi.org/10.1016/j.lwt.2005.03.015]

Özbek, Ç & Güzeler, N (2017) Yoğurt Peyniri Üretiminde Kullanılan Yoğurt Miktarının Peynirin Bazı Özelliklerine Etkisi. *Adnan Menderes Üniv Ziraat Fak Derg,* 14, 35-5. [http://dx.doi.org/10.25308/aduziraat.291988]

Özmen Toğay, S, Ay, M, Sandikçi Altunatmaz, S, Yilmaz Aksu, F, Erol Tinaztepe, Ö, İssa, G & Büyükünal,

SK (2016) 'Bazı Geleneksel Türk Peynirlerinden İzole Edilen Enterococcus spp.'nin Antimikrobiyal Aktivite Potansiyeli'. *Kafkas Univ Vet Fak Derg,* 22, 765-70.

Ozturkoglu Budak, S (2016) Dynamics of complex microbiota and enzymes in Divle Cave cheese and their biochemical consequences PhD thesis, Utrecht University.

Pedersen, TB, Vogensen, FK & Ardö, Y (2016) Effect of heterofermentative lactic acid bacteria of DL-starters in initial ripening of semi-hard cheese. *Int Dairy J,* 57, 72-9.
[http://dx.doi.org/10.1016/j.idairyj.2016.02.041]

Pereira-Dias, S, Potes, ME, Marinho, A, Malfeito-Ferreira, M & Loureiro, V (2000) Characterisation of yeast flora isolated from an artisanal Portuguese ewes' cheese. *Int J Food Microbiol,* 60, 55-63.
[http://dx.doi.org/10.1016/S0168-1605(00)00323-8] [PMID: 11014522]

Proust, L, Loux, V, Martin, V, Magnabosco, C, Pedersen, M, Monnet, V & Juillard, V (2018) Complete Genome Sequence of the Industrial Fast-Acidifying Strain *Streptococcus thermophilus* N4L. *Microbiol Resour Announc,* 7, e01029-18.
[http://dx.doi.org/10.1128/MRA.01029-18] [PMID: 30533920]

Psoni, L, Kotzamanides, C, Andrighetto, C, Lombardi, A, Tzanetakis, N & Litopoulou-Tzanetaki, E (2006) Genotypic and phenotypic heterogeneity in Enterococcus isolates from Batzos, a raw goat milk cheese. *Int J Food Microbiol,* 109, 109-20.
[http://dx.doi.org/10.1016/j.ijfoodmicro.2006.01.027] [PMID: 16529834]

Psoni, L, Kotzamanidis, C, Yiangou, M, Tzanetakis, N & Litopoulou-Tzanetaki, E (2007) Genotypic and phenotypic diversity of *Lactococcus lactis* isolates from Batzos, a Greek PDO raw goat milk cheese. *Int J Food Microbiol,* 114, 211-20.
[http://dx.doi.org/10.1016/j.ijfoodmicro.2006.09.020] [PMID: 17241681]

Ribeiro, SC, Stanton, C, Yang, B, Ross, RP & Silva, CCG (2018) Conjugated linoleic acid production and probiotic assessment of Lactobacillus plantarum isolated from Pico cheese. *Lebensm Wiss Technol,* 90, 403-11.
[http://dx.doi.org/10.1016/j.lwt.2017.12.065]

Romano, A, Casaregola, S, Torre, P & Gaillardin, C (1996) Use of RAPD and mitochondrial DNA RFLP for typing of *Candida zeylanoides* and *Debaryomyces hansenii* yeast strains isolated from cheese. *Syst Appl Microbiol,* 19, 255-64.
[http://dx.doi.org/10.1016/S0723-2020(96)80052-X]

Salum, P, Govce, G, Kendirci, P, Bas, D & Erbay, Z (2018) Composition, proteolysis, lipolysis, volatile compound profile and sensory characteristics of ripened white cheeses manufactured in different geographical regions of Turkey. *Int Dairy J,* 87, 26-36.
[http://dx.doi.org/10.1016/j.idairyj.2018.07.011]

Sofu, A & Ekinci, FY (2016) Bacterial diversity dynamics of traditional Turkish Ezine Cheese as evaluated by PCR-DGGE and SSCP analysis. *Int J Dairy Technol,* 69, 592-600.
[http://dx.doi.org/10.1111/1471-0307.12311]

Sönmez, Ş, Önal Darilmaz, D & Beyatli, Y (2018) Determination of the relationship between oxalate degradation and exopolysaccharide production by different Lactobacillus probiotic strains. *Int J Dairy Technol,* 71, 741-52.
[http://dx.doi.org/10.1111/1471-0307.12513]

Şimşek, Ö, Gürsoy, O, Dalca, SH & Yılmaz, Y (2016) Laktik asit bakterilerinde otoliz ve peynir teknolojisindeki önemi. *Akademik Gıda,* 14, 293-301.

Tarrah, A, Noal, V, Treu, L, Giaretta, S, da Silva Duarte, V, Corich, V & Giacomini, A (2018) Short communication: Comparison of growth kinetics at different temperatures of *Streptococcus* macedonicus and *Streptococcus thermophilus* strains of dairy origin. *J Dairy Sci,* 101, 7812-6.
[http://dx.doi.org/10.3168/jds.2018-14731] [PMID: 29960771]

Temiz, H, Aykut, U & Hurşit, A (2009) Shelf life of Turkish whey cheese (Lor) under modified atmosphere

packaging. *Int J Dairy Technol,* 62, 378-86.
[http://dx.doi.org/10.1111/j.1471-0307.2009.00511.x]

Temiz, H & Kiliç, S (2016) A survey of the chemical, biochemical, microbiological and sensorial quality of Aho cheese, a traditional cheese from Eastern Black Sea Region, Turkey. *Int J Dairy Technol,* 69, 209-16.
[http://dx.doi.org/10.1111/1471-0307.12237]

Torkar, KG & Vengušt, A (2008) The presence of yeasts, moulds and aflatoxin M1in raw milk and cheese in Slovenia. *Food Control,* 19, 570-7.
[http://dx.doi.org/10.1016/j.foodcont.2007.06.008]

Çelik, Ş & Uysal, Ş (2009) 'Beyaz Peynirin Bileşim, Kalite, Mikroflora ve Olgunlaşması, *Ankara Üniv. Ziraat Fak Derg,* 40, 141-51.

Vandera, E, Kakouri, A, Koukkou, AI & Samelis, J (2019) Major ecological shifts within the dominant nonstarter lactic acid bacteria in mature Greek Graviera cheese as affected by the starter culture type. *Int J Food Microbiol,* 290, 15-26.
[http://dx.doi.org/10.1016/j.ijfoodmicro.2018.09.014] [PMID: 30291917]

Welthagen, JJ & Viljoen, BC (1998) Yeast profile in Gouda cheese during processing and ripening. *Int J Food Microbiol,* 41, 185-94.
[http://dx.doi.org/10.1016/S0168-1605(98)00042-7] [PMID: 9706786]

Wyder, MT & Puhan, Z (1999) Investigation of the yeast flora in smear ripened cheese. *Milchwissenschaft,* 54, 330-3.

Yerlikaya, O (2014) Laktik asit bakterilerinin tanılanmasında kullanılan başlıca fenotipik ve moleküler yöntemler *Journal of Food and Feed Science - Technology,* 8-22.

Yuksel, FN, Akcelik, N & Akcelik, M (2015) Incidence of antibiotic resistance and virulence determinants in Enterococcus faecium and Enterococcus faecalis strains, isolated from traditional cheeses in Turkey. *Mol Gen Microbiol Virol,* 30, 206-15.
[http://dx.doi.org/10.3103/S089141681504014X]

Yuvaşen, A, Macit, E & Dertli, E (2018) Microbial species playing roles for the production of traditional Kasar cheese during pre-maturation period. *Lebensm Wiss Technol,* 91, 406-13.
[http://dx.doi.org/10.1016/j.lwt.2018.01.075]

Yüksek, FN, Akçelik, N & Akçelik, M (2015) Incidence of antibiotic resistance and virulence determinants in *Enterococcus faecium* and *Enterococcus faecalis* Strains, Isolated from Traditional Cheeses in Turkey. *Mol Gen Microbiol Virol,* 30, 206-15.
[http://dx.doi.org/10.3103/S089141681504014X]

Zago, M, Fornasari, ME, Carminati, D, Burns, P, Suàrez, V, Vinderola, G, Reinheimer, J & Giraffa, G (2011) Characterization and probiotic potential of Lactobacillus plantarum strains isolated from cheeses. *Food Microbiol,* 28, 1033-40.
[http://dx.doi.org/10.1016/j.fm.2011.02.009] [PMID: 21569949]

SUBJECT INDEX

Celile Aylin Oluk & Oya Berkay Karaca (Eds.)

B

C

L

M

N

O

P

Q

R

U

V

W

Y

www.ingramcontent.com/pod-product-compliance
Lightning Source LLC
LaVergne TN
LVHW070945210926
848482LV00002B/97
9789811432354